Best
of
Luck.
Gary Slick

Justin -
I have enjoyed working with
you very much. May you find
much intellectual challenges
(at your non-liberal arts school ☺) and
enjoyment in college.
Best - Leary

Justin,
What a great addition
you have been to this
school! Good luck
with all of your pursuits.
Jean Baugh

luck
Justin,
Linda
DeGraen

JUSTIN
ENJOY THE
NEXT STEP!

Taming HAL

Designing Interfaces Beyond 2001

Asaf Degani

First published 2004 by PALGRAVE MACMILLAN™
175 Fifth Avenue, New York, N.Y. 10010 and
Houndmills, Basingstoke, Hampshire, England RG21 6XS.
Companies and representatives throughout the world.

PALGRAVE MACMILLAN is the global academic imprint of the Palgrave Macmillan division of St. Martin's Press, LLC and of Palgrave Macmillan Ltd. Macmillan® is a registered trademark in the United States, United Kingdom and other countries. Palgrave is a registered trademark in the European Union and other countries.

ISBN 0-312-29574-X

Library of Congress Cataloging-in-Publication Data

Degani, Asaf.
    Taming Hal : designing interfaces beyond 2001 / Asaf Degani.
        p. cm.
    Includes bibliographical references and index.
    ISBN 0-312-29574-X
    1. User interfaces (Computer systems) I. Title.

    QA76.9.U83D45 2004
    005.4'3—dc21

                                                                        2003054934

A catalogue record for this book is available from the British Library.

Design by Autobookcomp

First edition: February, 2004
10   9   8   7   6   5   4   3   2   1

Printed in the United States of America.

Contents

To
Earl
and
Gail

Acknowledgements

Writing is a lonely process. The dark and fluorescent early mornings, the long dim nights, the foregone weekends, and the undulating pressure to get it out. That's what it was like for me—yet I did not do it all alone. These acknowledgements mark both the beginning and end of this book.

This book could not have been written without the help and support of my good friend and colleague, Michael Heymann. Many of the underlying concepts and ideas presented here are the results of our long and close collaboration; the last eight chapters in this book are the product of our mutual work on formal aspects of human-machine interaction. Michael wrote many sections in this book and had seminal impact on the approach and ideas presented. He has been with me throughout the development of this book—from first proposal to final draft, helping me with the details of each chapter and providing insights, encouragement, and invaluable support.

I am grateful to Earl Wiener for his mentorship and friendship, and for giving me the necessary foundations to do this work. With his wisdom and mindfulness, he helped me grow and set me on a path of scientific research. I learned a lot from Earl and received much support and generosity through many years. The groundwork for the concepts described in this book, as well as many of the devices and examples presented in chapters 5-12, comes from my dissertation work under the tutelage of Professor Alex Kirlik. With his engineering and ecological approach to understanding and describing human-machine systems, Alex introduced me to his unique way of thinking. There, at Georgia Tech, I was also influenced by Christine Mitchell and her approach to modeling human-machine systems, as well as by the work of Al Miller. These ideas, concepts, and modeling approaches run as a thread throughout this book.

I would like to thank Michael Shafto for continual and gracious support during many years of work, as well as for his guidance and foresight in laying out the scope of my research plans. The initial ideas and some portions of this book were written while I was a researcher with the San Jose University Foundation. Kevin Jordan provided nourishment, comradeship, and support for many years.

I further developed my modeling approach and learned to present it in a simplified way while teaching a seminar on modeling at U.C. Berkeley. I thank the enthusiastic students of that graduate class for helping me develop these ideas; their comments and involvement allowed me to refine my ideas and use them in chapters 1-4. Throughout the years, I have had the good fortune of teaching and working with many students on some of the topics presented in this book. Jim Lockhart, Ramak Asgari, Nancy Johnson, David Austin, Maria Romera, and Meeko Oishi gave me an opportunity to better understand my own work, and my interaction with each and every one of them has taught me valuable lessons. Maria Romero did the initial work on the visitor registration system in chapter 9 and suggested the example of the ceiling fan in chapter 1. David Austin was instrumental in doing the analysis on the VCR system described in chapter 7 and has helped me document some of technical details of the avionics systems described in chapters 13, 15, 16, and 17. The traffic light and autoland examples in chapter 17 were done in collaboration with Meeko Oishi and her advisor, Claire Tomlin.

I would like to thank Rowena Morrison for helping me from the very beginning to shape and reshape the message and scope of this book and then coming back toward the end and helping me pull it together. The title of the book is her idea, and she provided me with much substance and encouragement along the way. Doug Tamkin also provided much needed editorial support. I am grateful to Smadar Shiffman for painstakingly going over the entire manuscript and helping me see places that needed further elaboration and expansion. Tony Andre gave me important insights and reviewed the initial draft of this book.

I want to thank Kathleen Danielson for teaching me how to write from the heart. Cai Baker helped me find my own voice and reviewed early chapters. Mark Igler and Jennifer Cray reviewed early versions of the manuscript and gave me the benefit of their journalistic experiences. Martha Amarm, a dear friend, provided much needed support and helped me navigate the unchartered waters of agents, publishers, and the book writing process. I want to thank my literary agent, Ed Knappman, for his encouragement, especially at times when the stack of 20 some rejection letters seemed as tall and cold as Mount Everest. I am grateful to Toby Wahl, my editor at Palgrave, for his interest in human-automation interaction and user interface design and for holding the book together as a whole while I was busy with the details. He has helped me immensely to set the tone and feel of this book and provided necessary guidance as well as detailed editorial work. Alan Bradshaw took the manuscript and turned it into a real book.

Robert Mindelzun helped me along the way with medical topics, book-writing dilemmas, and reviewed the final page proofs of the book. John Moses wrote several sections in the maritime chapter (8), and shared with me his experiences in piloting ships around Cape Cod and through the Boston Traffic lanes. My appreciation to Tracy Golden for continual support, an illuminating review of the maritime chapter (8), and sharp insights that helped me in the process of writing the Internet chapters (9 and 10). Michael Dornheim gave me further insights into the cruise control system discussed in chapter 11. I would like to thank David Gaba for guiding me during the analysis of the blood pressure machine in chapter 12; and Assaf Morag, Alex Targ, and Oren Abramson for reviewing that chapter and sharing with me their medical expertise. Immanuel Barshi provided aeronautical information that was used in the procedures chapter (13), and was a companion for many long walks and fruitful discussions. I owe thanks to David Simmons for insights about the topic of priming, which is discussed in chapter 13. Loukia Loukopoulos provided a detailed review of chapter 13. Randy Mumaw, Mike Moody, and Curt Graeber were of great assistance and support in my research into cockpit automation and verification of interfaces which is summarized in chapters 16 and 17. Lance Sherry offered me valuable insights about the inner workings of the automated cockpit that is described in chapter 14 and the test flight accident of chapter 15. Jimmy Holmes and David Austin provided me with their flying expertise that helped me refine many of the aviation chapters. John Lee reviewed the discussion on trust in automation, and Ben Berman shared with me his familiarity with the aviation accidents described in chapter 17. Bill Sundstrom provided insights into the history of the Luddites.

I owe much to Kathy Abbott for introducing me to the world of aircraft certification, and the members of the joint FAA/JAA Flight Guidance Systems Harmonization Working Group for giving me a place to develop, test, and present some of the analytical methodologies described in this book. I wish to thank Ehud Krauss, Lisa Burnett, and the entire Zohar crew for taking me in and giving me an environment to dance away, at least for a while, from the all-consuming process of writing.

I would also like to thank the following individuals for their support, generosity, and encouragement along the way—Yehuda Manor, Ev Palmer, Vic Lebacqz, Todd Callantine, George Meyer, Helene Vencill, Alan Price, Rose Ashford, Lloyd Sauls, Mark Tuschman, and Helen Tait.

Last but far from least is the acknowledgement which is due to my family: My dad for his love of the written word and my mom for her love and touch of the paint brush; I cherish both. The Twelfth century Sufi poet Farid A-Din Attar, in his *Congregation of the Birds,* wrote that "whoever undertakes this journey should have a thousand hearts so that he can sacrifice one at every moment." It has not been an easy undertaking for me, and I thank my family for journeying with me along this toll road, accepting and enduring the many sacrifices that it entailed. My companion in life, Hila, has been a true light source, from the buds of an idea, through the fervor, upheavals, and difficulties that this process demands, all the way to completion and arriving at shore. Her understanding and (sometimes therapeutic) advice sustained me. Acknowledgement is also due my two daughters, Nitsan and Gal, for observing and tolerating this process that at times deprived them of attention and robbed them of precious time. Their inner strength and continual presence gave me indispensable encouragement and balance.

In writing this book I had to collect, assemble, draw, and describe hundreds of pieces of information. Knowing myself, and despite many reviews and inspections by others, I'm sure that a few errors have crept in. In this book I analyze many automated systems that either had problems or failed. I believe it is extremely beneficial to examine and understand design flaws, errors, mistakes, and resulting accidents—and that if we look deeper we can learn important lessons that will prevent them in the future. In general, I have avoided naming manufacturers and those involved in the design of systems that have problems.

With the exception of one or two examples, all the information used and presented in this book is publicly available and can be found in the academic and technical literature. In the endnotes for each chapter, I provide the source of the information as well as books, reports, articles, and historical facts pertaining to the topic. Finally, many of the topics discussed in this book, especially when it comes to the future of human-automation interaction and the contributing factors to some of the accidents described, are and will be debatable for years to come. I hope this book will help those who are interested in the topic of human-automation interaction by providing some light on an everyday, yet rather complex, problem. To this end, the descriptions, analysis, interpretations, and opinions presented here are solely mine and do not represent the views of any group, professional society, or any federal agency.

Introduction

"What does that mean—tame"?
"It is an act too often neglected," said the fox. "It means to establish ties."

—Antoine de Saint-Exupéry, *The Little Prince*, chapter xxi

Machines and computers are everywhere, progressively changing the way we go about our lives: We "interface" with ATMs, not bank tellers. At the gas station, we quickly swipe the card, pump, and press "YES" for a receipt. We buy airline tickets on line, use automated check-in machines at the airport, and drive rental cars along unfamiliar streets using an onboard navigation system. Slowly in some cases, overnight in others, almost every aspect of our work, leisure, and consumer world has come to involve automated systems. We interact with these sophisticated devices daily, and the only face we see is the interface.

Look around you now and you no doubt will see more than a few: your cellular phone, answering machine, personal digital assistant (PDA), laptop computer, car—all have built-in automation. These automated devices make our lives more efficient and give us impressive functionality and unprecedented access to information. We rely on them now, and we will rely on them even more in the future.

Automation is also embedded in medical systems, power generation facilities, ships, and aircraft. Automated control systems provide tremendous benefits in terms of accuracy and efficiency—physicians rely on them during surgery, bridge officers use them to navigate treacherous waters, and airline pilots depend on them to land safely in bad weather. Even more astounding benefits are just around the corner: intelligent highway systems that will aid the driver and eventually guide our cars; refrigerators that order groceries automatically; and "smart" homes that are connected to the Internet for security, lighting, heating, maintenance, and overall energy efficiency. Automated systems, our electronic servants, are here to stay.

Yet, we all have had those maddening experiences when automated devices seem to act as if they have a mind of their own: when you can't program the

VCR, when the clock-radio fails to wake you up, when the entertainment system won't play a movie. In most cases, the consequences are annoying and frustrating—yet trivial (so what if your VCR recorded the Home Shopping Network instead of the Super Bowl).

But just as we get frustrated when we cannot set up the VCR to record a favorite show, users and operators of highly complex automated systems also experience confusion and frustration with their machines. Take commercial aircraft, for example. Modern airliners are fully equipped with autopilots, automated navigation systems, and sophisticated information systems. Airline pilots hardly fly the planes with their own hands anymore; they rely on automated control systems to guide the aircraft from after takeoff to just before landing. And in really bad weather, when the visibility is 0, the system can land the aircraft automatically. While these automated systems are reliable and accurate, their interfaces have caused confusion, and in some cases, mishaps. Worse, during the 1980s and 90s, there were more than two dozen aviation disasters in which pilot confusion about the state of the automatic system contributed to the tragedy. And although pilots have learned to work around these systems and the rate of such accidents has since dramatically decreased, the root problem has not gone away.

When it comes to human interaction with automation, whether it is in the home, office, or in a more complex and high-risk environment, several important issues and potential problems exist. The first has to do with the sheer complexity of modern computer systems. As computers and automated control systems become increasingly complex, designers and engineers are unable to fully understand *all* the possible ways the system can behave. The implication, from our perspective, is that interfaces to these complex machines may not be a fully accurate representation of what is going on underneath. Second, automated control systems such as autopilots, automated trains, and even your car's cruise control are powerful; when in error they can cause havoc, even disaster. Third, slowly but surely, we see that many human functions are becoming automated in both everyday systems (e.g., antilock braking systems) and in high-end systems (e.g., envelope protection systems in aircraft). Furthermore, as technology evolves and becomes more available, we are giving machines more and more authority to take over—and sometimes override—our own commands in order to keep the systems safe. The last thing for us to remember is that there will be no end to increased automation. It is growing exponentially and will be with us for the long haul.

HAL's Legacy

Literature and film have exposed our hidden fears and anxieties about man-made creations that become unpredictable, overpowering, and ultimately unbearable. From Jewish folklore comes the legend of the colossal Golem that, after fulfilling its civil defense duties in sixteenth-century Prague, came back to hunt its creator, the chief rabbi, along with the entire community. Three centuries later, Mary Shelley created Dr. Frankenstein's "monster"—a creature of grotesque appearance that possessed human intelligence and feeling.

In *2001: A Space Odyssey*, Arthur C. Clarke and Stanley Kubrick provide us with an important message about the potential future of humans and automated machines. HAL 9000 is a highly advanced super-computer and the nervous system of a Jupiter-bound spaceship. HAL manages everything onboard—from meals and daily schedules to navigation and life-support systems. There isn't a single aspect of the spaceship's operation that's not under his control; his red "eye" is omnipresent. The astronauts onboard the spaceship *Discovery* do not fully understand how HAL does "his" work. Why should they? HAL seems intelligent, trustworthy, and always works as expected. He interacts harmoniously with the crew, plays chess, appreciates art, and his soothing and patient voice is reassuring. The astronauts are generally hesitant about intervening and challenging the computer's authority.

But when HAL goes quietly berserk and starts killing the astronauts, the situation is no longer ideal. In one chilling scene from the movie, HAL locks the sole remaining astronaut, Dave, outside the large spacecraft. Dave repeatedly commands HAL to "open the pod bay doors." The computer refuses: "This conversation can serve no purpose, Dave. Good bye," leaving the astronaut locked outside in the cold, dark void of space. HAL's soothing voice is a stark contrast to the life-and-death struggle that unfolds. The only way for the remaining astronaut to survive is to struggle with the machine. What follows is a harrowing ordeal that has the ingredients of a mythological man-God confrontation. (See the endnotes for additional information.)

Today, we don't have to look to the heavens or go to the movies to see powerful automated machines go bad. There are numerous examples: ATMs for Chemical Bank in New York City that billed twice the amount withdrawn to about 100,000 customers; the Therac-25, a computer-controlled radiation machine that massively overdosed patients and caused painful death; and an autopilot on a large commuter aircraft that did not clearly indicate to the pilots that it was working to its limits (countering a right turn due to accumulation of ice on the wing), nor give a warning that it was about to give up—it just quit suddenly, and as a consequence, the flight crews found themselves in a situation from which they could not recover. The aircraft crashed. Fortunately, these costly failures and resulting human casualties are the exception. Yet when they do occur, we are reminded of how powerful these machine are, how much we rely on them, and how they can bring disaster.

In all of the cases above, from the ATM to the autopilot system, the machine or the computer did not change its mind or become sinister. Modern machines cannot perform differently than programmed, and their interfaces do not turn malicious. What happens is that in some situations they start working differently than what the users expect. And many times the problem is further compounded by an interface that hides more than it reveals.

The inherent complexity of modern automated systems, and the fact that even designers cannot always account for every possible system outcome, make it difficult for users to understand, supervise, and interact with automated systems properly and safely. As we take to the sky in sophisticated and highly automated aircraft, many pilots themselves do not fully understand what the computer is doing, and some are hesitant about intervening and challenging the computer's control during critical maneuvers. According to one captain of a modern airliner, "You never know exactly what will be the result of flipping a switch or setting in a new parameter, so we don't interact with [the automation] during automatic landing. We simply don't know what it will do." The fact is that all of us have had similar apprehensive experiences when it comes to interacting with automated devices.

Objective

This book is written to address the problem of human interaction with automation by showing and illustrating how these automated systems work, the kinds of problems that people experience while interacting with them, and practical methods that can improve the situation. To this end, this book will take you on a lively journey. Together we will encounter more than 20 off-the-shelf devices and automated control systems, ranging from light switches, thermostats, VCRs, digital alarm clocks, phones, Internet applications, automotive and medical devices, all the way up to airplanes and autopilots. For each example, we begin with the human story, consider the resulting human–machine interaction, and proceed to examine the internals of the machine. Then we combine it all with the intent of pinpointing and revealing why some interfaces do not work and what can be done to improve them and put the human back in control.

In this book, therefore, we are less concerned with color-coding, layout, and other graphical aspects of user interface design. Instead, we focus our attention on the information conveyed to the user. Specifically, we will examine whether the information provided to the user through the interface is correct: Does it allow us to interact with the machine properly? Can we identify the current mode of the machine? Can we anticipate what will happen next? This is the foundation of many user interface problems that we encounter all too often while interacting with everyday automated devices (e.g., digital watches, VCRs, entertainment systems) and those we indirectly experience as patients going through surgery or passengers on cruise ships and aircraft.

The tour begins with simple machines and automated devices and then moves on to automated control systems such as cruise controls and autopilots. We look at how these systems work and what kind of functions they fulfill. We then look at humans—users and operators, naïfs and experts—to see how they work, the kind of tasks they need to do, and, ultimately, what types of information they need. Then we look at the two together. What will emerge is the fascinating and extremely fragile interplay between humans and their machines—revealing to us situations where the interaction is smooth and efficient and situations where the interaction is problematic, frustrating, and at times unsafe.

Why Read This Book?

Today, there are billions of user interfaces around the globe, and thousands more are developed each day. Computers, information systems, and their user interfaces are critical for the work we do, they are part of our day-to-day activities (even bathroom scales now have menu-driven interfaces), and are involved in almost any business transaction that we perform (banking, purchase orders, shipping). They are also an integral part of the service infrastructure (power generation, transportation, health care) that we all rely on.

The insights, concepts, and methods provided in this book will allow you to better understand and appreciate the delicate interplay between humans, interfaces, and machines; recognize when the interaction goes sour; and identify interface design problems before they contribute to mishaps. If you are involved in systems design, you will be able to use this knowledge to *improve* the design of user interfaces.

The first goal of this book is to open the reader's mind to how automation works and how it affects the user. You will come to see how the interfaces of remote controls, TVs, and phones work and why sometimes one gets frustrated with them. It will allow you to use interfaces more efficiently and avoid confusion and resulting errors. You will start asking yourself, "What's the problem with this device and why is it not working the way I expect it to?" instead of "What's the problem with me?"

This book will give you a better appreciation for the kind of problems that plague user interaction with automation and learn what triggers them. Through precise explanations and graphical depictions, you will actually witness interface problems and design deficiencies pop-up in front of your eyes, and be able to identify the problematic interface structures that cause human error. You will find that the same interface design problems you encounter in everyday life also happen to highly trained and experienced groups of people: those we trust to fly us in modern airliners, navigate our cruise ships and oil tankers, and operate on us with sophisticated medical equipment.

This book is also intended to help designers, engineers, and others involved in developing systems and interfaces. The approach and methods provided here can improve the design of interfaces and products, making them more efficient, reliable, and safe. When applied to interfaces of safety-critical systems, these methods can save lives.

Finally, the book is also written so that all of us will have an educated opinion about the design of current and future automated systems, and not be subject to the whims of technology gurus. Because at this point, it is no longer a question whether we can automate or not, but rather how it is done. "Technology," writes Peter Hancock, the former president of the American Human Factors and Ergonomics Society, "is the most powerful force shaping our world today, and those who mediate between humans and technology hold the key to our future, for better or worse."

Road Map

Part one of this book introduces the reader to new ways of looking at machines, automated devices, and user interaction with them. The first two chapters take us through light switches, ceiling fans, elevators, and climate control systems with the intent of showing the kind of design deficiencies that exist in these simple machines and the resulting frustration they create. These two introductory chapters also highlight the topic of abstraction, which is at the center of the interface design problem. Chapter 3 goes into non-determinism in user interaction and illustrates a hidden interface problem that plagues a digital watch. Slowly but surely we begin to see the basic structure of what drives us crazy when we interact with machines. This part ends with the story of why and how, in 1983, Korean Air Lines Flight 007, a Boeing 747 aircraft, deviated from its navigation path and was subsequently shot down over the Sea of Japan by Soviet fighter jets.

Part two focuses on several important characteristics of machines and principles of user interaction. We begin with a thermostat, an air conditioning unit, and three-way light switches to understand and describe fundamental characteristics of everyday devices (hierarchy, concurrency, and synchronization). We then proceed to analyze a design feature in a cordless phone that causes users to hang up repeatedly on the caller. Chapter 7 tackles the problem of the household VCR, the epitome of poor user interaction design, and shows why most of us become so frustrated while "programming" it. We then wake up with an alarm clock and go to sea to learn about the grounding of the cruise ship *Royal Majesty* and how a flaw in the design of its global positioning system (GPS) "tricked" the ship's officers into believing they were on course, when in fact the ship had strayed 17 miles off course. This part ends with the World Wide Web, and explores several Internet applications to better understand user interaction with "walk in and use" interfaces and information systems.

Part three explores user interaction with automated control systems. It begins with an interface problem in many of today's cruise control systems, and shows how this and other similar problems can be identified. We then investigate a problem with a blood pressure machine that contributed to a medical incident. Chapter 13 addresses the kinds of problems operators, such as pilots and technicians, encounter when using checklists and procedures to manage a complex and dynamic system. Although the example in this chapter comes from the world of aviation, the concepts that are discussed and developed throughout are common to medical and manufacturing settings, nuclear power, space flight, and many other domains where procedures are a necessity. The last two chapters describe new approaches and methods for identifying design errors in interfaces and automated systems before they manifest in a mishap. Both chapters 16 and 17 open with an automotive example with which we are all familiar and then proceed to a more complex aviation example. In these two chapters I will show how one can analyze interfaces and how to verify that the underlying system is efficient and safe. I end with several concluding remarks and a set of 35 interface design and usability guidelines that are based on the observations and insights presented throughout this book.

Part One

Foundations

Chapter 1

Machines and Maps

Forget about computers, the Internet, and automated systems for a while and let's talk about behaviors—human behaviors first. When we hear a loud, unfamiliar squealing sound, we turn our heads; when a person enters our office, we look up. We may blush while we listen to a very naughty story and turn pale at a horror one. We all respond in one way or another to external events that affect us. We also react to internal events, such as thoughts and memories. These responses are manifestations of our being. Each person is different; no two people are alike.

In the same vein, a machine—a light fixture, a digital watch, a kitchen gadget, a VCR, a medical device, a car, an aircraft autopilot—acts in response to external and internal events. The way a device responds and acts is called the machine's behavior. Take the simple light bulb, for example. It has two states: OFF and ON. When it's OFF, it's dead—no light, no heat, nothing. When we flip the switch, the bulb transitions to ON and light comes on. A thermostat senses that the room temperature has reached 72 degrees Fahrenheit, and (automatically) shuts off the heat source; a microwave stops cooking the popcorn when the two minutes and thirty seconds have elapsed. A light switch, a thermostat, and a microwave are all machines that respond to external and internal events.

But what is a machine?

Traditionally, machines were considered as a source of energy that provides some mechanical advantage. A car is a machine, and so is your dishwasher, oven, lawn mower, an industrial robot, and so on. Over the years the term has been extended to include mechanisms (such as clocks and typewriters) as well as computerized systems such as personal computers, answering machines, and CD players. In this book we are interested in the control aspect of these machines and computerized systems—the internal mechanism that controls the machine's behavior—because it is the key to understanding the impact of modern machines on our lives. Every machine accepts some input event, changes its state accordingly, and then produces an output. Whether the resulting outputs turn on the porch light, stop the gas heater, shut off the oven, pop up a screen window, initiate a database query, or make an automatic landing depend on what the machine's purpose is. The underlying mechanism of what makes it work is very much the same.

Modern machines exhibit complicated behaviors. A modern fax machine, for example, is no longer just a plain facsimile machine—it also functions as a telephone, voice mail system, scanner, and printer. All these functions and modes of the fax machine collectively define a kind of a map—a blueprint of the machine's behavior. The functions and modes of the machine are accessed through user interface and described in the user manual. But it is important to note that the interface and user manual are only a limited reflection of what goes on inside the machine. The interface is only a faceplate and the user manual is just a story. The design of the interface and user manual is all about making sure that this faceplate and story are indeed correct and adequate for the user.

Desert Map

Figure 1.1 is a map of a desert. The map does not portray gray asphalt roads, red-brown dirt roads, the colors of elevations, or the dots of lone shacks as on any other desert map. This map is different. It tells the evolving story of nomadic people journeying between water-soaks and wells in the desert. The desert is in Australia and the people are the Aborigines. Circles represent a water source, a stationary place. The connecting lines describe paths from one hole to another.

Knowing the location of water holes and paths in-between is the key for survival in the desert. Understanding how machines behave is the key for designing better and safer interfaces between humans and the automated systems that we all depend on. A behavioral map of a given machine is not much different from the map of this desert tract. The water hole map brings to life the fundamental relationships between the land and its nomadic inhabitants. In the same way, the behavioral maps that we will be developing in the next section and following chapters bring to life the interactions between the machine and its users.

States, Transitions, and Events

This is where we begin our journey: Look at your reading lamp, for example. The one in figure 1.2(a) has two behaviors, either OFF or ON—each one is a state of being. Just as a circle represents a water hole on the desert map, figure 1.2(a) depicts a state, a stationary place, as a circle. To move from one state to the

Fig 1.1. Map of the Desert (© copyright courtesy Aboriginal Artists Agency, Sydney, Australia).

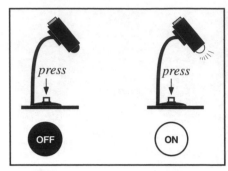 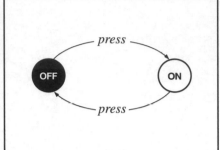

Figure 1.2 (a). Two states.   Figure 1.2 (b). And a transition.

other we *press* the little switch on the base of the reading lamp; it triggers a *transition* from the OFF state to the ON state. And just like the trails between water holes in the desert map, our map of the lamp portrays transitions as lines with arrows—snakes of change—telling us which direction to go, transforming the scene with their alchemy. Figure 1.2(b) is a behavioral map of the reading lamp. Each *press* on the switch counts as an *event*, and when the lamp's mechanism receives this event, it immediately transitions from OFF to ON. And such switching, from OFF to ON and so on, can continue forever.

In the map of figure 1.2(b) there is only one way to turn the light ON; it's an event called *press*. Clapping your hands or saying "light on" in a loud voice may work on other modern lighting systems—but not on this one. Our humble lamp accepts only the event that is in the description—it is deaf to any other events. The behaviors of the machine are completely predictable from the description: you press the knob and you get light; you first select FAX mode, then the "send" button, and the fax rolls down the telephone wire. Most machines have a limited, or finite, number of states and transitions. But just like the desert, finite is not necessarily small.

When we describe behaviors, we are in an abstract world. Our map of the machine is missing a lot. It's like a skeleton, with its lanky body, bony pelvis, and long dangling hands, telling us nothing about the flesh, and perhaps only hinting at the muscles. The behavioral description of the reading lamp in figure 1.2(b) tells us nothing about the size of the bulb, wattage, the filament, or the light's casing. But that's OK, we are interested here in the fundamentals: how the machine behaves and how humans will interact with it.

Elevators and User Information

Let's consider how this way of thinking can be used to understand more complicated machines, such as an elevator. But again, we are not interested in the details of the wood paneling or the tile pattern on the ceiling. We are interested in how the elevator responds to requests from the people who rely on it, how it processes requests, and how this information is presented.

Here is the scene: You arrive at a fine hotel—and drag your luggage to your 12th floor room. You shove the carry-on into the closet and close the door behind you. Hungry, all you can think of is a fast descent to the restaurant and bar on the 1st floor. Will the elevator take you there *directly*? No. Sometimes, actually more often than not, you will stop at other levels to pick up other guests. This time the elevators stops at levels 10, 9, and 5 before arriving at your destination.

An elevator is an example of a system that serves multiple clients. There are people inside the elevator who already "pressed" their requests, and there are guests waiting on various floors. The elevator's response depends on all these requests. Nevertheless, the elevator interface only shows you what you and the people inside the elevator requested. There is no information about the other guests waiting at various floors. Therefore, you will not be able to tell if the elevator is about to stop at the next floor, or cruise by.

By the way, the next time you go into a modern high-rise, take a look at how the high-speed elevators work. Watch where they go and where they stop. Some of them don't work as you would expect. You request the 12th floor and the elevator goes up. Then, to your astonishment, it passes the 12th floor, stops at the next floor, which is the 14th, lets off several somber-looking guests, and only then goes down to your modest room on the 12th floor. After almost walking out of this unpredictable elevator in favor of the staircase, you ask yourself, Why?

As it turns out, certain floors have higher priority (in this case, the 14th floor being the executive suite). The lobby has high priority, too. If the elevator is heavily loaded and speeding down toward the lobby, it will stop there first, then backtrack and release the sole person who wanted to check out the empty shops on the mezzanine floor above the lobby. Because of this logic, these sophisticated elevators are much more efficient in moving masses of people. Yet it still leaves elevator riders confused, anxiously pressing the button, demanding response. There are no signs, such as

> **This elevator may bypass**
> **your requested floor,**
> **please be patient, and it will**
> **bring you there.**

No, you have to carefully observe this behavior several times before you realize what is going on.

Three-way Light Bulb

Back in your hotel room, you find a vase-like lamp on the desk (figure 1.3[a]). You turn the knob and the bulb dimly lights. However, since you need to do some work-related reading , you turn the knob one more time and it is much better. A third twist makes the bulb even brighter. In a bit of a letdown, the fourth twist turns the light out. Figure 1.3(b) is a diagram of the states and transitions in this three-way lamp. The lamp has three lighting levels. By rotating the small aluminum knob on the base of the light fixture, you select among these three luminosity levels: 30-watt, 60-watt, and 90-watt. Simple enough.

But let's examine this vase-lamp system carefully. We have a single input event here, *rotate*, which triggers transitions among three luminance levels and one OFF state. Trivial, I know. But remember the elevator, and how the requests of people inside and outside the elevator affected the way the elevator made its way down? The same principle applies here: the new state of the machine depends on which state we came from—if we were initially in the OFF state, the event *rotate* will take us to 30 watts; if we were in the 90-watt state, the event *rotate* will take us to OFF. Notice what is going on here: an identical event, *rotate*, leads us to four different end states (30-watt, 60-watt, 90-watt, and OFF).

The general scheme is as follows: an event (*rotate*), changes the current state of the machine (e.g., to 30 watts), and that, in turn, determines how the subsequent event (another *rotate*) is processed and which state we will land in (e.g., 60 watts). This fundamental mechanism allows us to design machines that do very complicated pieces of work. But as we shall see next, it is also a mechanism that makes it difficult for users to anticipate what the machine is doing.

Bring in the User

This is the point where the user—me, you, and the rest of us—comes in. So first let's get the fundamentals covered: What do we want to do with the light fixture? For starters, we want to turn the light on, or perhaps off. And, if on, we also want to select the desired level. And yes, we also want to do these tasks efficiently and reliably; that is, we want to know exactly what will happen when we rotate the knob. No trial-and-error and no surprises, please.

Let us analyze these two user's tasks together. Anyone can easily determine when the light bulb is off, and predict that in one turn there will be light. But what about the 60- and 90-watt levels? First, of course, you need to know that this fixture has different light intensities, that this fixture is different from your

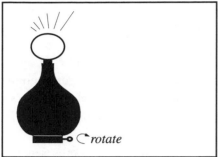

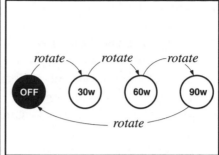

Figure 1.3 (a). Three-way lamp fixture.　　　　Figure 1.3 (b). And its behavior.

regular on/off reading light. There were no signs by the lamp and no directions anywhere. You have to know that this lamp has multiple settings and learn them by rotating the knob. That's one problem.

The second problem has to do with setting the desired light intensity. Have you watched people use three-way light fixtures? What most hotel patrons and unfamiliar users do is to rotate the knob as many times as it takes, until it is dark. Now they start counting the clicks: click, light; second click, more light; third click, enthusiastic light; another click—dark. What we all do in these situations is to try to establish some mental map of the fixture's behaviors. But it may take another pass, or even a third one, to finally establish the map in our memory. Only then do we click to the desired luminosity level and stop.

Why do we need one, two, or sometimes three passes through the light fixture's settings before we can be sure of the desired luminosity level? Why can't we do it in one shot, set the desired level, and go about doing things that are slightly more important? Well, this time-honored technique is in place because it is difficult to determine the current state of the light bulb. One noticeable state is OFF. But when it is ON, it's difficult for most people, especially the not-so-young and those of us without perfect vision, to determine right away whether the current level is 30 watts or 60 watts. And if we can't tell what the current state is, how can we predict what will happen when we rotate the knob?

Ceiling Fans and Time Delays

The problem becomes a bit more sticky when there's a time delay. Above your bed is an oscillating fan. A pull on the long dangling string starts it up. Now it's whirling and your stack of work-related papers are flying. You pull the piece of string to stop the fan, but it actually seems as if it is moving faster in a renewed fervor. You pull on the string again, but the fan doesn't stop. Angry yanks on the string only make the fan roar with vigor.

What's going on here?

This fan has several (speed) settings: OFF, LOW, and then HIGH. A *pull* on the string, as you can see in figure 1.4(a), is an input event that takes us to LOW. Another *pull*, and we are in HIGH. Once in HIGH, the fan whooshes to OFF by way of LOW.

Just like the three-way light fixture, there is no dedicated user interface. The fan blades themselves serve as the interface. So how does the user, who does not have figure 1.4(a) in front of his or her eyes, know what will result from pulling on the flimsy piece of string? Does it go OFF, ON, and then back to OFF? Or is it OFF, LOW, MEDIUM, HIGH, and only then OFF? The user has no way of knowing. The only way to find out is to cycle through—experiment is a more appropriate word—until you figure it out. Likewise, the fan is not very clear about enunciating its current state. Is it in LOW or is it in HIGH? Well, you have to figure that out by looking at the speed of the fan blades.

The above problems, however, are only part of the story. You see, the fan does not transition instantaneously from OFF to LOW and from LOW to HIGH; it takes several seconds for the fan to speed up and reach each one of these states. To describe this time lag, we add in figure 1.4(b) four transitory states: After we *pull* on the string in the OFF state, we enter a transitory state called REV-UP TO LOW. Here we remain until the fan speed reaches 90 revolutions per minute, and only then does the fan transition to LOW. When we decide to pull the string again, we enter REV-UP TO HIGH until the fan speed reaches 160 revolutions per minute. Only then do we transition to HIGH. The same story applies for revving down to LOW and then finally to OFF.

The common frustration with this time delay occurs at LOW on the way down to OFF: Pulling on the string to stop the rotation and not recognizing that the fan *is* indeed winding down. Thinking that the pull did not register or wasn't hard enough, you give it another yank, which, as we can see in figure 1.4(b), only sends the fan to LOW. Now you have to cycle it up and then down again before you can put this whirlpool to rest.

Finally, there is another nuance here that is beyond the hotel patron who is unfamiliar with this fan. Even when an "experienced" user enters the room when the fan is at LOW speed, he or she can't really tell if a *pull* will send it to HIGH, or a pull will wind it down to OFF. Why? Because as figure 1.4(b) shows, we kind of have two LOWs here: one on the way to HIGH and one on the way to OFF. There is simply no way to know in which LOW you are in (even if you have trusty figure 1.4 in your hand). You must pull on the string, and patiently wait to see what happens next.

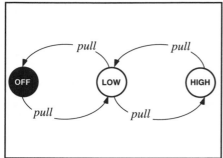

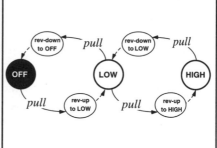

Figure 1.4 (a). The behavior of the fan.

Figure 1.4 (b). The behavior of the fan with time delay.

Transoceanic Lights

Time delays, especially when they come together with a not-so-clear interface, can really confuse us, just like some hotel showers in which the hot water handle works counter-intuitively and confuses us, and the time delay in getting the desired temperature seals our fate. Such shower arrangements are not only annoying, but can also cause minor burns. On the topic of time delay, the next time you take a flight on an airplane, take a look at how people turn on their reading lights. You know, that little rounded pushbutton on the armrest alongside the audio channels and the flight attendant call button. It turns out that in many aircraft, there is a short time delay between pressing the button and when the light comes on. It's about two or three seconds—certainly not instantaneous like our light switches at home. Maybe the switch has an air-driven connection, or maybe there is another reason for the delayed response. But let's put the technical reason aside, and look at the following scene.

A decent-looking gentleman enters a dim cabin and sits down in an aisle seat. He opens up the London *Times* and then pushes the button to turn on the light. Nothing. Eyes still on the newspaper, his fingers crawl to the button and another push. Just then, the light comes on, but then shuts off immediately! Then he tries it again, this time looking at the button, and it shuts off again. Now a third time with caution, slower, waiting for a response; the light comes on and he gets the hang of it.

Let's think about this for a moment. The only interface is the light shining from above; the pushbutton has no indicator light. What happens is that when we don't get an instantaneous response, we immediately assume that we did not press hard enough, or that the connection is loose, or something else went wrong. But actually the lighting system has already registered the first press, moved out of OFF, and in two seconds the light *was* about to come on. If our passenger had not intervened with another button press, the light would have come on just fine. But we are all experienced "light" users and when we see no light, we hit it again, which sends the system, promptly, back to OFF.

Now here is the underlying principle again: pressing the "light" button has different consequences depending on the current state. And when you cannot really tell which state you are in, then you will be surprised. Always! This is the first rule of life in the age of machines.

So what's the big deal you say? We have been using these three-way light bulbs for years and no major design change is expected in the future. Ceiling fans are not going to change tomorrow, and pushbutton reading lights on aircraft are here to stay. The point, however, is that it's annoying. Some fumbling is required before you get it right. This is no big deal for lights and fans, but what about alarm clocks? What if because of such a design the alarm won't ring for you in the morning? Well, that would be discouraging. What if something like this unexpectedly engages the cruise control of your car, accelerating while in a tight turn? Well, that would be a bit on the dangerous side. Medical devices? Now we are in harm's way. And what about misleading interfaces in the bridges of cruise ships and missing indications in cockpits of commercial aircraft? Well, that's definitely something to worry about. Welcome to the world of looking in detail at machines and user interaction.

Chapter 2

Users and Maps

Your rental car is waiting for you down in the parking lot. It has one of those new climate control systems that has several modes and an LCD display (figure 2.1). Let us examine this system, because it will help us to better understand the kind of difficulties we frequently encounter with modern devices in which the old set of knobs, dials, and switches, which were previously provided to the user, are now squeezed into a small display with multiple modes. There are many advantages to this trendy shift to modes and LCD displays—it saves space, is easier to clean, reduces wear and tear on bulky controls, and, of course, looks more modern. But is it always good for the user?

To answer the question, we need to evaluate the interface. And the interface here is rather simple and straightforward. But how do we go about doing this evaluation? Where do we even start? Well, we first need to understand how the machine works, and only then analyze user interaction with it.

Figure 2.1 shows the climate control system. To select the fan speed, you have a large rotary knob, on the left, with OFF, LOW, MEDIUM, and HIGH settings (see figure 2.2[a]). To select the air source, you press the "mode" button, and switch from one source to another (see figure 2.2[b]). The mode button accepts the touch and springs back, and a little symbol comes on the LCD

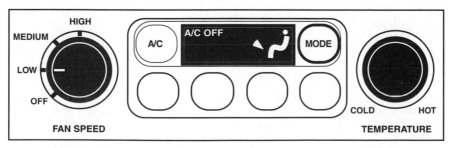

Figure 2.1. The interface to the climate control system.

display, letting you know what mode it is in. As long as the fan speed is in LOW, MEDIUM, or HIGH the LCD displays the setting. However, if the fan speed is OFF, the LCD is blank, just like your car radio's display.

When the fan is working we can tell the current state of the air source by looking at the LCD display. But what about the *next* mode? Can you tell, sitting in your rental car, what will be the consequence of pressing the "mode" button? You immediately recognize that the interface here is not different from the three-way-lamp and the fan. Unless you have the user manual in your hand, you can't tell what will be the next mode. You need to scroll through the modes to find what you need.

To see if there are additional problems here, we first need to describe this climate control system. When we consider this system, we realize that the climate control is made up of two interacting sub-systems, the fan speed and the air source. One way to describe the whole system is to combine the 4 fan settings with the 5 air source modes for a total of 20 different combinations. Each one of these 20 combinations represents a possible configuration of the climate control system. So in front of us in figure 2.3 is a full description of what this system can do. And how do we get from one configuration to another? Well, we need to add the snaky transition lines. Figure 2.4 now includes all the possible transitions in the system. On top of each line is the user-initiated event (*mode, to-low, to-medium*, etc.) that triggers the transition. (Note that for the sake of brevity, only a representative sub-set of all the events are listed in the figure.) Finally, before moving on, you may find it interesting to

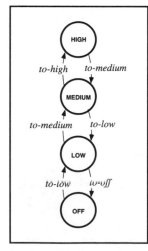

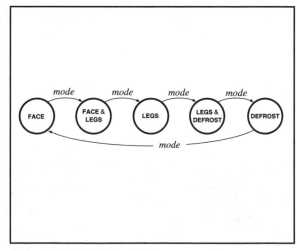

Figure 2.2 (a). Fan speed settings.

Figure 2.2 (b). Air Source modes.

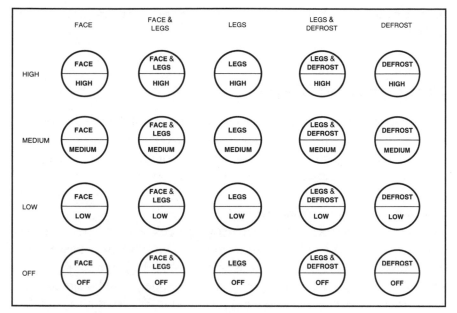

Figure 2.3. All 20 configurations of the climate-control system.

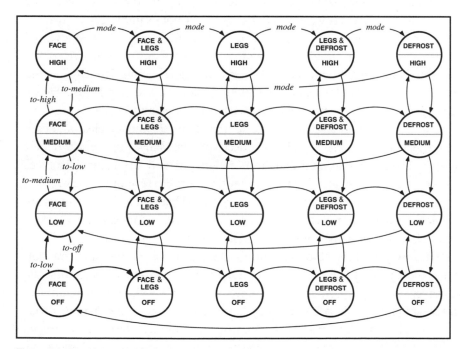

Figure 2.4. Machine model of the climate control system.

compare figure 2.4 with the Aborigine map of the desert (figure 1.1). See the similarity?

Underground Maps

What we have in figure 2.4 is a behavioral description of the climate control system. Not only does it show us what this system can do (i.e., all possible fan and air source configurations), but also how it transitions from one configuration to another. In many respects, behavioral descriptions are like modern maps of subways and airline networks, showing passengers how they can move from one location to another.

Figure 2.5 is a map of a well-known network. Look familiar? It is a diagram of the train system in London, the Underground, from 1933. Since then, tracks were added, stations opened, stationed closed—but the graphical presentation has not changed much. The purpose of the Underground diagram is to help travelers move, between stations, under London. We can think about the stations as *states* and the lines that connect the stations as *transitions*. The transition lines show us how to move from one station to another; where we can go direct and where we need to change trains. For example, say we want to go from Bank station to Mansion House station (figure 2.6[a]). We would leisurely board a Central Line train to Liverpool Street, transfer to the Metropolitan and after one stop arrive at Mark Lane station. There we would switch to the District Line, and travel three stations to Mansion House. Figure 2.6(b) shows the same route as a state transition diagram.

The Underground map and the state transition diagram are clear and easy to navigate because they are abstract descriptions. Many unnecessary details are removed: the size of the stations is ignored and the actual distances between stations are not all that important. What we care about is the order of stations and how we move from one to another. That is, which transitions are available and which transitions are not (it is impossible, for example, to take a train *directly* from Bank station to Mansion House).

The designer of the Underground diagram used this notion of abstraction with the intention of helping commuters and travelers move about the complex network of stations and lines. There is hardly any geographical information in the diagram; the only landmark you see is the Thames River. Look carefully at the map and note the strict diagrammatic format—the tracks follow a horizontal, vertical, or 45-degree path, and stations are depicted as nodes—almost as if this description came out of some blueprint of an electrical system. Everything is neatly arranged and readable.

But this diagram, with which we are so familiar, is actually a very sophisticated fake!

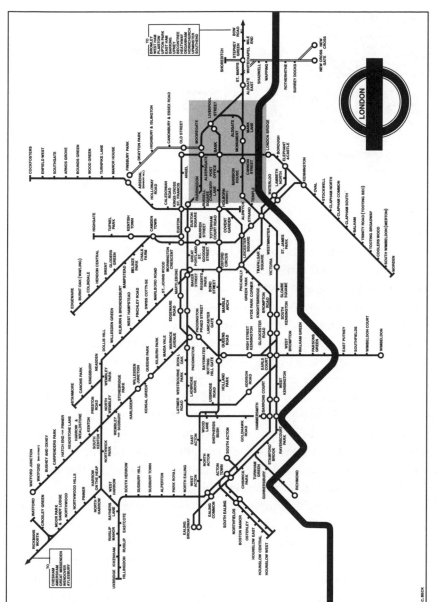

Figure 2.5. Diagram of the London Underground (from 1933).

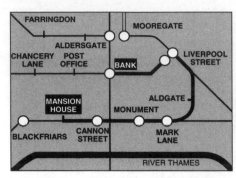

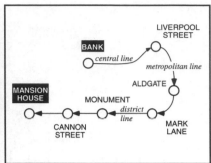

Figure 2.6 (a). From Bank to Mansion House.   Figure 2.6 (b). From Bank to Mansion House in a state machine description.

The actual map of the Underground is what you see in figure 2.7. Here you can see all the twists and turns of the train tracks and the real curves of the Thames. But is it necessary? Not really. In fact, commuters had a hard time using this geographical map; it was confusing and had too much superfluous detail. When the "abstracted" diagram was introduced in 1933, it was a big hit with Londoners, because it catered to the commuter's information needs— and not to the geographical details. But the ingenuity of the London Underground diagram, which is considered one of the graphical marvels of the twentieth century, goes beyond the schematic clarity: the central area in downtown London is enlarged, while the suburb areas are compressed. The distortion makes it easy to locate stations in the downtown area and understand how to move about from one line to another. In the suburbs, were there is usually one available track, we care only about the order of the stations (and not about the geographical distance between them).

Abstraction is a fascinating topic, very primitive and ancient on the one hand, and very sophisticated and futuristic on the other. Understanding abstraction and knowing how to use it to reduce the complexity of automated systems and their interfaces is a key technology for our future. Coming up with abstraction is far from trivial, because not every proposed abstraction is reliable and efficient (see the endnotes for this chapter for some Underground examples). We will further discuss this topic of abstraction, and its implications for interface design, many times throughout this book. But for now, let's return to the car.

The User's Perspective

Our description of the climate control system has 20 states and 50 transitions (figure 2.8). Let us now consider what the driver can actually see while

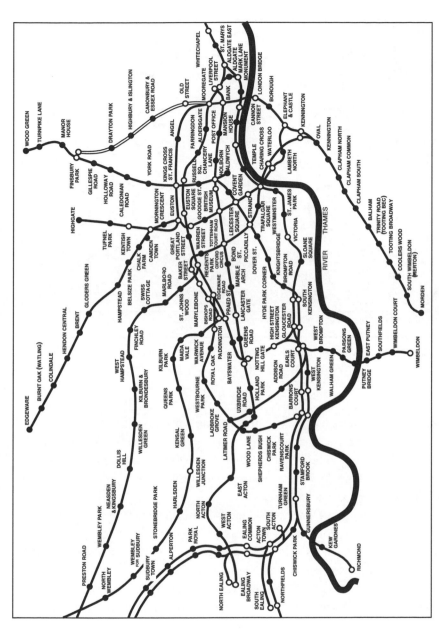

Figure 2.7. Geographical map of the London Underground (from 1932).

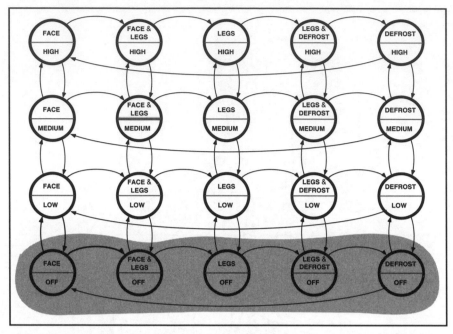

Figure 2.8. Model of the system from the user's perspective.

operating the system. First things first: can a driver distinguish among all the 20 states of the system? Look at the bottom of figure 2.8. There are five configurations with OFF inside the gray region. Is it possible to distinguish among these five OFF states? Can our driver, by simply looking at the system, tell in which OFF state the system resides? The answer is no; there are no LCD indications when the fan is off.

But does this matter? Do we really need to know which is the current mode when the fan is OFF? Not really. An OFF is an OFF, and who cares which is the current mode?—no air is coming out anyway. OK, but what about the transitions? Will there be any effect on our ability to anticipate the next configuration of the system? Does it matter that we can't distinguish among the five OFF states? The answers to these questions are a little bit more difficult to answer intuitively.

We are starting to see that there are differences between what the user can see and what the underlying system is doing, and that these differences are important when it comes to designing interfaces. We already have a model of the machine's behavior, so why don't we try to build a description of the system from the user's perspective. From this perspective, all five OFF configurations are just one big blur; the user is unable to distinguish among them. Therefore, we corral them in figure 2.9 with a rounded box and write OFF on it with big letters. The rest of the "user model" is just like the machine model (we

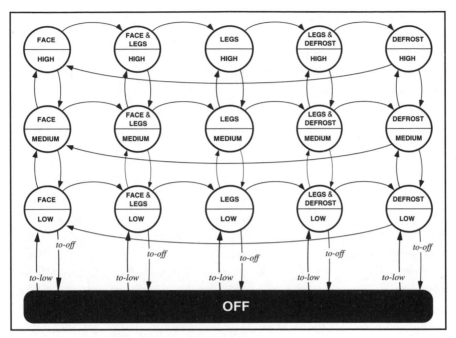

Figure 2.9. User's model of the climate control system.

assume, for the sake of this discussion, that the driver has memorized the sequence of modes and therefore knows what will be the next mode). Figure 2.9 describes what the user can see about the state of the system, and what the user knows about the system.

Comparing Models

By now we have two models: The first is a "machine model" (figure 2.4) that describes all the configurations and transitions of the underlying system, and the second is a "user model" (figure 2.9) that describes what the user sees and knows about the system. The two models share the same overall pattern of states and transitions. But there are differences, and these differences hold the key for identifying potential user interface problems.

Naturally, we all want an interface that is "good." "User friendly" is another common term. But what's a good and user-friendly interface? Well, with respect to the information provided to the user, we want an interface that does not *hide* information from the user (nor does it *lie* to the user). We call such an interface a "correct interface."

But in order to judge an interface as correct, we need criteria—just like when buying a car, where price, for example, is one criterion and engine size is another. We do the same thing here. The correctness criteria we will be using to judge that an interface doesn't hide information are twofold: (1) Can we reliably tell the current state of the machine?; and (2) Can we reliably anticipate the next state of the machine? (We will deal with interfaces that lie later on).

We do not want a system in which our user must fumble, flipping between modes and settings before reaching the desired configuration. We want our user to know exactly the outcome of his or her action. We talked about the consequences of such user confusion earlier. We want an anesthesiologist, after getting a request from a surgeon to increase the rate at which blood pressure is measured, to know exactly what the consequence will be of changing the setting of the blood pressure machine. We want the pilot to be able to anticipate reliably the consequences of commanding the aircraft to change its altitude, not a trial-and-error ordeal.

Does failure to do these sound far-fetched? Not at all! These two deficiencies, one in a modern surgery room, and one on board a commercial aircraft, contributed to mishaps that we will discuss, in detail, in chapters 12 and 16.

But before we step too far ahead, we need to be sure that we first understand this correctness issue and know how to identify interfaces that are incorrect. So let's continue with the car. Look at figure 2.9, say we are in the OFF state and we want to DEFROST the window. We rotate the fan-speed knob to LOW and get? Sometimes FACE, other times LEGS, and if we are lucky—DEFROST. Why? Because it depends on what mode was previously used. If we were driving earlier with the air source at our FACE, then that's what we will get when we turn it back on. So how can the user reliably anticipate what mode shows up? The answer is that we simply cannot. There is just not enough information on the interface to help us determine the previous mode. We have to trial and err until we get it right.

Here is the general structure of this problem. It is important to understand it because this is the fundamental problem that plagues interfaces and causes much confusion to users. Look at figure 2.9 again and note the five to-low transitions emanating from the big OFF state. The problem we just discussed relates to these transitions. The event that trigger these transition is the same, but look at the five different modes to which these transitions take us (FACE, FACE & LEGS, LEGS only, LEGS & DEFROST, or only DEFROST). Where you end up after you rotate the turn knob to-low depends on which of the five hidden OFF states you started from. But of course you can't tell these apart—and therefore you can't tell where you will end up.

There are also five transitions that go into the big OFF state. We know from figure 2.5, the machine model, that they all go to their respective OFF configuration (FACE and OFF, FACE & LEGS and OFF, and so on). And we also know that these individual OFF states are replaced by a big OFF state in figure 2.9. OK, but does the user care? I am sure you will agree that we don't really care about knowing which of the five OFF states we end up in after we turn off the air. All the user wants to do is to turn it off. From this point of view, this ambiguity has no bearing on the user.

By now we have accounted for all the differences (states, transitions out, and transitions in) between the two models and identified two deficiencies with this system. The first, which we quickly identified in the very beginning, has to do with not knowing what will be the next mode of the air source. Unless you memorize the sequence of modes, you have to keep pushing the button until you find your desired setting. The second problem has to do with transition out of OFF. You just can't tell which mode will be there when you turn ON the climate control system; there is just no way to tell. Can we work around this problem? Sure. But that's the point; it's a design that requires a work-around. And as we all know from our daily interaction with machines (think about setting the time on your car's radio with its small LCD display and all those modes), we need to do a considerable amount of trial-and-error and much work-around to get things done.

Modifying the Interface

We can't meet our criteria of a reliable and efficient user interaction unless, of course, we modify the interface. So what should we do? Let's deal first with the problem of not knowing the next air source mode that we identified early on. One thing to do is to provide the user with small indications (text or graphics) of all the possible air source modes, just like weekday indications in some digital watches. That way the user, in addition to knowing the current mode, can see the full spread and sequence of all other modes.

And what about the problem of anticipating the next mode after the system is OFF? One solution is to display the modes on the LCD even when the system is off. That way, at least you will know where you are coming from. You may have another idea for modification, but the solution, whatever it is, must allow the user to distinguish among the five OFF states prior to rotating the fan knob.

Note, however, that the redesign does not have to account for the ambiguity of not knowing which of the five OFF states we end up in after shutting off the system. With respect to this specific interaction, the abstracted interface is OK. And here is an important advantage of doing this type of evaluation: we can precisely judge any design, and, of course, any modification idea. We do this in a rigorous, systematic, and traceable way. Finally, the point of this discussion was to show that every interface is a reduced description, or abstraction, of the underlying system. Knowing what to abstract and how to abstract it is the foundation of any user interface design.

Remote Control

You are back in the hotel room. Bored and tired, you look around the room. The VCR is here, TV is here, the remote is here, and apparently some audio system is somewhere, because you hear music when you push the "audio" button. There is also cable TV (CTV) on the remote. Figure 2.10 is a line drawing of the device—a "universal" remote control.

You flip among TV stations only to find poor entertainment. Some videos are on the shelf—all second- or third-rate movies—but you stick a tape in the VCR anyway. You switch between VCR, TV, and C-TV several times in hopes of something better, before settling back to the video. All right, it will be a video night.

Then the telephone rings. You pick up the phone and hear a faint voice. The noise from the movie is too loud, so you search for the remote control and fumble your way to the "mute" button. You press it once. Nothing. Then a second time, but still no mute. You push the "mute" harder. Nothing; "this remote is broken," you say to yourself while getting up from the bed in search of the TV volume control.

Is it really broken? Not really. It turns out that when you are in the VCR mode, the remote control does not respond to TV commands, "mute" being one of them. In order to mute the volume, you must first press the TV mode button and then press the "mute" button. But how should you know this? Well, unless you had diligently read the manual (which of course does not exist in

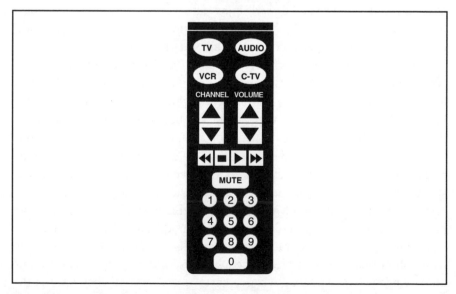

Figure 2.10. A "Universal" remote control.

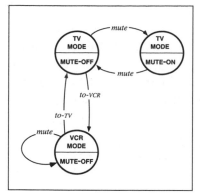

Figure 2.11 (a). Machine model.

Figure 2.11 (b). User model.

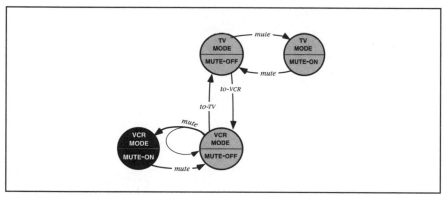

Figure 2.11 (c). Composite model.

this hotel room), you really have no clue. And even if you knew when MUTE works and not, it may not be easy to determine, with all the flipping and switching, which mode (TV or VCR) you're in. The buttons on this remote say nothing about whether the active mode is TV or VCR, nor is there any display to look at. The TV screen does flash VCR when you first select this mode, but then the indication goes away after a few seconds. Can you remember which of the four buttons you *last* pressed?

Figure 2.11(a) is a machine model of the TV and VCR modes and the mute function. You switch between TV and VCR modes by pressing the respective mode button. When you are in TV mode and hit the "mute" button, it silences the TV audio. When you are in VCR mode, the command (or event) *mute* is ignored. The event does take place, however (you did in fact push the button), and the event registered—but the machine does not respond to it when in VCR mode. Figure 2.11(b) is the user model of the "unknowledgeable" person. The user thinks that the "mute" button works in both TV and VCR modes.

To see what happens when our user interacts with the remote, let's superimpose the user model on top of the machine model and look at them together. This is our composite model of Figure 2.11(c). Everything matches— bubble-to-bubble, transition-to-transition—except the transition out of the VCR and MUTE-OFF bubble. On pressing the "mute" button, the user believes that the VCR will mute (VCR and MUTE-ON), but the machine goes nowhere. At this point, the two models are no longer in sync. Look carefully at this structure in which the same user action (pressing "mute") leads to two different end states (MUTE-OFF, MUTE-ON). This is an architecture that *always* breeds unwanted surprises and confusion. We will discuss this in more detail in the next chapter.

Chapter 3

Non-determinism in User-Machine Interaction

Non-determinism

T here is a basic and fundamental notion in the world of automated machines and software design that is used to characterize the type of user interaction problem that we identified in the remote control example from the previous chapter. It is called *non-determinism,* and refers to a system that behaves in a way that cannot be determined. The machine's responses, from the user's point of view, become unpredictable. They confuse us, and therefore, at times, are quite dangerous.

There is a basic human fear, rooted deep in our psyche, about things that we create coming back to haunt us. I mentioned in the introduction stories about man-made systems that have gone out of control: Golems, creatures, and futuristic super-computers that end up behaving in an unpredictable and disastrous way. Interestingly enough, they all start with good intentions. Even the structure of the stories is the same: man and his needs, creation and power, and finally disaster, which always results in a struggle—man against machine, man against his own creation, man against himself.

Our inability, in certain scenarios, to fully control automated systems is a reality. A horrible accident that resulted from a struggle between the pilots of a large modern airliner and the aircraft's autopilot is one chilling example. It took place several years ago over Nagoya Airport, Japan. During the approach to the runway, the copilot, who was flying the aircraft manually, mistakenly engaged the GO-AROUND mode. The autopilot immediately commanded the aircraft to climb and go around. The copilot, however, wanted to continue the landing and was pushing the control wheel down. The more the copilot pushed down, the more the autopilot countered and adjusted the aircraft's control surfaces for climb. In a struggle between man and machine, the autopilot eventually won; it had more control authority. But at the end of this duel everyone lost. The aircraft stalled and crashed on the runway.

The Nagoya accident and other automation accidents, some of which will be discussed in detail in the coming chapters, are the ugly buds of an emerging problem, which, if we are not attentive to it and fail to address it now, will come back to haunt us. These are situations in which we rely on powerful machines and automation, but the interfaces as well as the entire human-automation interaction is problematic. The pilots in the Nagoya accident did not understand how the machine worked in GO-AROUND mode, the autopilot interface did not indicate that the autopilot was working against them, and the aircraft manual was not detailed and specific enough about how the GO-AROUND mode worked. The immediate consequences were that the machine and automation, as it appeared to the pilots, behaved in an unexpected and unpredictable way. And when they do so, we all pay the price. This is one of the dangers of our technological future, and this book, at least in part, is written to highlight these problems.

But before we deal with these daunting life-or-death situations, let's try to understand this problem at its roots. We begin with unpredictable machines of the simple kind: You pick up the phone to call your mother, dial in the number—and wait. Maybe you get a ring tone, a busy tone, or, if you call overseas, perhaps "all lines are busy to the country you are calling, please try again later." Can you tell, before picking up the phone and calling, which of the three states you will end up in?

From the caller's point of view, the system is ambiguous: you simply won't be able to tell until you try. You dial, you wait, you listen—and whatever happens, happens. Most of the time we accept this non-deterministic behavior of the telephone system. But just try when it's 4:55 P.M. on a Friday afternoon, the office closes at 5:00 P.M., *all* lines are busy, and you have an urgent message to pass along. You frantically call, time after time, hoping that a line will open up. Non-determinism can drive us nuts.

You dial up your local Internet provider. Will you get connected? Who knows! 9:00 A.M. on a business day—most of the time, forget it (at least with my provider). Or maybe they will disconnect and throw you out just when you are about to make an important transaction. "Your connection has been terminated, please try again." Another example is the airport ATM machine. You ask for $500 in $50 bills. But when you hear the clack-cluck-clack going on for a long time you realize that you'll be getting 50 $10 bills. A fat wallet in a place of known pickpockets is not a good omen. And how many times did you insert carefully collected coins into vending machines, only to be surprised and hungry?

So you see, all of us are constantly bombarded with non-determinism while interacting with machines. The above machines behave in a way that is unpredictable. Do we like it? Not really. But we accept it as part of our modern times. Nevertheless, the telephone, your Internet server, and the ATM machine are, for the most part, completely deterministic when you only look at them from the inside. It is just that in many cases, when you look from the outside— from the interface side, that is—you realize that you are not provided with *all* the necessary information. You are blind to what goes on inside the machine, and therefore the machine "appears" non-deterministic. Capricious is a term some people use.

But does it have to be that way? In the case of the phone call to your mother, for example, can you envision a display with all your frequently dialed phone numbers and an indicator beside each number informing you, ahead of time, when the line is available? Why not? Some ATMs already display a message telling you, before the transaction, what bill denominations are available.

Vending Machines and Juices

For a more refreshing and illustrative example, let's consider the vending machine down the hallway. You walk up to it and 75 cents later the machine says "MAKE A SELECTION." You press the "orange juice" button, the can rolls down like a log, you pick it up and it is root beer. Most of the time this vending machine works just fine, but on some rare occasions, like now, it fails and gives you the wrong can. It happens—and we really don't care if this happens once in 5 attempts or once in 500. The plain fact is that sometimes, in this vending machine, you can get root beer instead of orange juice. (The unseen reason for these unexpected surprises is that the service guy who re-stocks this machine is occasionally absent-minded and misplaces the cans.) Figure 3.1(a) is a schematic description of this thirsty game of chance. Figure 3.1(b) is an identical, albeit more general, description of this non-deterministic structure.

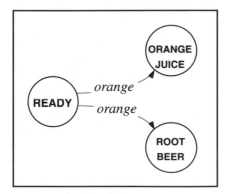

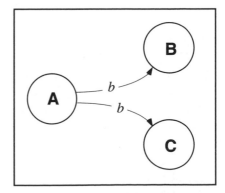

Figure 3.1 (a). Orange juice.

Figure 3.1 (b). Non-determinism, again.

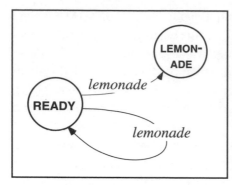
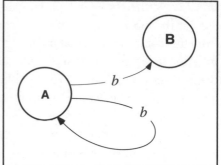

Figure 3.2 (a). Lemonade.

Figure 3.2 (b). An abstract and general description of non-determinism.

Annoyed, but still thirsty, you try again. Slide in a one-dollar bill, hit the "lemonade" button, and wait for the rolling can. It doesn't come. Most of the time it does, but today it doesn't. Not a good day for vending machines, as user frustration turns to heavy clobbering. The lemonade story is in figure 3.2(a) and the more general description of this non-deterministic structure is in figure 3.2(b): we start at some initial state A, and initiate request (b) expecting to get B. Some days we will get A and sometimes nothing. The same input event b takes us either to B or back to where we started, which means that nothing really happened. Non-determinism can be bitter.

Actually, if we think about these two non-deterministic structures in a practical sense, figures 3.1 and 3.2 are not much different. Why? Because in both cases the same event b leads to two different outcomes. Whether the lower transition goes to state B or returns to initial state A is, basically, the same bad thing. We simply cannot reliably tell what will be the consequence of pressing b. This ambiguity will be there, time after time, just like in the remote control from chapter 2. It will always lead to confusion, frustration, and at times errors.

The problem of non-determinism in human interaction with machines is not only limited to a system that has two end states. Remember the problem with the climate controls in the rental car (figure 3.3[a]) where we were in the big OFF state and wanted to select DEFROST? We agreed that it was impossible to perform this task reliably. There was just not enough information for us to anticipate what would be the new mode. Most of the time you have to fumble your way through other modes to get to DEFROST. The same event (to-low) leads to five different outcomes.

Now look at figure 3.3(b). It has an initial state A and five end states: A, B, C, D, and E. The same event (b) may take us to five different end states. Does this sound familiar? The underlying problem is that the interface of the climate

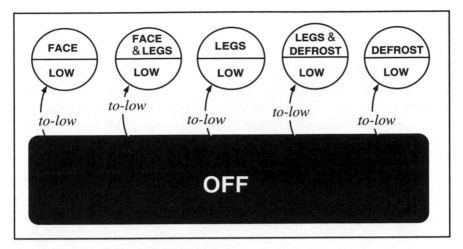

Figure 3.3 (a). Non-determinism with five end states.

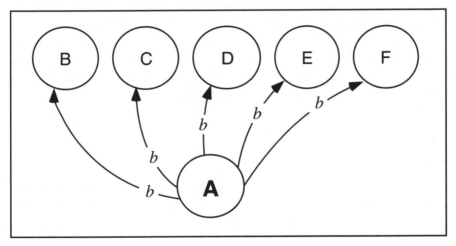

Figure 3.3 (b). Non-determinism.

control is non-deterministic. It is the same as the vending machine example, but with more end states. Cars and vending machines aside, we now have a principle and a formal structure to understand, and also explain, one of the main sources of confusion in human–machine interaction.

Night Watch

Now let's move on and talk about time. Watches, to be exact. Not fancy watches, but a rather ordinary digital watch manufactured by a reputable watchmaker. We will use this watch example to understand and actually see the kind of non-deterministic structures that are embedded in everyday devices. This digital watch has four basic modes: TIME, DATE, ALARM, and STOP WATCH. Figure 3.4(a) is a line drawing of this watch in TIME mode—showing the hours, minutes, and seconds as well as the alarm enable symbol (shockwave) on the far right. There is also a traditional three-handed clock display. You change modes by pressing the "mode" button (the lower-left button), which triggers transitions among the four modes (figure 3.4[b]).

The watch brightens up by a flick of a button (in the lower right corner of the watch). After seconds the light goes off automatically. The watch can also wake you up: in ALARM mode, you set the wakeup time and then select "alarm on" or "alarm off."

Now imagine that you are lying in bed. You have to catch a plane early in the morning and you want to set the alarm on the watch. You press on the "mode" button and settle into ALARM mode. After setting up the alarm for an early 5:30 A.M. wakeup, you look carefully for the shockwave symbol to ensure the alarm is indeed enabled. It is. You turn off the bedside light and fall to sleep.

You wake up in the middle of the night with a fright. Is it time to wake up yet? You hit the "light" button to see and it's only 3 o'clock or so. You turn on your side and go back to sleep. Several hours later, you wake up to a bright sunny morning. Looking at the watch with consternation, you see it is past 8 A.M. You missed the flight!

Machine Model of the Watch

So what happened? Why not 5:30 A.M. as planned? Did the alarm sound and you were so sleepy that it fell on deaf ears, or was it something else? Sadly enough, the alarm never did ring.

Why?

Figure 3.5 is a machine model of the watch. We are viewing the behavior of the watch with respect to the four basic modes (TIME, DATE, ALARM, STOP WATCH). Each bubble describes one configuration of the watch. Look at the top-left bubble: we are in TIME MODE, and the alarm setting is enabled (ALARM-ON). Below this time bubble are the DATE, ALARM, and STOP WATCH configurations. This is the first column of mode configuration. In each mode, when you press the "light" button, a transition labeled *light-on* takes place and four little diodes shine on the display. The second column of states represents all the configurations in which the LIGHT is working. As such, all the other configurations in which the light is off are marked as DARK. So far so good.

Just like any light, this one doesn't last forever. Three seconds after you press the "light" button, the light goes off automatically, and we are sent back to the first column via the *light-off* transition. *Light-off* is an internal transition,

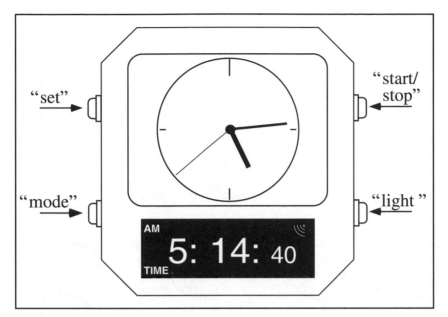

Figure 3.4 (a). Digital watch.

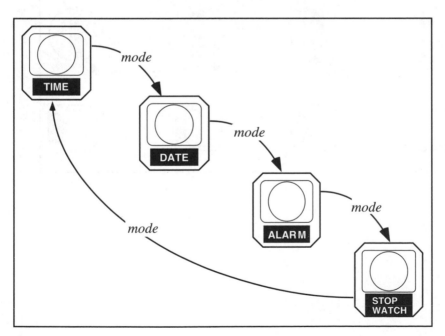

Figure 3.4 (b). The four modes of the watch.

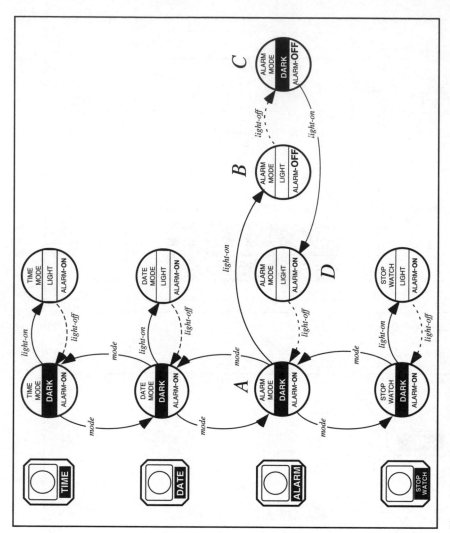

Figure 3.5. Machine model of the watch.

triggered by the watch itself, and belongs to a class of transitions that are automatic, in the sense that they take place without any direct user action. Such automatic events are depicted in figure 3.5 with a broken line.

In all modes of the watch, the events *light-on* and *light-off* do the same thing: shine light on the watch face and then shut it off. However, in ALARM mode, when you hit the "light" button you not only turn on the light, but you also disable the alarm. Look again at figure 3.5, third row down. The transition out of the "ALARM-MODE & DARK & ALARM-ON" (state *A*), takes us to a new state *B* in which the watch shines, but the alarm is disabled (ALARM-OFF). And after three seconds the light goes off automatically and we jump to state *C*. Press the light button again, and the alarm is enabled (state *D*).

This architecture, in which an event (*light-on*) triggers more than one configuration change (e.g., the watch is *lit* and the alarm *disabled*), is very common in computer-based and automated control systems. It is sometimes called *side effect, coupling,* or *linkage,* but all it means is that there are additional consequences, perhaps in other modes or in another machine altogether, following the initial event. Sometimes, as is the case here, it is just for the sake of not adding an additional button. Other times, it is for simplifying the interaction and reducing burden on the user. Propagation of data and automatic updating is one example, as in a case of electronic forms and reports, where one change of your address prompts a global change to all such fields in tax preparation software. Side effect is an extremely powerful feature of any automated system (and we will see more examples of this later in the book in chapter 5 and 14). It can be employed with considerable benefit, eliminating arduous labor and making the system easier to use. But it can also lead to a lot of "labor-saving" complexity, if the user's tasks and expectations are not thoroughly considered by designers.

User-Model of the Watch

Going back to the watch, let's consider your plight in the dark. But to see what happened we need to build a user model of the watch. In particular, let's build one for the night: you can't see anything in the dark, let alone your wristwatch. From this nocturnal point of view, you can't see any modes whatsoever. All the configurations in figure 3.5 that have the word DARK in them are hidden from us and cannot be seen. We therefore encompass the four dark modes of the first column, as well as the other dark-alarm configuration (the one with ALARM-OFF and DARK) into one state. The large ALL DARK state in figure 3.6 masks from view all the configurations that are dark.

And what about mode transitions in the dark? Is it possible to change between TIME, DATE, ALARM, and STOP WATCH modes in the dark? Sure it is; you won't know what mode is active, and you won't know which mode you landed

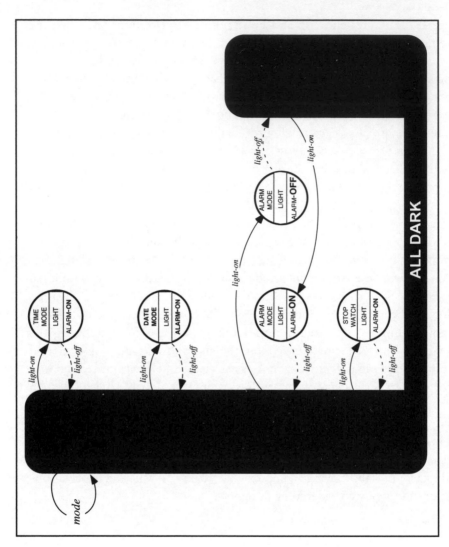

Figure 3.6. User model of the watch.

on—but one can cycle among the modes in the dark, no problem. We represent this way of interacting—cycling among the modes but without knowing what is really going on—as a transition that looks like a loop. You find it there at the top of the big ALL DARK state in figure 3.6. And it should come as no surprise to you that such a transition is called a "loop transition"; it loops around without saying anything about what really goes on underneath. From a user's point of view, all we know is that the mode-switching event called *mode* took place—but we have no idea what change actually resulted.

The rest of the user model is the same as the machine model. Once the "light" button is pressed, the light shines for three seconds and we can see the lit configurations. No problem. And at this time we are done with the user model.

Evaluation of the Watch Design

To fully understand and see what went wrong here, let's superimpose the user model on top of the machine model (just as we did with the remote control from the previous chapter). The machine model, the most complete description, is our foundation. It's always there, complete, but underground. The user model, a reduced description, is what we, as users, actually see. Figure 3.7 shows the two models one on top of the other. We call this stacked model the "composite model."

What we are trying to do here is to analyze whether the interface really supports what we wish to do with the machine. We can see that at night the user cannot distinguish among the modes and that only after hitting the "light" button can you tell which mode you are in. We've been here before, right? Just like the climate control, you have to fumble your way to get to the desired mode. The non-deterministic structure, where we have the same *light-on* event leading to five different watch modes, emerges in figure 3.7. That's one problem.

Now consider the transitions in and out of ALARM mode. Some of them just bring out the light, but two, marked with heavy lines in figure 3.7, do more than that. The first one (from A to B), turns the ALARM-OFF, and the second transition (from C to D) makes the ALARM come ON.

Note that both transitions emanate from the ALL-DARK state, and each one delivers us to a markedly different configuration. Let's consider this in an intuitive way first. When we are in the ALL-DARK state and hit the "light" button, sometimes we just see the light, but at other times we also change the alarm status (from ALARM-ON to ALARM-OFF, or vice versa). But since we can't see in the dark which mode we started from, we can't tell which one will take place. Specifically, sometimes when you turn the light on you disable the alarm, and sometimes you don't!

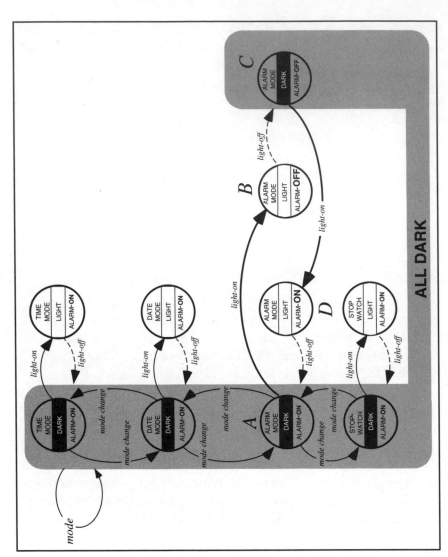

Figure 3.7. Composite model of the watch.

Non-deterministic Wakeup

To trace the sequence of actions and see what really happened here, we need to go back to bed. When you set the alarm before going to sleep, you made sure that the ALARM was indeed ON. At that point the system was in state A. In the middle of the night you woke up to check the time. After pressing the "light" button, the system transitioned to state B. You read the time and then the light went off. As soon as you entered state B the alarm was turned OFF. And of course you knew nothing about it!

Can a diligent and responsible user recover from this predicament? Hypothetically, yes. Had you looked carefully at the alphanumeric display, you perhaps might have noted that the little shockwave (ALARM-OFF) symbol disappeared. But practically, who would notice—half asleep—that a tiny symbol went away? And all of this in three seconds (before the light automatically extinguished)?

At this point we must conclude that in this watch, the design of the light and the ALARM ON/OFF function is inadequate and for all practical purposes incorrect. The problem we uncovered cannot be solved by a work-around; it only breeds confusion, error, and no wakeups. This is a flaw that requires a design change.

So before we leave the watch and move on, let's consider some possible design solutions to the problem. There are a few: one is to use a different button (e.g., "set") to turn the alarm ON and OFF. But this solution, as you can imagine, may also be just as vulnerable to error if the user accidentally hits the "set" button. Another solution is to incorporate the ALARM ON/OFF function into the sequence of setting the alarm time.

The Structure of Non-deterministic Interaction

Now, let's use the watch example to see the generic structure that is responsible for this type of problem. Look at figure 3.8(a). We start in ALL-DARK state and press the "light" button (event *light-on*). On the one hand, the resulting transition can simply light the display and deliver us to either TIME, DATE, or STOP WATCH mode, which, for illustration purpose, are bunched together in the figure. On the other hand, the resulting transition can turn OFF the alarm.

Looks familiar? It should; it goes back to our discussion of non-determinism, where the same event can bring about two different outcomes. And there's no way to tell, a priori, which outcome will take place. Figure 3.8(b) is identical to figure 3.1(b). It is the same non-deterministic story.

Non-determinism is a structure that designers of computer hardware and software are extremely concerned about. It is the kind of design flaw that

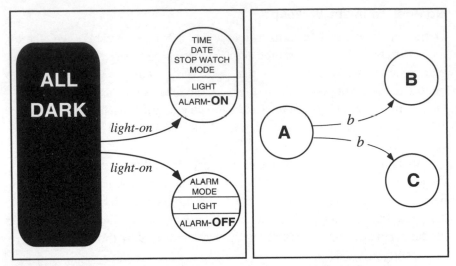

Figure 3.8 (a). Light-on transition out of ALL DARK. Figure 3.8 (b). Non-determinism, again.

causes your computer to crash, the printer to stop, and the Internet connection to fail. In the same way that it has been known to wreak havoc in computer operating systems and communication protocols, non-determinism *can* also create serious problems in user interaction. The key word here is *can*.

Some kinds of non-determinism are intentionally built into the interface. For example, some modern cars have a CHECK ENGINE indication. It may come on due to a fault in the ignition computer, a faulty emission control system, or any number of engine-related problems. Yet this specific information is purposefully hidden from the driver. What the driver should do is to take the car to a service station for repair. Other consequences of non-determinism we do care about, but we can also work around them. It may be inefficient and annoying, but we can live with it. The climate control is one example. But when we *do* care about which mode we land upon first, when it is difficult or impossible to "undo," or when the action is irreversible and the environment is unforgiving—well, in these instances, non-deterministic behavior is problematic. Here is where gizmos, devices, and machines confuse us and sometimes drive us nutty.

Interacting with devices and interfaces that behave erratically and respond in a non-deterministic way becomes an unnecessary game of chance. Moreover, of course, there are more dangerous situations, when we consider high-risk systems such as missiles and fire control systems, where the consequence of an improper mode change is hazardous and, at times, irreversible. When that happens, we lose control of these powerful man-made systems and find ourselves at the whim of the Golem-machine.

Chapter 4

The Crash of Korean Air Lines Flight 007

For want of the nail the horse-shoe was lost;
For want of the shoe the horse was lost;
For want of the horse the rider was lost;
For want of the rider the battle was lost;
For want of the battle the kingdom was lost;
And all for the want of a horse-shoe nail.

— George Herbert (1593-1633), "Horse Shoe Nail"

We are now ready to make a leap. Quite a big one, from watches to high-flying aircraft, showing that the same underlying structures that render the watch confusing also exist in the world of complex safety-critical systems. In fact, this is one of the central messages of this book. But please don't be intimidated by the perceived complexity of the aircraft systems, autopilots, and navigation systems—you will see that, basically, they are not different from fans, climate control systems, and digital watches. We will be using the same descriptive language with which we are already familiar to look at and understand the inherent problems that cause confusion and error.

Flight 007

One of the most tragic and perplexing civil aviation disasters in the twentieth century was the loss of a Korean jetliner in 1983. Korean Air Lines Flight 007 (KAL 007), a Boeing 747 jumbo jet flying from Anchorage, Alaska, to Seoul, South Korea, deviated more than 200 miles into Soviet territory and was subsequently shot down. There were no survivors; all 240 passengers and 29 crewmembers perished in the crash.

When we hear about such a horrific disaster, all of us ask, Why? Why would a well-equipped aircraft piloted by an experienced crew veer more than 200 miles off course? And why was a harmless passenger aircraft shot from the sky? This set of questions was, for many years, a puzzling mystery that tormented the victims' kin, baffled the aviation industry, haunted many, and gave rise to more than seven books and hundreds of articles, ranging from technical hypotheses and covert spy missions, to a handful of conspiracy theory offerings. But as you will come to see, the truth, as always, is cold, merciless, and needs no embellishment.

Alaska

The date: August 31, 1983. Time: 4:00 A.M. Location: Anchorage Airport, Alaska. Flight 007, which originated in New York City, was now ready and fueled up for the long transpacific flight to Seoul. After a long takeoff roll, the heavy aircraft, full of cargo, fuel, and passengers, pitched up and climbed slowly into the gloomy morning sky. After reaching 1,000 feet, the white aircraft rolled gently to the left and began its westbound flight. Leaving Anchorage behind, Korean Air Lines Flight 007 was given the following air-traffic-control instruction: fly directly to the *Bethel* navigational waypoint and then follow transoceanic track R-20 all the way to Seoul (see figure 4.1).

The aircraft followed the instructions and changed its heading accordingly. But as minutes passed, the aircraft slowly deviated to the right (north) of its intended route, flying somewhat parallel to it, but not actually on it. It first passed five miles north of *Carin Mountain*, a navigation point on the way to *Bethel*. Then, instead of flying straight over *Bethel*, a small fishing hamlet in Western Alaska, it passed 12 miles north of it. With every consecutive mile, the rift between the actual location of the aircraft and the intended route increased. Two hours later, when the aircraft reached *Nabie*—an oceanic waypoint about 200 miles west of the Alaska coast—it was already about 100 miles off track (see figure 4.1). In the early morning darkness, the weary passengers on this westbound flight were fast asleep. In the cockpit, the flight crew reported to air traffic control that they were on track, flying toward *Nukks, Neeva, Ninno, Nippi, Nytim, Nokka, Noho*—a sequence of navigation waypoints on route to Seoul.

Everything looked normal.

But it wasn't. As the flight progressed and the divergence between the actual aircraft path and the intended flight route increased, the lone jumbo jet was no longer flying to Seoul. Instead, it was heading toward Siberia. An hour later, Flight 007, still over international waters, entered into an airspace that was closely monitored by the Soviets. In the very same area, a United States Air

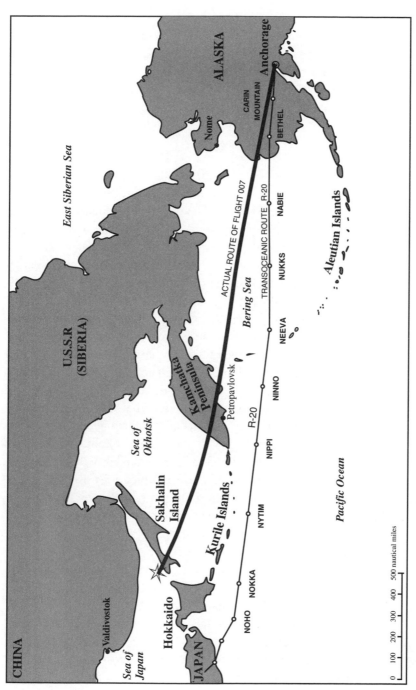

Figure 4.1. The flight path of Korean Air Lines Flight 007.

Force Boeing RC-135, the military version of the commercial Boeing 707 aircraft, was flying a military reconnaissance mission, code named "Cobra Dane." Crammed with sophisticated listening devices, its mission was to tickle and probe the Soviet air defense system, monitoring their responses and communications. The U.S. Air Force aircraft was flying in wide circles, 350 miles east of the Kamchatka Peninsula.

Kamchatka

Kamchatka is a narrow and mountainous peninsula that extends from Siberia to the Bering Sea. It has many active volcanoes and pristine blue, yet acid-filled, lakes. In those mountains, the Soviets installed several military radars and command centers. Their purpose was to track U.S. and international flight activities over the Bering Sea. At the height of the Cold War, under the leadership of Yuri Andropov, the General Secretary of the Communist Party, the Soviets were obsessed with keeping their borders secured and tightly sealed. As the U.S. Air Force Boeing 707 aircraft was circling over the frigid water in the dark of the night, purposely coming in and out of radar range, Soviet radar operators were monitoring and marking its moves. And then, during one of the temporary disappearances of the reconnaissance aircraft from the radar screen, the Korean airliner came in. The geographical proximity between the two large aircraft led the Soviet air-defense personnel sitting in front of their radar screens to assume that the target that reappeared was the military reconnaissance aircraft. They designated it as an *unidentified* target.

The Korean airliner continued its steady flight toward Kamchatka Peninsula. In the port town of Petropavlovsk, on the southern edge of the peninsula, the Soviets had a large naval base with nuclear submarines. KAL 007 was heading straight toward it. But the pilots could not see Kamchatka, because although the night sky above them was clear, everything below them was pitch dark. When the jetliner was about 80 miles from the Kamchatka coast, four MiG-23 fighters were scrambled to intercept it. The fighter formation first flew east for the intercept, then turned west and started a dog chase to reach the fast and high-flying Boeing 747. Shortly after, low on fuel, the fighters were instructed to return to base. The Korean jetliner, now 185 miles off track, crossed over the Kamchatka Peninsula and continued into the Sea of Okhotsk. Over international waters, safe for the moment, the large aircraft was heading, unfortunately, toward another Soviet territory—Sakhalin Island a narrow, 500-mile-long island off the Siberian coast, just north of Japan.

Sakhalin

On the radar screen inside a military control center on Sakhalin Island, the approaching blip was now designated as a *military* target, most likely an

American RC-135 on an intrusion mission. Because of this military designation, the rules for identification and engagement were those reserved for military action against Soviet territory (and not the international rules for civil aircraft straying into sovereign airspace). Apparently, there had been more than a few such airborne intelligence-gathering intrusions by U.S. aircraft into Soviet airspace in the preceding months, not to the liking of Soviet air-defense commanders.

As the target approached Sakhalin Island from the northeast, two Soviet Su-15 fighters, on night alert, were scrambled from a local military airbase toward the aircraft. The Soviet air defense system had fumbled in its first encounter with the intruding aircraft, but was unlikely to miss the second chance. A direct order was given to local air-defense commanders that the aircraft was a *combat* target. It must be destroyed if it violated state borders.

About 20 minutes later, Flight 007 crossed into Sakhalin Island. The flight crew—sitting in their womb-like cockpit, warm and well fed—had no idea they were flying into the hornet's nest. At 33,000 feet, while the pilots were engaged in a casual conversation in the lit cockpit, monitoring the health of the four large engines, and exchanging greetings and casual chat with another Korean Air Lines aircraft also bound for Seoul—a gray fighter was screaming behind them to a dark intercept. The fighter pilot made visual contact, throttled back, and was now trailing about four miles behind the large passenger aircraft. The fighter pilot saw the aircraft's three white navigation lights and one flickering strobe, but because of the darkness was unable to identify what kind of an aircraft it was.

The identification of the target was a source of confusing messages between the fighter pilot, his ground controller, and the entire chain of air-defense command. When asked by his ground controller, the pilot responded that there were four (engine) contrails. This information matched the air-defense commanders' assumption that this was an American RC-135, also with four engines, on a deliberate intrusion mission into Soviet territory. The Soviets tried to hail the coming aircraft on a radio frequency that is reserved only for distress and emergency calls. But nobody was listening to that frequency in the Boeing 747 cockpit. Several air-defense commanders had concerns about the identification of the target, but the time pressure gave little room to think.

Completely unaware of their actual geographical location, the crew of KAL 007 were performing their regular duties and establishing routine radio contact with air traffic controllers in Japan. Since leaving Anchorage they were out of any civilian radar coverage. After making radio contact with Tokyo Control, they made a request to climb from 33,000 feet to a new altitude of 35,000 feet. Now they were waiting for Tokyo's reply.

Meanwhile, a Soviet air-defense commander ordered the fighter pilot to flash his lights and fire a burst of 200 bullets to the side of the aircraft. This was intended as a warning sign, with the goal of forcing the aircraft to land at Sakhalin. The round of bullets did not include tracer bullets; and in the vast and empty darkness the bullets were not seen or heard by the crew of KAL 007. The four-engine aircraft continued straight ahead. Flying over the southern tip of Sakhalin Island, Soviet air defense controllers were engaged in stressful communications with their supervisors about what to do. The aircraft was about to coast out of Soviet territory back into the safety of international waters; the target was about to escape clean for the second time. The Sea of Japan lay ahead—and 300 miles beyond it, mainland Russia and the naval base of Vladivostok, the home of the Soviet Pacific fleet.

Sea of Japan

The air-defense commander asked the fighter pilot if the enemy target was descending in response to the burst of bullets; the pilot responded that the target was still flying level. By a rare and fateful coincidence, just as the aircraft was about to cross into the sea, KAL 007 received instructions from Tokyo air traffic control to "Climb and maintain 35,000 feet." As the airliner began to climb, its airspeed dropped somewhat, and this caused the pursuing fighter to slightly overpass. Shortly afterward, the large aircraft was climbing on its way to the newly assigned altitude. The fighter pilot reported that he was falling behind the ascending target and losing his attack position. This otherwise routine maneuver sealed the fate of KAL 007. It convinced the Soviets that the intruding aircraft was engaging in evasive maneuvers.

"Engage afterburners and destroy the target!"

The fighter pilot moved in to attack. He climbed back to behind the target and confirmed that his missiles were locked on and ready to fire. He could see the target on his green radar screen and also visually; the steady white light from the tail of the aircraft was shining through and the rotating strobe flickered against the cruel arctic darkness. With the morning's first light behind him and the lone aircraft ahead of him, he was in a perfect attack position against his foe. The only problem, of course, was that the high-flying aircraft, with 269 innocent souls onboard, was neither a foe nor a military target. But the fighter pilot, as well as the entire chain of air-defense command, did not know that, and if any officer, including the fighter pilot, had any doubts about the target's designation, they certainly did not voice them.

Seconds later, the fighter aircraft launched two air-to-air missiles toward the target. One missile exploded near the vulnerable jet. The aircraft first pitched up. The blast burst a hole in the aircraft skin and caused a loss of pressure inside the large cabin. The public address speakers gave repeated orders to don the yellow oxygen masks, as the aircraft started to fall while rolling to the left. The captain and his copilot tried helplessly to control the aircraft and arrest its downward spiral. Two minutes later, the aircraft stalled out of control and then plummeted down into the sea. It impacted the water about 30 miles off the Sakhalin coast.

The Many Whys of This Accident

Following the downing of the aircraft, Russian military vessels searched for the aircraft wreckage. Two month later, deep-sea divers operating out of a commercial oil-drilling ship brought up the aircraft's "black boxes"—the cockpit's voice recorder and the digital flight data recorder. The Soviets analyzed the recorders to find clues about the accident, but kept their findings secret. Ten years later, after the Iron Curtain fell, the tapes were handed over to the International Civil Aviation Organization (the civil aviation arm of the United Nations Organization). The accident investigation was completed in 1993.

With that, let's look into the first "why" of this disaster. Why did the aircraft deviate from its intended flight route? To answer this question, we need to trace the crew's interaction with the aircraft's autopilot and navigation systems from the very beginning of the flight. Two minutes and ten seconds after takeoff, according to the flight data recorder, the pilots engaged the autopilot. When the pilot engaged the autopilot, the initial mode was HEADING. In this mode, the autopilot simply maintains the heading that is dialed in by the crew. The initial heading was 220 degrees (southwesterly). After receiving a directive to fly to *Bethel* waypoint, the pilot rotated the "heading select knob" and dialed in a new heading of 245 degrees (see figure 4.2). The autopilot maintained this heading until the aircraft was shot down hours later.

Using HEADING mode is not how one is supposed to fly a large and modern aircraft from one continent to another. In HEADING mode, the autopilot maintains a heading according to the *magnetic* compass, which, at these near arctic latitudes, can vary as much as 15 degrees from the actual (*true*) heading (not to mention that this variation changes and fluctuates as an aircraft proceeds west). Furthermore, simply maintaining a constant heading does not take into account the effect of the strong high-altitude winds so prevalent in these regions, which can veer the aircraft away from its intended route. The aircraft was also equipped with an Inertial Navigation System (INS). This highly accurate system encompasses three separate computers linked to a sophisticated gyro-stabilized platform. Every movement of the aircraft is sensed by the gyros and then each of the three computers separately calculates the corresponding change in latitude and longitude. The three computers

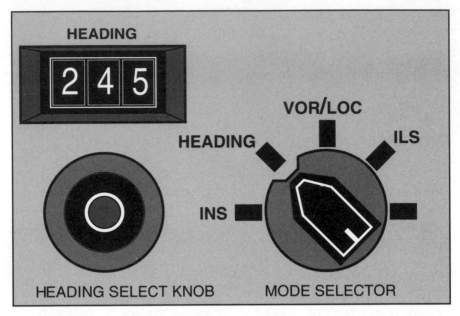

Figure 4.2 Heading indication and autopilot mode selector switch. The autopilot is in HEADING mode.

constantly crosscheck one another, making sure that their calculations agree. The system thus provides triple redundancy and is usually accurate within a mile of actual location. Throughout the flight, the pilots can see their latitude/longitude location on the digital readout of each computer, and know their exact location. In addition, the pilots can enter their entire flight route into the inertial navigation computers, and let them give steering commands to the autopilot, thereby creating a fully automatic navigation system. The pilots select inertial navigation (INS) mode by rotating the autopilot mode selector to the left (see figure 4.2).

On this flight, the route of flight was transoceanic track R-20 with some 10 waypoints (*Bethel, Carin Mountain, Nabie, Nukks, Neeva, Ninno, Nippi, Nytim, Nokka, Noho*), which passes very close to, yet shy of, Soviet-monitored airspace. To comfortably fly the aircraft along R-20 to Seoul, the pilots should have engaged the autopilot in INERTIAL NAVIGATION mode and let the system do the rest. That's what they were trained to do and that's what they have done in every transoceanic crossing with this aircraft. But they did not do it on this flight.

Why?

To answer this question it is first necessary to understand how the autopilot works and how pilots interact with it. The initial autopilot mode was HEADING. Then the pilot rotates the switch and selects INERTIAL NAVIGATION mode, and the autopilot should guide the aircraft along the route of flight (e.g., R-20 to Seoul).

Sounds simple enough. But actually, it's a bit more complicated, and this was part of the problem. It turns out that several things have to take place before the autopilot *really* goes to INERTIAL NAVIGATION mode: one is that the aircraft must be close to the flight route. Specifically, the distance between the aircraft's position and the flight route must be within 7.5 miles for the INERTIAL NAVIGATION mode to become active. That's one condition; necessary, but not sufficient. The other condition is that the aircraft must be flying *toward* the direction of the flight route (and not, for example, 180 degrees away from it). Only when these two conditions are met, or become True, will the autopilot engage the INERTIAL NAVIGATION mode.

Guards

There is a software routine that constantly checks for these two conditions. It is called a guard. A guard is a logical statement that must be evaluated as True before a transition takes place. Sound complicated? Not so. You and your friend go into a downtown bar, and there is a beefy guy there, leaning to the side and blocking the entrance with his wide body, checking young barflies' ages—no ID, no entry. Automatic teller machines, or ATMs, same story. You can't make an ATM transaction unless you punch in your personal identification number (PIN). All said and done, the PIN, just like the beefy guy, is a guard. Unless you supply an age-valid ID or the correct PIN, you are locked out.

Back to the autopilot: the first condition says that the lateral distance between the aircraft and the nearest point on the route must be within 7.5 miles. What is meant by this is that the distance should be "less than or equal to" 7.5 miles. Right? As long as the aircraft is 1, 2, 3, 4, 5, 6, 7, or 7.5 miles from the route, then we are allowed in. We write this as,

The distance between the aircraft and the route is "less than or equal to" 7.5 *miles.*

This logical statement, once evaluated, will produce either a True or False result. The second condition says

aircraft is flying toward the direction of the route.

You can see these two conditions written on the transition in figure 4.3.

The guard—the software routine that evaluates whether this transition will fire or not—cares not only about the first and second condition, but also about the relationship between them. In our case, the relationship is the logical

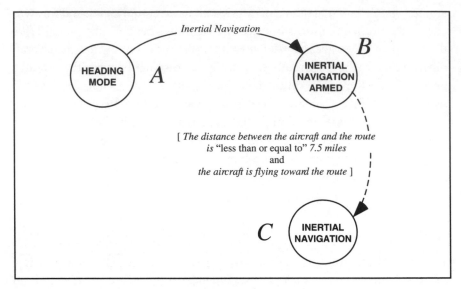

Figure 4.3. Machine model (of the autopilot's logic).

"and," which means that both conditions must be evaluated True for the transition to fire. In figure 4.3, the guard on the transition to the INERTIAL NAVIGATION constantly checks that *the distance between the aircraft and the route is "less than or equal to" 7.5 miles* and that the *"aircraft is flying toward the direction of the route."* Unless the two conditions are True at the same time, the guard will hold us back and no transition will take place.

So let's consider this scenario for engaging the INERTIAL NAVIGATION mode. Look again at the machine model in figure 4.3. Initially we are in HEADING mode. This is state *A*. We then decide to engage INERTIAL NAVIGATION. Very well—we reach forward and rotate the mode selector, and initiate event *Inertial Navigation*. This fires the transition to state *B*, ARMED, which is like a waiting room. We wait there until the two conditions are True before we can move on. If they are, we automatically transition to state *C*, which is what we wanted in the first place. Kind of awkward, isn't it?

This business of a waiting room, where we await patiently while the guard checks our entry credentials, is indeed awkward because if any one of the conditions is False, we have to wait until it turns True. But this begs an interesting question: What will the system do while we wait? According to this system's internal logic, the autopilot will stay in an awaiting state called INERTIAL NAVIGATION ARMED. What will happen from now on is that, each second, the guards evaluate the conditions. True, you're in, False, stay out. This will happen over and over (and stop only when the transition to INERTIAL NAVIGATION finally takes place).

All right, we are stuck in this ARMED state, but which autopilot mode is now active and flying the aircraft? This problem of what to do when conditions have not been met or when conditions are no longer valid is a common concern in the design of automated systems (and is discussed in detail in chapter 9). In our case, the designers of this autopilot chose to stay in a mode in which the autopilot keeps on maintaining the current heading of the aircraft. When in INERTIAL NAVIGATION ARMED, the autopilot is actually in HEADING mode. Note what is happening here: the mode is INERTIAL NAVIGATION ARMED, the display says "INS" (with an amber light), but the autopilot is in HEADING mode. Was all this a factor in the KAL accident? We'll soon see.

Automatic Transitions

By now we have seen two kinds of transitions that exist in automated systems. The majority were *user-initiated*. The user made the system move from one state to another in a very direct and deliberate way; the user pushed a button and the machine changed its mode. We have also seen *timed* transitions, such as the transition from LIGHT-ON to LIGHT-OFF in the digital watch. After three seconds in LIGHT-ON, the system automatically switched to LIGHT-OFF. The transition from ARMED to INERTIAL NAVIGATION mode is also *automatic*. But this one is not triggered by some predetermined (e.g., 3 second) timed interval; rather, it takes place when the guard becomes True. That is, it may stay in its current mode or switch depending on the two conditions (the distance between the aircraft's position and the route of flight). In a way, an automatic transition has a life of its own, switching and changing modes by itself. A guarded transition is the basic architecture of any automated system, and we shall therefore consider this structure very carefully in this book.

So who is involved in making an automatic transition take place? It's not only the pilot, is it? No, the transition happens without any direct user involvement; no buttons are pushed and no levers are moved. So who is it? For one, it is whoever wrote the software routine for the autopilot—this is where it all begins. And then there are the events themselves, some of which are only indirectly controlled by the pilot (such as flying toward the route) and some that are triggered by events that may be beyond the pilot's immediate control. More complex automated systems, with ever more sophisticated routines and capabilities, have authority to override the pilot's commands, and in some extreme cases the automated system will *not* give the pilot what he or she asked for.

And who bears the responsibility when something goes wrong? The legal responsibility of operating a system has resided traditionally with the user— the aircraft's captain in this case. But is that fair? The answer is not so clear-cut as it used to be in the many centuries of maritime tradition and one century of powered, human-controlled flight. And this indeed is a serious legal matter that is still pending in several courts of law, all involving cases of faulty user interaction with complex aircraft automation.

User Model

For now, let's leave this problem of responsibility and legality aside. What we need to remember is that there are two things that govern automatic transitions: the *guards*, which are conditions that are written by the software designer, and the *events* themselves. Therefore, to track and monitor what the autopilot is doing, the pilot must have a model in his or her head of specific modes, transitions, conditions, and events. And we recognize that this model may at times be incomplete and perhaps also inaccurate because of poor training and lousy manuals, or simply because, over time, the pilot might forget the minute details, especially the numbers. So let us assume, for the sake of this discussion, that the pilots of KAL 007 had an inaccurate user model about the behavior of this autopilot. Say they knew about going toward the flight route, and that the aircraft needed to be west of Anchorage, but they only knew that the aircraft must be *near* the flight route for the autopilot to transition to INERTIAL NAVIGATION. And let us quantify this by saying that "near" is a distance of up to 20 miles.

Our user model of figure 4.4 is quite similar to the machine model. There are three modes: HEADING, ARMED, INERTIAL-NAVIGATION. Transition from HEADING

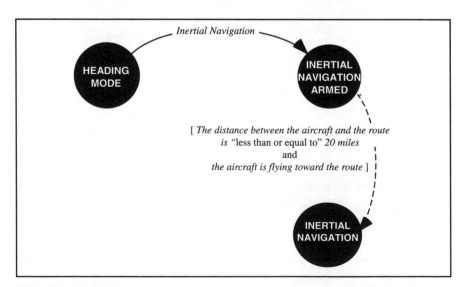

Figure 4.4. User model.

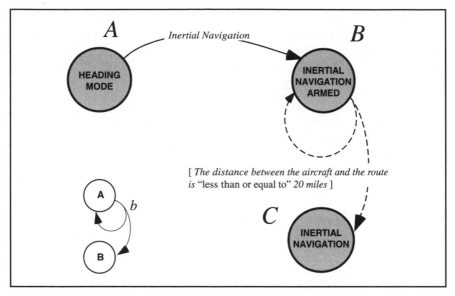

Figure 4.5. The composite model (and a non-deterministic structure in the insert).

to INERTIAL NAVIGATION ARMED is accomplished by rotating the mode selector. But according to this model, the transition from ARMED to INERTIAL NAVIGATION occurs when the aircraft's distance is equal to or less than 20 miles away from the route and the aircraft is flying toward it.

Now we have a machine model (figure 4.3) and a user model (figure 4.4). Sounds familiar? Just as we did with the remote control of chapter 2 and the watch in chapter 3, we combine the two models to create a composite model (figure 4.5). Quick observation: this composite model looks different from either the machine model or the user model of the autopilot. Let's look at the differences one at a time. First, note that the *aircraft is flying toward the route* condition is omitted. I did this on purpose, because we want to focus our attention on the distance condition only (which is where the problem resides). With that out of the way, you can see in figure 4.5 that there are two transitions out of INERTIAL NAVIGATION ARMED (state *B*): one is going to INERTIAL NAVIGATION (state *C*) and the other one loops back to state *B*.

Why?

We know for a fact that if the aircraft's distance from the route is up to 7.5 miles, the transition to INERTIAL NAVIGATION will take place. The pilot, however, thinks that if the aircraft is up to 20 miles from the route, the transition to INERTIAL NAVIGATION will take place. As long as the aircraft is up to 7.5 miles from the route, the pilot model will not cause any problem. However, if the aircraft is more than 7.5 miles from the route, the pilot thinks that the autopilot will transition to INERTIAL NAVIGATION—but actually the autopilot will stay in ARMED.

From the pilot's point of view, the autopilot is very confusing; sometimes the autopilot will transition to INERTIAL NAVIGATION and sometimes not. Think about it this way: say that we do repeated flights and try to engage the inertial navigation system. On the first flight, the distance between the aircraft and the route is 5 miles. On the second flight it's 10 miles, and on the third flight it's 15 miles. Our pilot, who has the "up to 20 miles" model, will get completely confused. The pilot expects that the transition to INERTIAL NAVIGATION will take place in all three cases, when in fact it will only engage on the first flight (5 miles).

The source of this problem, as you already know, is non-determinism. In the composite model (figure 4.5), the same event generates two different outcomes. The structure of the composite model is the same as the non-deterministic structure of the vending machine. You can see the generic structure at the lower left corner of the figure. It's the same old story.

Back to the Air

Now that we better understand the essence of automatic transitions, we can return to the tragic saga of KAL 007 and finish the story, although from here on we are navigating in uncharted territory, because what really happened in that cockpit will never be known completely. From the data recovered from the black box, we know that the crew engaged the autopilot two minutes and nine seconds after takeoff. It was initially in HEADING mode. Afterward, two possible sequences of events could have occurred: one, that the crew completely forgot to select the inertial navigation system; two, that the crew did select inertial navigation, but it never activated.

We'll follow the more probable scenario, the second one. The crew of Korean Air Lines Flight 007 most probably selected INERTIAL NAVIGATION on the mode selector panel once the aircraft was *near* the route. This, after all, was the standard operating procedure on every transoceanic flight. They anticipated that the transition to INERTIAL NAVIGATION mode would take place and the autopilot would fly route R-20 all the way to Seoul.

Based on radar plots recorded while the aircraft was in the vicinity of Anchorage, we now know that the actual distance between the aircraft and the flight route, at the time the pilots could have switched to INERTIAL NAVIGATION mode, was already greater than 7.5 miles (see figure 4.6). If that was indeed the case, the first condition (*the aircraft's distance from the route is* "equal or less than" *7.5 miles*), was evaluated False. As the aircraft started to deviate from the intended flight route, the distance between the aircraft and the route only grew and grew. Every second the guards evaluated the condition, and every time it

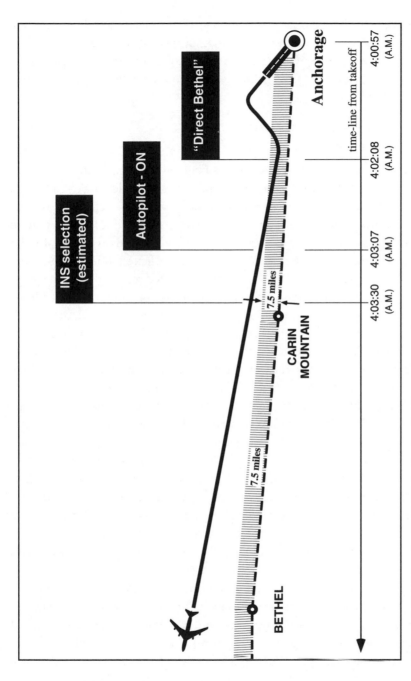

Figure 4-6. The flight path of KAL 007 shortly after takeoff.

came back False. The system was perpetually stuck in ARMED and the autopilot was on a straight, 245 magnetic heading, all the way to Siberia. The pilots probably had a very different mind-set. According to their model, the autopilot had indeed transitioned to INERTIAL NAVIGATION and was following transoceanic route R-20 to Korea.

Autopilot Displays

But why did the crew of Korean Air Lines Flight 007, sitting in the cockpit for five long hours, fail to detect that the autopilot was not following the intended flight route and that it wasn't in INERTIAL NAVIGATION mode? Fatigue, boredom, complacency? Many factors may have come into play, but it is a fact that on their way home, the crew was weary and tired after a long five-day trip. For centuries, captains and navigators have always feared homeward-bound carelessness on the final leg of a long and protracted journey. But lest we get lost in speculations, let's stay on our primary topic, pilot interaction with the autopilot. You see, the only hope of recovering from such a divergence between what the mind imagines and what the autopilot is actually doing is by getting some kind of feedback from the machine about its active mode—something that will alert the pilot that there is a problem. Were there indications about the active mode in the cockpit? Yes and No.

The inertial navigation system had a dedicated display to show the pilot the active mode. There is an indicator labeled INS. If the inertial navigation system is ARMED, the indicator lights up with an amber color; if the inertial navigation is engaged, the indicator is green. This is the "Yes" part of our answer. However, there was no indication anywhere in the cockpit that the autopilot was in HEADING mode. Why? It's hard to tell, but we are talking here about the early generation of automated flight guidance systems, when the importance of human factors was only remotely considered, if at all. Fact: when the autopilot was in INERTIAL NAVIGATION ARMED, there was no indication that the autopilot was actually maintaining HEADING. This is the "No" part of our answer.

Did this lack of mode indication play a role here? Most probably, yes. Did the pilots mistake INERTIAL NAVIGATION ARMED for INERTIAL NAVIGATION engaged? Perhaps; after all, it's the same indicator, and who knows what were the exact lighting conditions in the cockpit, and we all know how the mind sometimes sees what it wants to see. Did the pilots understand the subtle difference between these two modes? Did they know about guards and triggering events and how automatic transitions take place? Maybe, but we'll never know.

What we do know is that the lack of indication about the autopilot's active mode deprived the crew of an important cue. Such a cue might have drawn their attention to the fact that the INERTIAL NAVIGATION was *not* engaged and that the aircraft was actually flying on HEADING mode. Following the accident, all autopilots were modified to include this information. In today's autopilot systems, the display also shows the currently engaged mode as well as the armed mode; this is now a certification requirement. But this, as is always the case in accidents, is the blessing of hindsight. This design change came too late to help the crew and passengers of Flight 007.

Conclusion

The story behind Korean Air Lines Flight 007 was one of the greatest mysteries of the Cold War era (see endnotes). It is unnerving to consider that the sequence of actions and coincidences that finally led to the chilling shoot-down of a commercial aircraft in which 269 innocent people died, began with something that *didn't* happen—a tiny mode transition from HEADING to INERTIAL NAVIGATION. The autopilot was probably stuck in INERTIAL NAVIGATION ARMED mode, and that prompted the deviation over the initial waypoint (*Bethel*) that, for whatever reason, was not queried by Anchorage air traffic controllers. Then we have the similarity, at least initially, between the aircraft heading and the intended flight route. That heading, 245 degrees, took the aircraft toward the orbiting military RC-135 plane, which, coincidentally, was leaving the Soviet radar screen just as the 747 was coming in. Flight 007 was first labeled as *unidentified* and then, when it crossed into Soviet territory, it was designated as a *military* and then *combat* target because of the stringent rules of engagement that the Soviets employed at the height of the Cold War. The early morning darkness, which may have prevented any real visual identification of the white aircraft with the red Korean Air Lines logo on its large tail as an innocent civilian airliner, was another factor in this complex tragedy. Finally, what sealed the fate of KAL 007 was the climb from 33,000 feet to 35,000 feet, just after the fighter aircraft fired a warning burst in order to force the airplane to descend and land on Sakhalin. The climb was mistaken for an evasive maneuver and the aircraft was shot out of the sky and crashed into the sea. And all for the want of a little automatic transition.

Part Two

Characteristics of Machines and Principles of User Interaction

Chapter 5

Characteristics of Machines and Frugal Air-Conditioners

P art one of this book (chapters 1-4) laid the groundwork for understanding how humans and machines interact, what kind of problems emerge, and how it is possible to understand and identify some of these problems. We have seen how time delays may confuse users. We have seen how the underlying structure of machines, where the same event can lead to several outcomes, produces errors. I introduced the concept of abstraction and we have discussed examples (e.g., the diagram of the London Underground) that highlight the use and benefits of this concept for interface design. The basic concept of non-determinism was introduced and we have come to see its problematic implication (e.g., in the digital watch) for interface design. Hopefully, you began to intuitively feel and realize that there is a very fine balance between simplification of the interface (by abstracting-out superfluous information) and over-simplification of interface to the point that it becomes non-deterministic. This is a key issue for interface design, and we will expand on it in later chapters. But most of all, chapters 1-4 gave us a foundation to organize the way we describe user-machine interaction (by using a *machine model, user model,* and *composite model*) and, given what we already know about abstraction and non-deterministic structures, identify potential design problems.

In part two of this book, we will use all that we have learned so far to describe and discuss many principles that govern efficient (as well as problematic) human-machine interaction. We will focus our attention on consumer electronics and identify the kind of user-interface design problems embedded in many "everyday" devices. But before we move on, we need to discuss three important characteristics of machines that will help us to better understand automated devices and user interaction with them.

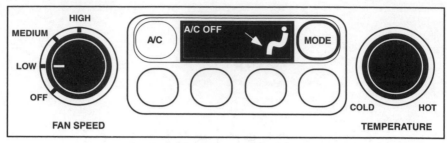

Figure 5.1 (a). The interface to the climate control system.

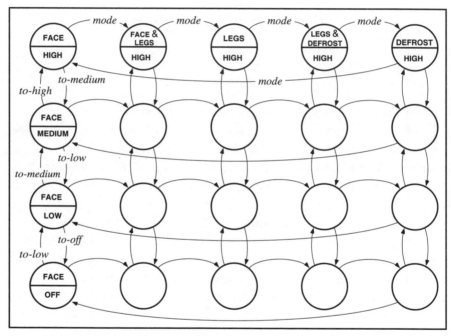

Figure 5.1 (b). Model of the climate control system

Concurrency

Remember the climate control system from chapter 3? You can see it in figure 5.1(a), and all its 20 separate configurations and transitions in figure 5.1(b). It is really a rather simple system; we only had five modes and four fan settings, but look at the muddle of bubbles and transition lines in the figure. The mess, which arises from the multitude of configurations needed to describe the system, hinders our ability to understand the machine, let alone evaluate user interaction with it.

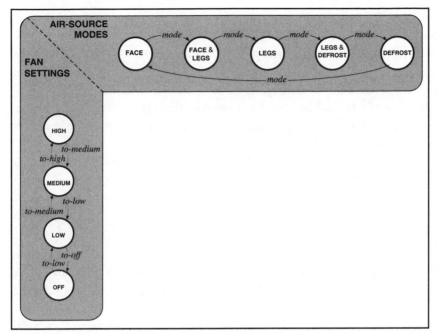

Figure 5.2 (a). Abbreviated model of the climate control

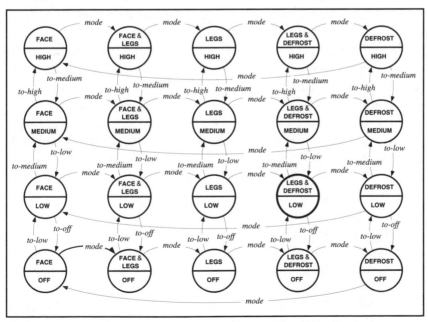

Figure 5.2 (b). Full model of the climate control system (configurations and transitions).

So, the first concept that we will explore here is concurrency, which will allow us to reduce much of the complexity in the description. Concurrency means that several things are happening together in the same time and in the same space. Look at it this way: we all live and work, concurrently, with other people, and we all do more than one task at the same time—driving and using a cell phone is an all too common example, though maybe not the safest. Concurrency not only exists in human activities—machines and computers also run concurrently.

Back to our climate system. When we think about it with the idea of concurrency in mind, we quickly realize that this system has two concurrent components. The fan setting is one component and the air source mode is another. There are four *fan settings* and five *air source* modes. Is there a way to capture this concurrency in the way we describe the system?

Here is one way. Note that in figure 5.1(b) the names of most of the configurations as well as the events are omitted. Does this matter? Actually, we have been doing this all along as a way of reducing unnecessary clutter in the figures of our climate control system. We could get away with it, because it is possible to use the names of the configurations in the upper row and left-most column to determine what should be the name of every configuration. Along the same lines, we can also get away with not naming many of the events. So here is a radical suggestion: why not just keep the upper row and left column, and do away with the muddle in the middle?

Figure 5.2(a) shows such an abbreviated description (in gray) of the climate control system; but just to be sure, the full description is provided in figure 5.2(b). So let's employ it to check and see if the abbreviated description can be used to describe all the configurations and transitions in the full model. On the checkerboard of figure 5.2 we start with the left index finger on fan and say it's OFF. The right hand is on the modes, and the first one is FACE. The corresponding configuration in the full model is FACE and OFF. Now we change to "low fan" by sliding the left finger up to LOW. The corresponding configuration in figure 5.2(b) is FACE and LOW. Now we slide to FACE & LEGS, and then to LEGS with our right finger (but our left index finger is still on LOW). The current configuration is now LEGS and LOW. By moving your two fingers along the two side components, one finger on the fan and the other on the modes, you travel, like a faithful pilgrim, along the corresponding states and transitions of the full model.

Now it's time to let the full description go. Figure 5.3 is an abbreviated description of the climate control system. The broken line indicates that there exist concurrency between the FAN settings and the AIR SOURCE modes. Note how simple and clear is the description. Moreover, the savings in size are immense. Think about it: there were (4 x 5) = 20 states in the original model and only (4 +5) = 9 states in the abbreviated model. We had 50 transition arcs

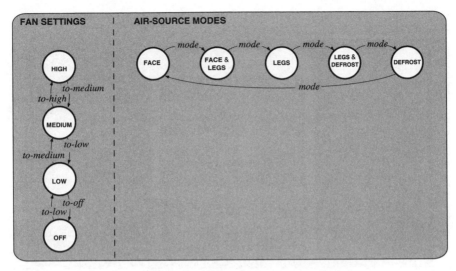

Figure 5.3. Abbreviated model of the climate control system

in the full dress-up description and only 11 in the abbreviated description. (See the endnotes of this chapter for a short discussion and reference to the *Statecharts* description language.)

Here is another example: Say that we had two computers that control a robotic arm. One computer (A), controls the vertical axis and the other (B) the horizontal. In computer A we need to describe 40 states, and in computer B we describe 50 states. For the full description we will end up with (40 x 50) = 2,000 states altogether. However, when we use the abbreviated description, the number of configurations would be only (40 + 50) = 90. 2000 or 90—a magnitude of more than 10! So we are happy with the concept of concurrency because it can really make things simpler. Now let us consider the concept of hierarchy.

Hierarchy

If you look carefully at any man-made system, you will see a hierarchy in the way it is organized and built. As an example, consider the air-conditioner unit detailed in figure 5.4(a). The unit has several modes and a simple display. The user interface is in figure 5.4(b): there are "on" and "off" buttons, a mode switch, buttons for switching between "inside" or "outside" air source, and two buttons for setting the desired temperature. Let's start with the modes first.

The behavior of the unit is shown in figure 5.5(a). We start at IDLE, which is the initial state (indicated by the little arrow). Press the "on" button, and the

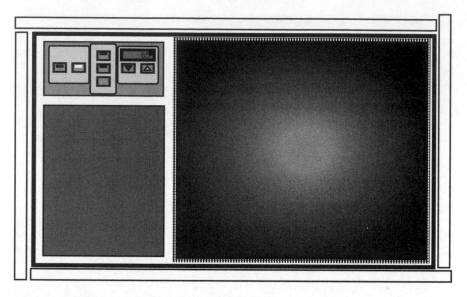

Figure 5.4 (a). The air-conditioning unit.

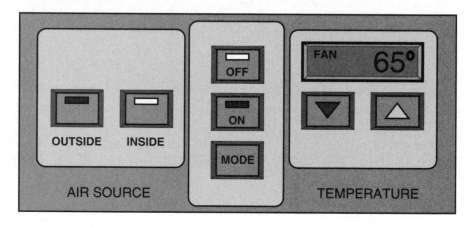

Figure 5.4 (b). The unit's control panel.

unit's fan starts to spin. Press the "mode" button, and you are in AIR CONDITION-1 mode. If it's not cool enough, press "mode" again, and now it's AIR CONDITION-2. If you hit the "mode" button another time, you reach HEAT. Mode it again, and you are back in FAN mode. Finally, note that the *off* event triggers a departure from all modes back to the IDLE state.

Now look at figure 5.5(a) from a distance. Do you see a certain repetition with respect to all the *off* transitions into IDLE? We already know that the event

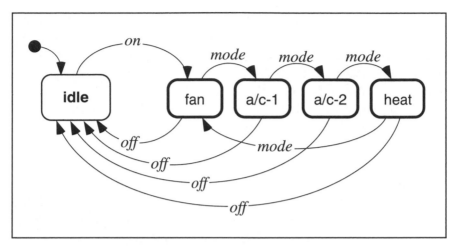

Figure 5.5 (a). Full description of the unit's modes.

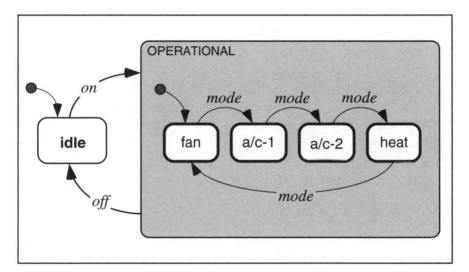

Figure 5.5 (b). Abbreviated description of the unit's modes.

off works the same for all four operational modes. Therefore, one way to simplify the description is to cluster all four modes into a newly created mother state that we shall call OPERATIONAL. You can see this in figure 5.5(b). Once we established this overarching super state, it is possible to substitute the four *off* transition lines with a single outgoing transition (from OPERATIONAL to IDLE). This *off* transition comes out of the boundary of the super state, and means that regardless of which mode we are currently in, an *off* event always takes us back to IDLE.

We have created a hierarchy. OPERATIONAL is the maternal super state and embedded in it are four baby modes (FAN, AIR CONDITION-1, AIR CONDITION-2, and HEAT). But the embedded states are not all that equal—FAN is always the start, or initial, mode when we enter OPERATIONAL. This is what the little arrow pointing at FAN tells us. And with that, all the behavioral particularities of the system in figure 5.5(a) are preserved in the abbreviated description of figure 5.5(b).

So it is evident that the usefulness of employing concurrency and hierarchy provide for a description that is more concise and compact. But this is not the only reason. By observing and capturing the subtle relationships between states and events in the system, we are able to reveal the underlying structure of the system and its modes, and you can readily see this in figure 5.5(b). As we shall see soon, this way of revealing the underlying structure is quite a benefit when it comes to evaluation and design of user interaction with automated machines.

That said, it is important to note that hierarchy is an innate structure for humans. It is one of the prevailing methods for organizing information and it is also entrenched in the way we think, design, and build systems. Military and corporate organizations are obvious examples, but look around you and you will see hierarchical structures almost everywhere: in the menu arrangement of your word processor, in the arrangement of modes and functions on your cell phone, in the way you move through a voice-activated system, and in many other devices that you use every day. It is a human trait to find comfort in hierarchy.

Synchronization

As mentioned earlier, almost all modern machines contain several components that work concurrently. We have two components in our air-conditioner: the modes, which we already discussed, and the air source setting. There are two "air source" states in our unit, OUTSIDE and INSIDE, which determine whether the air comes from the outside or only circulates inside. You select "outside" or "inside" air by pressing a button on the interface (see figure 5.4[b]).

So consider this scene: you are in a small roadside motel, by the ocean. It has been a long, hot, and humid night. You wake up early in the morning with a headache. The air conditioner has been working dutifully all night. The room is cool but the air is extremely stale. You wonder about this because the air outside is fresh and a gentle sea breeze is blowing. With one small window that does not open, this air conditioner unit is your only source of fresh air, but nothing enters to revive the room. Something went wrong for sure.

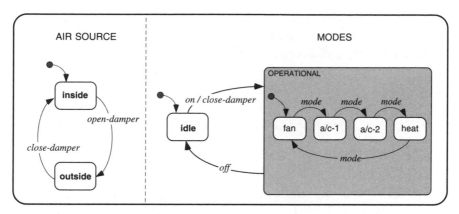

Figure 5.6. Air source and modes.

You bend down and gaze at the "outside" and "inside" buttons. The "outside" button opens a damper that allows fresh air from the outside to enter in; the "inside" button closes the damper. And, indeed, the unit is set on "inside" air. Ah; you hit the "outside" button, inhaling from the vents, making sure that it is really fresh. But wait a minute—you vividly recall that you set it to "outside" the previous night just before you turned on the air-conditioner.

So why did the setting change?

You sit on the floor by the unit, pressing the buttons and switching the air source from OUTSIDE to INSIDE and vice versa (see figure 5.6), trying to figure out what's up with this system. Everything looks fine. You start playing with the modes and turning the unit on and off—and then you notice the trick: the damper closes up immediately after the air conditioner is turned on. It does this automatically every time you press the "on" button. Therefore, you always end up with "inside" air. Now you can always push the "outside" button to get fresh air, nobody is limiting freedom here, but you have to remember to do it.

But why is all this trickery necessary in the first place? Why should the damper close automatically when you turn the machine on? Well, it has to do with the "bottom line." Inside air is cheaper to treat, because it takes less time for the air conditioner (or the heater) to reach and maintain the selected temperature, and this of course is the management's preference. The quality of the inside air is not a concern so long as the electric bill is low.

You can see all this in figure 5.6, where the event *on* in the mode component triggers the event *close-damper* in the air source component. This synchronized link between the two components is written like this

on / close-damper

where the left-most event (*on*) triggers the following event (*close-damper*).

Now that we understand how the trick works let's see how it plays out in figure 5.6. The initial states are IDLE and OUTSIDE. Now you can play with the air source control as much as you want; inside, outside—whatever. But as soon as we turn the unit on and transition from IDLE to OPERATIONAL, a signal is sent to close the damper. If the damper is already set on INSIDE, the event *close-damper* falls on deaf ears. But if the damper is set to OUTSIDE, then of course, transition *close-damper* takes place, and now the air conditioner only circulates inside air. That's how this little frugal trick is done.

The point of this discussion, as you can imagine, is beyond motel air conditioners and proprietors' frugality. It is about how multiple components that make up a system work together. In every computer system, certain events (e.g., *close-damper*) that occur in one component get picked up by other components. These events synchronize the system and provide the glue that makes it all come together. Synchronization is an important characteristic of any machine, computer, and automated system.

From the user's perspective, these synchronized events are called *side-effects* (see also chapter 3). Side effects can be extremely helpful for the user (minimizing button pushing and other interaction), yet they can be sometimes elusive. And the reason for that is because they "appear" to the user as automatic—coming out of nowhere. Yet as we know, side effect events are synchronized and triggered by an event somewhere else in the system; they are not automatic per se. We will see more examples of side effects in chapter 6 and also in chapter 14.

Chapter 6

Cordless Phones and Population Stereotypes

W e remain in the motel by the sea where the climate is under control and the inside is now fresh and cool. We now direct our attention to the bowl-like light fixture on the wall. There are two switches for this light: a switch by the door and a switch by the bed. This is the all too familiar dual light switch arrangement—two switches that operate the same light fixture. This arrangement confuses many, and we shall use it as a stage to consider the kinds of problems that exist in concurrent systems. You can see the layout of the motel room with the light fixture and the two switches in figure 6.1.

Dual Light Switch

To consider the dual light switch system, we first need to understand how it works. So let us go beyond appearances and abstract away the context and all the architectural details. Figure 6.2 is a stripped-down version of the two switches and the light fixture. We start with the light fixture. It has only two states—either it's ON or OFF. Next are the light switches. The light switch can be either in the DOWN or UP position. The transition from DOWN to UP takes place because of an event, *flip-up*, which is triggered by the user. Once in UP, the event *flip-down* takes us back to DOWN. When the event *flip-up* or *flip-down* takes place, it generates another event called *change*. And this change is what makes the light transition from ON to OFF. So there you have it—the behavior of the switches and their internal mechanism, standing naked in front of you.

The initial configuration of the light system is that both switches are in the DOWN position and the light fixture is OFF. This is how the electrician that installed the switches rigged the system. In figure 6.2 we denote the initial configuration with a little arrow. But don't take this business of an initial

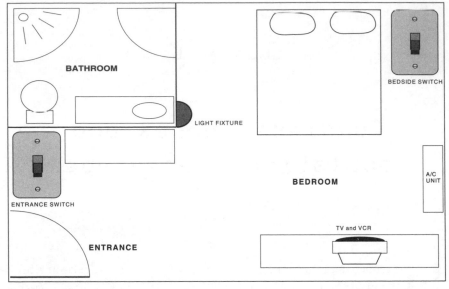

Figure 6.1. The motel room.

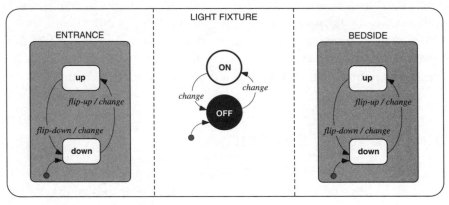

Figure 6.2. Model of the dual light switch system.

configuration lightly, because as we will see soon it is a very important factor and reveals much about the potential confusion for user interaction.

Night Scenario

Now that we understand how the dual switch system works, let's observe user interaction with it. You walk into your room late at night and it is completely dark. If you look at figure 6.3, the corresponding configuration is *A*; everything is down and the light is off. You enter the room, *flip-up* the switch by the door, and the light comes on. This is configuration *B* in figure 6.3. You do what you

CONFIGURATIONS entrance bedside	ENTRANCE SWITCH	BEDSIDE SWITCH	THE LIGHT IS
A	down	down	OFF
B	up	down	ON
C	up	up	OFF
D	down	up	ON

Figure 6.3. Table of configurations.

have to do, roll into bed, and then *flip-up* the bedroom switch to turn off the light and go to sleep. Now both switches are UP, the fixture is off, and the room is dark. This is configuration *C*.

Hours later, you wake up in the middle of the night. You stumble to find the bathroom door and end up in front of the entrance door. Your meandering hand finds the switch, *flips* it *down*, and light up the system in configuration four *D*. After finding your way to the bathroom and back to bed, you roll to the side, *flip-down* the switch, and fall asleep. Now both the entrance switch and the bedside switch are in the DOWN position and the room is dark again. We are back, in a circular way, to our initial configuration *A*.

User Model

By now we have a description of the machine (figure 6.2) and we know all the possible user interactions (figure 6.3). The next step is to explore what the user needs to know in order to operate this system effectively. For starters, we can ask our motel patrons to memorize the table of figure 6.3 so that they will be able to work the lights properly. By noting the status of both light switches, and recalling the table, any guest should be able to predict the consequence of flipping a switch, always knowing whether the lights "should" be on or off— even if the light bulb is broken or burned out. But this solution, and I'm sure you will agree, is a bit too demanding.

Or perhaps there is something more simple. Take a look at figure 6.3 for a hint. Note that in the dark (configurations *A* and *C*), the switches are in the same position (either DOWN & DOWN or UP & UP). For the light to come on, the switches must be in opposition (either UP & DOWN or DOWN & UP) as in configuration *B* and *D*. Would this revelation help us simplify the operation of the system?

It sure would. We can summarize our findings, and inscribe them on the wall by the switch, as follows:

> If both switches are UP or both are DOWN, the light is OFF. Otherwise, the light is ON.

This is indeed a simpler user model, but it leaves us with two little problems. The first is that the user must be able to view *all* the light switches and their respective positions in order to use the inscription. The second problem, which I will elaborate on in the endnotes for this chapter, is, What if we had three switches—or perhaps four, five, or more?

Is there a better way?

From the description of the system and also from our collective experience, we all know that all it takes to change the light from ON to OFF (or vice versa) is a flip of the switch. True. And with that in mind, the instruction on the wall can read:

> "To turn the Light OFF or ON, flip any switch"

This is by far most simple instruction yet. But note what we just did here—and be aware of the consequences. We incorporated the interface, the light fixture, into the user model. When we were dealing with the two previous instructions, we did not care about the interface; we could just rely on switch position. Tell me the switch configuration and I can tell you if there *should* be light or not. But in the case of the "flip any switch" instruction, the rule holds true only as long as the interface is fully reliable. If the bulb is burned out or otherwise defunct, the rule collapses immediately. Agree?

Generalizations and Rules of Thumb

This business of observing patterns and making rules of thumb is something that we all do. Mariners, for example, have more than a few. "Red sky in the morning, sailors take warning; red sky at night, sailors' delight" is a well-known rule of thumb. But in making such simplifications and generalizations we must proceed with caution, because they usually invite exceptions, meteorological errors of sorts that sometimes render the rule inadequate. The mariners rule, for example, works only in the mid–Northern Hemisphere. It fails miserably if you sail in the Southern Hemisphere. Likewise, the simple statement about the dual switch system works only if the light fixture is functional and is not burned out or defunct.

Whether we like it or not, we all tend to create such simplistic and general explanations to deal with things that are beyond us. With respect to the dual switch, if you ask people to explain to you how it works, you will get a wide range of explanations. Try it. A common one is that there is some central controller that sets the switches. Actually, and we mentioned it earlier, it is all in the initial configuration that is set by the electrician who wired the switches in the room. Once that initialization is done, flipping switches will always follow the same pattern. And if our loyal electrician had initially rigged the system differently, i.e., DOWN & UP is dark, then, of course, UP & UP and DOWN & DOWN would always produce light. Fortunately, it is a common practice that the initial configuration of switch positions is DOWN & DOWN (although you may find other rigging in older homes). Another folk tale is that there is a master-slave relationship between the switches. That is, one switch location is the master and it always works UP for light and DOWN for dark, and the other is the slave and just works to change the current setting. False again.

Population Stereotypes

And if we are already in the business of folklore, let us venture a bit into the psychology of user interaction with the dual light switch. Most people are confused by the arrangement and find it annoying at times. Why? One important factor is the lack of consistency in the direction of the flipping: In one situation flipping up the entrance switch will turn the light on (from configuration A to B in figure 6.3), and in another situation a flip-up will turn the light off (from configuration D to C). But since we usually expect that flipping up a switch turns the light on (and not off), we get confused sometimes.

You see, we all have an internal model of how things ought to work—a fixed and conventional set of expectations telling us what our interaction with a machine "should" result in. We expect that when we flip-up the switch the light will come on. And indeed, some of these expectations are so uniform as to constitute a "stereotype." "Population stereotype" is the term to describe these expectations. When it comes to light switches, the world as we know it is split in two. Europeans, for example, will not find flipping down a switch, and seeing the light, a violation of their "stereotypical" expectations. To Europeans, this is *the* way light switches "should" and are expected to work. (Nevertheless, when it comes to the dual-light switch, the setup described in figure 6.3 breeds confusion for European and North American populations insofar as it contains *both* stereotypes.)

Reach Out and Call Someone

Let us leave the lights behind and turn our attention to telephones. In our room by the sea there is a phone—a rather popular cordless phone made by a reputable manufacturer of consumer electronics (see figure 6.4). The phone consists of two units: a base and a handset. The base is a cradle for the handset. And the handset itself is not different in any way from any other cordless or cellular phone that you are familiar with. A large 'talk' button, a tiny light indicator telling you when the phone is on, and some 12 keys are the main features here.

As in most phones, the user presses the "talk" button to activate the phone (either for answering incoming calls or for dialing an outgoing call), but this phone has an additional feature: when the handset is on the base and an incoming call arrives, the listener lifts up the handset from the base and the handset is automatically turned on. The little red light is on and the phone is ready for use. There is no need to press the "talk" button. But when the handset is off-base, you *must* press the "talk" button to listen and talk. If you have ever used such a phone, you know very well that it breeds confusion and results in errors. Let's find out why.

Machine and Interface

To understand the problem, we first need to describe how this device works. We begin with the machine: There are three components that we care about (see figure 6.5). The PHONE BEHAVIOR contains two main states, OFF and ON—the phone can be off (and standing by for an incoming call) or on and actively connected. That's all we need to know about the phone's behavior. The interface contains a little indicator light, telling you when the phone is active; and then there is the talk button, our agent of change.

The talk button is initially in IDLE, doing nothing. Pressing the "talk" button generates an event called *change* (and when you let go of the talk button, it springs back to IDLE). The event *change* is picked-up in the PHONE BEHAVIOR component and changes the state there from OFF to ON (or vice versa). It's the same thing as the lowly light in the motel room: flip a switch and you change the state of the light. But here we also have a dedicated interface; a little red light beacon, informing us when the phone enters ON mode and turning dark when the phone enters OFF mode.

Figure 6.4. Line drawing of the cordless phone.

Note that in the PHONE BEHAVIOR component of figure 6.5, the transition from OFF to ON has two elements to it (*change* or *lift-handset*), telling us that there are two ways to get ON: One way is by pressing the "talk" button and generating a *change* from OFF to ON. The other is by lifting the handset from its base (*lift-handset*). The "or" tells us that either way will work.

The phone begins to ring. You rush to answer. You reach for the cradle, pick up the handset, press the "talk" button, and listen

—to a long dial tone!

Guess what happened? You just hung up on the caller!

Analysis of User Interaction

Let us analyze this quite frustrating scenario so that we can understand its roots and not blame and shame ourselves for being dozy and inattentive. Look at figure 6.5 again. If we are walking with the phone in hand, away from the cradle, we can transition from OFF to ON and then back to OFF and so on by repeatedly pushing the "talk" button. The event *change* triggers these circular transitions, and there is no hanging up on people.

The problematic situation arises only when the handset is sitting on the cradle and the phone rings. By lifting the phone, you generate a (side effect) transition to ON (with event *lift-handset*). And if you press "talk" at this point, you trigger an immediate transition back to OFF. In blunt words, you just hung up on the caller. You see, pressing on the "talk" button leads to two different outcomes—OFF or ON—depending on the previous state of the phone. And if

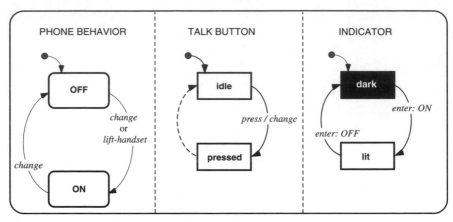

Figure 6.5. The phone, talk button, and indicator.

you can't discern between them, then you will always be surprised with the outcome. The source of the problem is that there are two different initial states here when it comes to OFF: "OFF on the cradle" and "OFF away from the cradle." To the user, however, after picking up the handset, it all seems the same.

But wait a minute, isn't there information on the phone itself to help the user? True, there is an indicator, albeit small, which signals that the phone is indeed ON after we lift the phone from the cradle. Another source of information is the place from which the handset was picked up (on-base or off-base). But unfortunately, it turns out that even users who have read the manual, can perfectly see the little red indicator, or have already been through this embarrassment before, still tend to hang up on their friends and loved ones.

Why?

Capture Errors

Remember those population stereotypes we mentioned earlier? There is a population stereotype, among all of us who use phones, to press the "talk" button to talk. The sequence of only lift and simply talk is indeed a violation of this convention. But there is something more general than just the cordless phone here. Think about driving for a moment: How many times has it happened to you that you found yourself at the entrance to your work place, when in fact you only wanted to go to a nearby establishment—a restaurant, a movie theater, or a store? Or conversely, the situation where you promised to stop along the way and get some milk but ended up at home empty-handed?

These embarrassing errors occur when a familiar and well-rehearsed sequence of actions competes against a unique sequence. It usually involves a common set of activities (for example, getting in the car, driving on the same freeway) and then a branching point. And at the branching point, we end up taking the exit toward work instead of the exit to the movie theater. The same happens with our phone. There is a familiar and well-rehearsed sequence of lifting the handset, and then a branching point with two pathways: one is the "press talk to talk," the other is simply "talk." Once we hear the ring and lift the phone, we easily sway toward the "press talk to talk" sequence. It is such a familiar sequence that it simply takes over.

This well-practiced sequence of actions that takes over, captures us, and prevents us from doing what we really wanted to do is called a capture error. There is also some, but not conclusive, evidence to suggest that the more "stressed" the user is, the more likely he or she is to be taken over by such capture errors. But please remember, these sequences are not wired into our brains and there is nothing in our genes about it. It is simply based on our deeply rooted experience with cordless and cell phones, just like our routine day-to-day experience of driving to work.

Final Notes

The problems we uncovered here go one step beyond pure interface issues, because technically speaking the interface here is just fine. There is a light to indicate the current state of the machine (OFF, ON), and of course the user can always see where the phone is lifted from. The real problem is a combination of the "lift to talk" sequence on the machine side and the population stereotype and resulting capture error on the human side. The moral here is that just having a correct interface, one that provides all the necessary information, is not enough. The design of the interaction between the user and the interface must also be suitable—taking into account users' expectations, population stereotypes, and other "human" factors.

It is clear that designers of phones and other devices should take advantage of these population stereotypes instead of violating them. If you survey other cordless phones on the market, you will find that several manufacturers have attempted to improve on this design. Here is one manufacturer's solution: after the phone is lifted from the cradle, the "talk" button is disabled for three seconds. In this design, the problem we uncovered here is eliminated (if you mistakenly hit the button within three seconds). But it introduces a timed transition that is hidden from the user and not mentioned in the manual. So guess what happens when you wish to hang up immediately on those annoying prerecorded sale pitches. . . .

Chapter 7

VCRs and Consistency Problems

The Video Cassette Recorder

B ack in the motel, we are now ready to deal with the video cassette recorder (VCR). Let us go together and explore this human-VCR relationship, something that has become the epitome of inefficient and at times frustrating user interaction. We want to understand what exactly are the problems here and why many of us shy away from using many of its modes and functions.

The VCR in front of you is manufactured by a well-known electronics company and found in many hotels and motels (figure 7.1). Oddly enough, looking at the VCR, nesting on top of the TV box, you note that the usual display, the necessity for programming the VCR, is absent; only a small "on/off" button, and the usual "play," "fast-forward," and "rewind" buttons

Figure 7.1. TV and VCR.

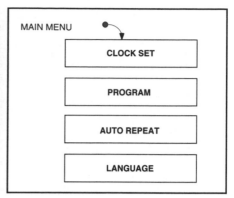

Figure 7.2 (a). TV and menu screen.

Figure 7.2 (b). Model of the main menu.

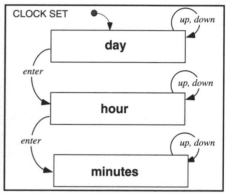

Figure 7.3 (a). Clock set screen.

Figure 7.3 (b). Model of CLOCK SET sub-menu.

are present. So how do you set up the VCR? How do you program it to record your favorite show?

In this VCR, everything, you soon realize, is done from the remote control and the TV screen. To set things like time or language, and to program for a future recording, you must first hit the "menu" button. And then the menu screen pops up, informing you that you can set and program the VCR along the four menu options in figure 7.2(a).

Sub-menus

There are four different settings on this menu: CLOCK SET, PROGRAM, AUTO REPEAT, and LANGUAGE. Clock set is the initial state as you can see in figure 7.2(b). On the remote control, just below the menu button lies another button called "enter." Press it, and you are in the clock set sub-menu itself (figure 7.3[a]). Press the "up arrow" or the "down arrow" buttons on the remote to change the

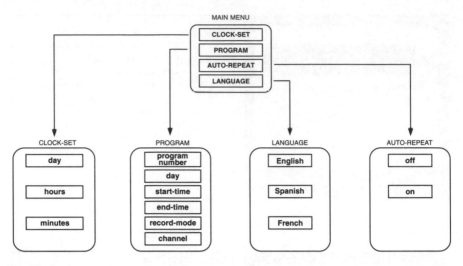

Figure 7.4. Hierarchy in the menu system.

day, "enter" to set the hours, and then another "enter" for minutes. Nothing new under the sun here, and this does start to read like user-manual language. But please hang on, because we will soon go beyond the placid appearance to discover hidden insights beneath the surface of this everyday device.

So let's continue with the sub-menus (figure 7.4): PROGRAM is where we set the VCR to record future shows, a task that many of us shy away from. But if we have to, we first assign a "program number" for the show we want to tape, set the "day" for recording, the "start time" and "end time" of the show, and so on. AUTO REPEAT tells the VCR to repeat this happening on a weekly basis, and LANGUAGE makes the entire escapade also appear in Spanish or French. So far so good.

Now let us focus on transitions going into and coming out of the four sub-menus, because this is where the problems sneak in. In figure 7.5, we have the entire system in front of our eyes. We already know that event "enter" takes us into the sub-menus—but getting out is a different story, as we shall soon see. So here we go: while in CLOCK SET and PROGRAM, "enter" takes us out of the menu screen into TV, while "menu" takes us back to the main menu. CLOCK SET and PROGRAM are identical in terms of sequences of interaction and we therefore cluster them together in a gray super-state.

Now watch what happens when we are in the LANGUAGE and AUTO REPEAT sub-menus. Here "enter" scrolls you inside the sub-menu, and "menu" (contrary to what the name implies) takes you out to the TV screen! Why? Well, we will get into this in a minute. But how do we get back to the menu screen if you need to "program" or set the clock? Well, you would have to hit "menu" again, and start from the beginning.

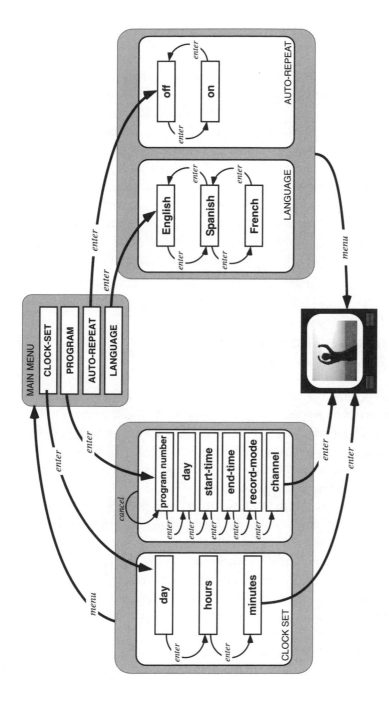

Figure 7.5. Description of the menu system (including transitions).

Consistency

What we see here is a problem in consistency among the sub-menus: "enter" takes us out to TV in CLOCK SET and PROGRAM, while in AUTO REPEAT and LANGUAGE it scrolls us indefinitely inside the sub-menu. Next we take issue with "menu." In CLOCK SET and PROGRAM, "menu" takes us back to the main menu, but in AUTO REPEAT and LANGUAGE, the "menu" works contrary to its name, and takes us completely out of menu and into the TV (or VCR-playing) screen. And if you look carefully in figure 7.5, you will note that while the "cancel" button in PROGRAM indeed cancels the setting, in all other sub-menus it does nothing.

To the casual user, all these inconsistencies are confusing. The user assumes that all sub-menus work the same. Since they are all nested under the same main menu, that's what we all expect. But reality is different. Arbitrary and inconsistent design decisions like this one make it difficult for us to understand, let alone remember, how to work a system.

Standards for User Interaction Design

What is desperately needed in the many devices that we all encounter, is some standardization of interaction. These standards should be intuitive and appealing—taking advantage of prevailing population stereotypes—and of course creating new populations stereotypes for the future. We want one way to go *down* the hierarchy and another way to go *up*. When the design of the VCR menu mixes the two, it confuses us.

The topic of standardization of user interaction is many times overlooked by manufacturers of devices and machines, sometimes only for the sake of providing us with a unique and "novel" design—something that makes their own design "different" from the competitors'. While this may be arguable for unique functionality and behavior, it defeats the purpose when it comes to the mechanics of how we interact with the devices. Why is it that we have to learn and relearn how to go up and down a menu in VCRs, cell phones, and almost every different device we use?

Consider automobiles for example. Display icons and indications (such as for fuel and low oil pressure) are standard across all manufacturers. It makes it easy to work the car, especially a rental, and know exactly what each indicator means. But still, you find some cars where the headlights switch is on the dashboard, and in others it is part of the turn signal on the steering wheel. The same is true of location and settings of wiper controls. Among American cars you will find many differences in the location and arrangements of controls and settings. And it should not come as a surprise to you that many accidents involving rental cars occur inside or just outside the rental-car lot. But check out different Japanese cars, and you will quickly note that their controls are very similar in arrangement, and that this standardization is across the board.

So why was this VCR, made by a reputable manufacturer, designed in this way? Perhaps it was a way to shorten the sequence of button presses or a way to eliminate an extra button. But what it actually does to users is to increase confusion and incite frustration. And indeed, such inconsistency throws many of us off, to the point of avoiding any VCR setups and programming altogether.

Can you see a solution to this problem? Start by looking carefully at figure 7.5 and I'm sure you can come up with a few. The goal is to eliminate inconsistency in entering, cycling within, and exiting a sub-menu. Here is one solution: change the interaction such that "enter" only scrolls us within the sub-menu and "menu" always takes us back to the main-menu screen, and another button (or an additional menu item) takes us out of the menu screens.

Universal Stereotypes

But we are not done yet, because there is an additional problem with this device—one that has to do with violation of a very common population stereotype. Look at figure 7.6(a). It is a description of the step transitions in the MAIN MENU screen. To move the cursor *down* the screen to PROGRAM, AUTO REPEAT, or LANGUAGE you need to press the . . . "up" arrow. And yes, you got it right, to scroll up you need to use the "down" arrow button. Up-arrow for down movement and down-arrow for up is a clear violation of a rather strong population stereotype. (You can visually see this in figure 7.6(a) by noting the discrepancy between the transition arrows and the button arrows).

For some additional confusion, look at figure 7.6(b) that now includes all the possible transitions involving the "up" and "down" arrows. Specifically, focus on the long transition from the CLOCK SET down to LANGUAGE. Here the arrows *do* match. You press the "down" arrow to go down to LANGUAGE, and the "up" arrow to CLOCK SET. I'm sure you immediately recognize the ambiguous consequences here: in some situations the arrow button works as expected and in some, they don't—and we already know quite a lot about how such problems cause confusion and frustration to users.

Furthermore, unlike the *flip-up* and *flip-down* light switches in chapter 6, which dealt with two different stereotypes (European vs. North American), the problem with the arrows is universal. It is a well-established principle of user-interface design that the direction of movement of a control should be compatible with the resulting movement on the screen. You know, the computer's mouse moves away from the body and so does the cursor. The joystick is pushed down and the cross hairs fly in that direction. So natural, so simple. But not here.

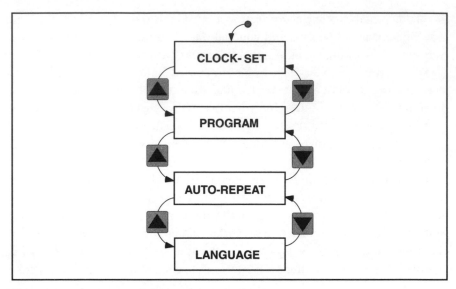

Figure 7.6 (a). Step movements of the cursor in main menu.

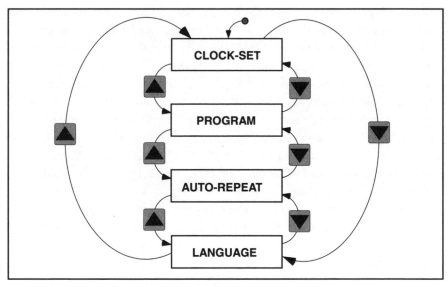

Figure 7.6 (b). All possible movements of the cursor.

The infamous VCR machine flashing 12:00 is often heralded as the symbol of user's frustrations with computerized devices. (And if you think that the auto-clock feature in some new TVs and VCRs solved the problem, be sure to read the endnotes for this chapter.) Many of us avoid setting our VCR clock, let alone programming a VCR. And don't even mention doing such "programming" in a hotel room or with a device that we are not familiar with. Setting up

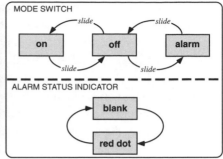

Figure 7.7 (a). Clock radio (photo has been enhanced).   Figure 7.7 (b). Mode selection.

a VCR is a case where many of us users simply accept defeat. But is this really because we are clumsy, stupid, and technologically challenged?

Clock Radio and Reference Values

It is once again time for us to set our alarm for tomorrow's wakeup. After our experience with the wristwatch in chapter 3, we learn not to trust our own time-keeping devices. A white clock radio with a slick interface is by the bedside (figure 7.7[a]). On top of the device, there is a series of buttons and switches to set up the clock time, the alarm time, radio, and modes. The mode switch shuttles between three positions: ON, OFF, and ALARM (see figure 7.7[b]). You *slide* the switch to the left to turn ON the radio and then two slides to the right for ALARM. Once in ALARM mode, there is a red dot, signaling that the alarm is ready and armed.

It's getting late and you need to set your wakeup time for tomorrow. You press and hold the buttons down until it reads 5:00 A.M. That's it, we are done with the alarm. The radio is next. You find satin-smooth jazz to listen to as you wait to fall asleep. You turn the volume down a bit. Ten minutes later you're still awake, so you turn the volume lower. Soon after, you slide the mode switch to ALARM, see the red dot appear, roll to the side and drift away. . . .

Machine Model

Now we can quietly tip toe in to explore the internal workings of this device, because as you can imagine, there are some hidden insights here beyond the casual-looking interface. Figure 7.8 details the clock: within the clock component there is the alarm mode. But in order for the clock and alarm to work, time, an ever-persistent cycle, must be counted. To the right of the clock component there is another (concurrent) component called CLOCK TIME. In it, time TICK is updated every second with an event called *tock*. Tick and tock count time for us and serve as a reference value for the device's internal states and modes. A reference value is usually a parameter (in this case, time) that is used by the device's internal logic to change its behavior (e.g., switch modes).

Now we can patiently examine the alarm component, which, as you can see in figure 7.8, is harbored inside the CLOCK component. Looking inside, you can see that initially, the ALARM is DISABLED. However, when we slide the mode switch (on the interface) to the right and enter alarm mode, the clock enters the (alarm) ENABLED super state. The alarm is ARMED and ready to wake us up. Now even if you wake up in the middle of the night you can immediately tell that everything is all right: the red dot, which signals that the alarm is enabled and ready, is in sight. And you can easily note that the alarm time is set right.

With the intended wakeup hours away, let's explore and understand how the wakeup will occur. Since the alarm is ARMED, once the time reaches 5:00 A.M., the alarm will ENGAGE and the radio should come to life. Music, news, or inane chatter between DJ's will go on for 60 minutes, after which the clock radio times-out with an automatic transition back to ARMED (for the next morning wakeup). No problems so far.

Finally, an additional note regarding reference values and their importance in making time work for us: just as the tick-tock is a reference value for the clock, alarm-time (5:00 A.M. in our case) is the reference value for the alarm component. Everything seems to hinge on it. But as we shall see soon, simply arming the alarm is only one condition for wakeup.

Our next stop is the radio, which can be either silent in OFF or playing in ON (figure 7.9). The radio plays either because we turn the mode switch ON *or* the alarm ENGAGES automatically (at the appropriate wakeup time). We write these two conditions like this:

enter (MODE SWITCH: ON)

or

enter (ALARM: ENGAGED)

As for going back, we exit RADIO-ON in the reverse way that we entered. But before we leave the radio, note that it also has its own reference values: one is the volume, which you *click* with the thumb wheel till you reach the desired audio level (from 1 to 10). The other one is the radio frequency. Both must be set properly for the radio to play and be heard.

Alarm Not

It is now past 5 A.M. The alarm, for whatever reason, did not activate, even though the red dot was there and everything looked shipshape.

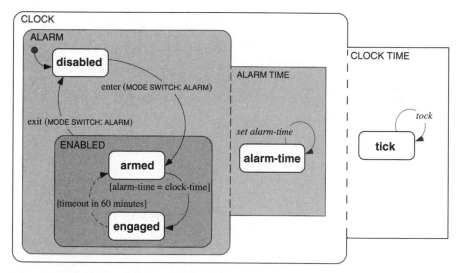

Figure 7.8 The clock (and alarm) component.

What happened?

Things went awry this early morning because although the alarm ENGAGED and the radio turned ON, the volume was left below hearing threshold from last night, and therefore the alarm did not activate. For the alarm to sound so that the user will wake up, several conditions must be well coordinated. One condition, which appears on the display, is that the alarm must ENGAGE. That's necessary but not sufficient. The radio can be on, but if any one of its two reference values, namely volume and radio frequency, are not set properly— the wakeup fails.

There is a subtle distinction here between *engaged* and *activated*. "Engage" means that a mode latches on; "activate" means that the mode (and all its reference values) produce the desired behavior. In later chapters, we will see other examples, from maritime electronics, medical devices, and aircraft, where the divergence between what is engaged and what gets activated leads to errors and results in unnecessary mishaps.

Implications for Design

In the context of designing a device for human use, there are two main issues, and we will address each one of them separately and then show how they intertwine in this example. The first one has to do with the task requirement. To consider any interface from the user's perspective we must begin by asking

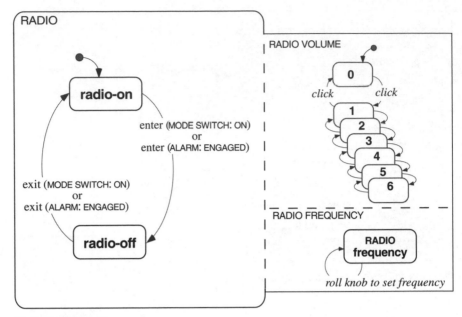

Figure 7.9. The radio component.

what are the user's tasks. With respect to this clock radio device, one fundamental task here is that the user should be able to differentiate, by looking at the display, whether the alarm is armed and ready for wakeup or not. We want the user to correctly and reliably predict the next state of the machine. But when we examine this specific display, we realize that it is not easy to distinguish between the two: if the volume is left at 0 (or below the user's hearing threshold), or the radio frequency is not set properly—it is not apparent to the user that the alarm will *not* activate.

The second issue has to do with the way we think about modes. This device and the many others that we will discuss in the following chapters have reference values that are associated with the modes. In this clock radio, time, alarm time, volume, and radio frequency are all reference values. The point is that when it comes to automated devices, we can no longer think of a mode as a separate entity from these reference values. The CLOCK uses time, ALARM needs the alarm time, and for the RADIO to activate and be heard, the radio frequency and volume must be properly set.

The relationship between the mode and its associated reference values must be carefully considered when it comes to designing the user interface. This is where the definition of the user's task is intertwined with the modes issue. The user wants to know whether the alarm is armed and ready to activate at 5:00 A.M. That is part of the user's task requirements. Therefore, the display must be designed such that the user will be able to determine if the alarm is armed and ready for wakeup. But if we consider the interface of this clock radio, we realize that it only indicated that alarm is armed (ALARM: ENABLED). The red dot, sad to say, is simply not enough. Worse, it gives the impression that all is well, when in fact it is not.

Some solutions

One way to try to solve this problem is to combine, through the indication, the mode status (ALARM: ENABLED) and the two reference values—volume and radio frequency. But there is a problem with this approach. While we can display the volume and the setting of the radio frequency knowing that they will not change on their own, we cannot guarantee that the radio station will not quit transmission at midnight only to resume at 6 A.M. Therefore, one design solution is to include a sensor to detect a reasonable level of audio at wakeup. If it is not present, then the device should automatically sound a buzzer.

There are several other ways to solve the problem, and perhaps you can think of a few. One thing to keep in mind is that functionality and the resulting interface must also be considered from the user's point of view (and not only from the machine's). After all, it is for the user. This clock radio fails because it was designed as two separate machines (radio and clock) glued together, as opposed to an integrated device for human use.

Chapter 8

The Grounding of
the *Royal Majesty*

*"There are three things which are a wonder to me: The way of an eagle in the air;
the way of a serpent upon a rock; the way of a ship in the midst of the sea . . ."*

—Proverbs 30: 18-19

With summer winds out of the northeast at 16 knots, any sailing ship would make good headway to the New England coast. But the *Royal Majesty*, a modern cruise ship, did not need sail power to slice the waves—she had two titanic diesel engines with hefty amounts of horsepower to propel her all the way to Boston Harbor. Built at the renowned Finnish shipyard Kvaener Masa, her slanted foremast and streamlined chimney stack gave her the sleek appearance of a stately cruise liner. Elegant looking and with a high superstructure, this modern leviathan of the sea was almost 600 feet long and 10 decks tall, well-built and powerful. But in her hold, she was hiding an old sea tale with a modern twist.

Only three years in service, she sported the state-of-the-art in navigation equipment: three radars, an integrated bridge system with an autopilot and a radar-map display, a sophisticated speed log, and an accurate gyrocompass. Unlike vessels of the past, which had to rely on hugging the coastline or sighting constellations to find their way, the *Royal Majesty* was equipped with a Global Positioning System (GPS). The GPS is a marvel in the world of navigation. If the invention of the magnetic compass was the first revolution in navigation, and finding longitude with chronometers the second, then the global positioning system with its artificial constellation of orbiting satellites is surely the third; no need to rely on distant stars, sun, moon, and no prayers for

Figure 8.1. The cruise ship *Royal Majesty* (source: National Transportation Safety Board).

a cloudless sky. The GPS displays the latitude and longitude with an accuracy of about 100 yards, anytime, anywhere.

Out to Sea

June 9, 1995, Bermuda. Departure time: 12:00 P.M.. As the 1,200 passengers were readying themselves for the last two nights of a seven-day cruise, the officers and staff were also preparing themselves for the return voyage. The navigator, perched up on the bridge an hour before departure, was testing and checking all his navigation systems, the pride of his job. They were in "perfect" condition, no anomalies were detected.

The bridge of the *Royal Majesty* was smart. With its huge front windows and sophisticated equipment, it looked like a set from a science fiction movie. In the front was a large crescent-shaped console planted with several radar screens, navigational displays, and ship alarm indicators. Behind it, separated by a partition, lay the long chart table. Mounted just above the table, within an arm's reach, were the GPS and the LORAN-C units (see figure 8.2). The LORAN-C is another navigation system, providing reasonably accurate position data by triangulating radio signals from onshore transmitters. Its major drawback is that it deteriorates in accuracy as the distance from shore increases. Naturally, the officers always preferred the more accurate GPS. On this departure, like every other, the navigator selected GPS as the primary source of position data.

While the GPS and LORAN-C display units were inside the bridge, their antennas were mounted on the roof to better facilitate the reception of radio

Figure 8.2. The chart table and the bridge. The left arrow points to the GPS unit, the right arrow points to the LORAN-C unit. Source: National Transportation Safety Board (the photo has been enhanced).

waves. The roof above the bridge of a vessel is called the "flying bridge." Passengers were never allowed to walk on the roof, and the officers hardly ever used it during a passage, but electricians and technicians would climb up there on a fairly routine basis to check the cables and maintain the GPS and LORAN-C antennas. The *Majesty*'s flying bridge was just like any other—a low railing all around, antennas of all sorts, satellite receiving domes, and bundles of connecting wires—except that the gray coaxial cable coming out from the closed bridge and leading to the antenna of the global positioning system was hanging loose; it wasn't sufficiently secured or strapped to the flying bridge's roof.

The departure from Bermuda was normal. On the bridge, the officers were busy bringing all the necessary systems on line, checking their critical functions and alarms, reading the weather charts, setting the watches, and examining the sailing plans and nautical charts. Just before noon, while the buffet lunch was being served in the dining halls, the ship departed the Ordnance Island terminal. The bridge was busy with the captain, harbor pilot, and officers conning the ship, taking orders, and maneuvering the ship in the small harbor. Outside, lookouts on both sides of the bridge were guarding her flanks for any unforeseen or unusual circumstances. Commands and directions were given with authority, and she began to move and turn smoothly with her bow thrusters and main engines. Now, with her stern to the terminal and her bow to the northeast, she slowly glided on her own power away from the pier and through the breakwaters into the blue Atlantic Ocean.

At 12:05 P.M., she was making 14 knots and every line, cable, flag, and unlashed cloth flapped in the wind. 16 knots by 12:20 P.M. A small launch came alongside to pick up the harbor pilot at 12:25 P.M. With engines revving up to cruise speed, the ship was approaching 20 knots. Bermuda was slowly disappearing aft until it was a low coastline behind the ship.

Inside the bridge, the tense port departure activities began to calm down. The focus was now on the 677 nautical miles that lay ahead to Boston Harbor. The ship's position (called a fix) was plotted on the chart and regular reports from the engine room followed. At 12:30 P.M., the navigator walked to the chart table and compared the GPS and LORAN-C positions—one-mile difference. "Not bad," he muttered to himself with pride. The GPS was computing the ship's position by receiving timed transmissions from three or more satellites orbiting the Earth, and then converting time to distance. Then the GPS went on to compute, based on the distance to the three satellites, the location of the ship. The LORAN-C was doing basically the same thing by receiving timed transmissions from three or more land-based transmitters on the East Coast of the United States. The accuracy of the LORAN-C deteriorates with distance, and therefore a mile difference in a location as far off the U.S. mainland as Bermuda was a testimony to the accuracy of the LORAN-C.

Satisfied with the accuracy of the fix he just laid on the map, the navigator walked away and joined the captain, who was standing by the large windows. Above them, on the flying bridge, the loose GPS cable was banging freely in the wind and slapping the flying bridge's deck; the connection between the GPS cable and the antenna was coming apart. At 12:52 P.M., interruptions in the global positioning data began: after a few seconds of no position signal, a one-second alarm chirp sounded, similar to a wristwatch alarm. Nobody heard it; it was barely audible. At 1 P.M., the captain commanded "change course to 336, take us up to Boston." Nine minutes later, the global positioning system signal was back again. But it did not last long, because at 1:10 P.M., the connection between the cable and the antenna broke away for good.

Around 2 P.M., the navigator set the autopilot in navigation (NAV) mode. In this mode, the autopilot would get position data from the GPS and then guide the ship along a pre-programmed route that would take the ship to Boston with little human intervention. On this run there were three legs: 500 miles from St. George, Bermuda to a point just south of Nantucket Island; then skirting the infamous Nantucket Shoals up to the northern part of Cape Cod; and finally around the Cape into Boston Harbor.

Figure 8.3 is a map of the route between Bermuda and Boston. But maps, as we already know, are only an abstraction. In the real ocean, there are winds, currents, tidal streams, and rolling waves. They all combine to push, or offset,

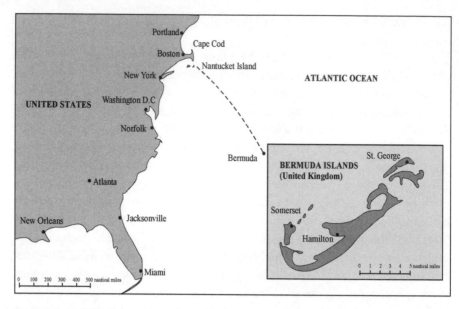

Figure 8.3. From St. George, Bermuda (see inset) to Boston, Massachusetts.

the ship away from its intended route. If the autopilot simply maintains the course plotted on the chart, it will never keep the ship on track. To compensate, the autopilot steers a slightly modified course.

But how does the autopilot know how much to compensate for these effects? Well, the autopilot calculates the difference between its current position and where it is supposed to be. Once you know where you are and how much you have deviated, you can determine what course changes are needed to get back on track. We all do this when we drive along an open road with high winds. We get blown to one side and then steer our way back to the center of the lane. After a few such episodes, we learn to oversteer a bit into the wind to counter its effect on the car. The autopilot does the same thing. But since the autopilot has no eyes to gauge how much to oversteer, it uses GPS position data to do this. Under normal sea conditions, these modified course changes are so little, they are hardly even noticeable—a degree to port here, a degree to starboard there. Yet without these minute course corrections, any vessel will wander off its intended track.

Change of Shift

By late afternoon Bermuda was well beyond the horizon and the passengers were resting after a luxurious lunch buffet, but the routine work in the bridge continued—there were fixes to make, engine room readings to record, information and events to be entered in the ship's log book, communications with other ships and vessels, and recording of sea and weather information. A decent formality has always been observed aboard ships at sea. Mariners do not think of work in terms of hours, but of watches: At 4 P.M. the chief officer takes the afternoon watch and relieves the navigator; at 8 P.M. the second officer is on duty; at midnight the navigator returns; at 4 P.M. of the next day, the chief officer stands again; second officer at 8 P.M.; navigator noon.

It was now 28 hours since leaving Bermuda and the monotonous work on the bridge continued. Sea duty is like a clock that never stops. Every hour, on the hour, the position of the ship in the midst of the ocean is fixed. There are many ways to do this: in coastal waters, one can find the ship's position by taking compass bearings from two separate landmarks. The intersection of the two lines *fixes* the position of the ship. Using radar, a fix can be obtained by taking the distance to two objects or three objects (such as radar beacons, buoys, and capes), and then plotting a distance arc around each one. Where the arcs intersect is where the position is fixed. When the vessel is far from shore, the ship's position can be fixed with reference to the stars, the moon, or the sun. And finally, there is also the GPS and the LORAN-C. On the *Royal Majesty*, the captain and officers relied solely on the GPS.

A few minutes before 4 P.M. on that Saturday, the chief officer came to relieve the weary navigator. The lookout was also being relieved by his mate. Shift change on a ship's bridge is a well-choreographed and time-honored ceremony that repeats itself every four hours. A fix is drawn on the map, logbooks are signed, reports are relayed, and relevant information such as changes in wind and current, objects or close-by ships, is exchanged. The chief officer conferred quietly with the navigator over the charts. The ship was about 80 miles away from shore and they were transitioning from deep-ocean navigation into the more hazardous coastal passage where there are many vessels and fishing boats to worry about and, of course, the vicinity of land. The point of transition was just south of Nantucket Island. They were scheduled to dock at Boston in the early morning of Sunday, June 11.

Nantucket, the capital of a nineteenth-century whaling empire that ranged from the Atlantic to the Pacific, is a small crescent-shaped island, actually a sand bank, located about 20 nautical miles south of Cape Cod. Spread around it is Nantucket Shoals—a moving set of shoals that insulted and bedeviled mariners for centuries. Ships of all sizes, some of them Nantucket's own, have grounded and sunk in those treacherous waters. To protect ships from the dangerous shoals and help separate traffic of vessels in and out of Boston Harbor and around Cape Cod, the U.S. Coast Guard created the Boston traffic lanes (figure 8.4). Just like a boulevard with two opposing lanes and a divider in between, there are lanes for northbound and southbound ships. Defining the separation zone is a series of six lighted buoys. The BA buoy marks the southern entrance into the lanes and the buoys that follow—BB, BC, BD, BE,

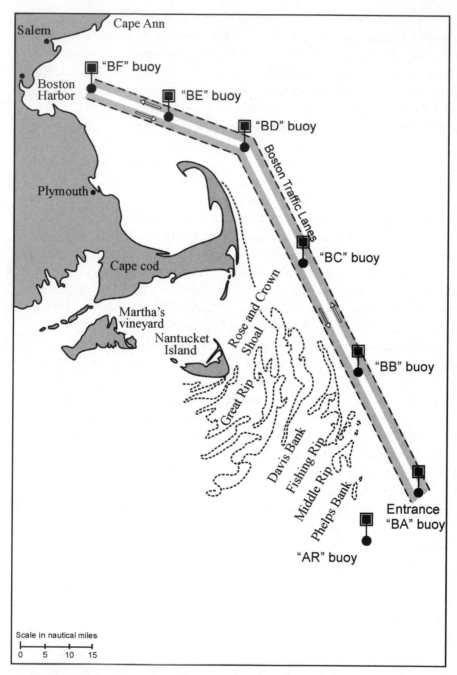

Figure 8.4. The Boston Traffic lanes, Cape Cod, and Boston Harbor.

and BF—mark the way to Boston Harbor. Each buoy has a radar reflector and a unique flashing characteristic.

Twenty minutes after 4 p.m., the captain entered the bridge. He inspected the sophisticated radar display on the spacious console and the logbook. Everything looked fine. Ready to leave, he walked by the chart table, followed by the chief officer. The chart was spread wide open, ready. Poised above the chart table, the GPS displayed in green glowing red alphanumeric digits the latitude and longitude of the ship (see figure 8.5[a]). The two officers looked at the latitude and longitude coordinates on the GPS display and compared it with the small pencil mark on the map.

"Here we are, and here are the traffic lanes," said the chief officer.

"Excellent," replied the captain.

But what neither of them noticed was that in addition to the latitude and longitude coordinates and all the usual data, the GPS display showed two small, innocent-looking text symbols—sol and dr (figure 8.5[b]). But these two symbols were hardly innocent—they portrayed a serious problem in the GPS. sol stands for solution, indicating that the GPS was *no longer* computing an accurate position solution. Actually, it had not been producing accurate position solutions for quite some time—ever since the connector between the cable and the GPS antenna had given up an hour after departure, the unit was unable to receive any satellite signals and update its true location.

So what information was the GPS displaying for the past 27 hours?

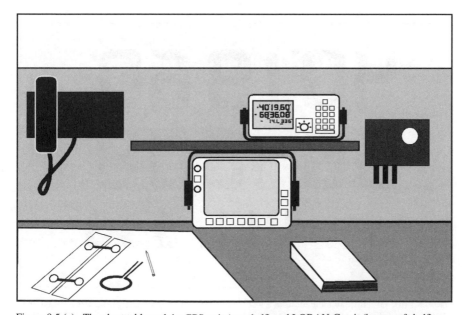

Figure 8.5 (a). The chart table and the GPS unit (top shelf) and LORAN-C unit (bottom of shelf).

Not receiving any satellite signals, unable to determine its position, the GPS defaulted to dead reckoning (DR) mode. Dead reckoning is an old method of navigation dating back to thirteenth-century Mediterranean mariners and the introduction of the magnetic compass. Dead reckoning—which should properly be written *d'ed* reckoning, an abbreviation for deduced reckoning—boils down to laying down a compass course and estimated distance on the chart. The navigator would start from the port of departure, plot the course line on the map, and then mark the distance traveled. With every course change, a new line was drawn and the new distance marked. With this simple geometry, the estimated position of the ship was calculated.

But how does one get the distance traveled? There were no odometers on thirteenth-century ships—but if one knows the speed of the ship one can compute distance. The early method was to throw flotsam in the water, note how long it took to travel from the bow to the stern (a known distance), and convert time and distance to speed. In the age of Spanish caravels and inaccurate hourglasses, sailors would throw the flotsam and sing a fast sea chantey—and when the flotsam reached the stern mark, the pilot would note the last syllable reached in the chant, and convert that syllable into speed. Columbus, they say, was a master in dead reckoning navigation, making amazingly accurate landfalls during his return trips from the New World.

In modern ships, dead reckoning is calculated by obtaining the ship's heading from the gyrocompass and distance traveled from the speed log—a sort of a ship's odometer. Using the same arithmetic as employed by navigators

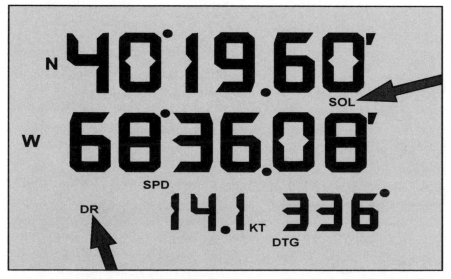

Figure 8.5 (b). GPS display indications at 8 P.M. on the night of the grounding. The arrows show the DR (dead reckoning) and SOL (solution) indications.

in the age of discovery, a fix is calculated. This is exactly what the GPS does in dead reckoning mode. And just like thirteenth-century dead reckoning navigation, it needed a starting point. In the case of the *Majesty*, it was a point some 30 miles northeast of Bermuda, corresponding with the last GPS fix, just before the antenna connector finally broke.

At first, the effects of sea, wind, tidal currents, ocean currents, compass and steering errors were negligible; the *Royal Majesty* was sailing almost parallel to route and the latitude/longitude was only slightly off. But as the hours passed and Bermuda fell behind, slowly and surely the gentle eastward swell rolled her aside and the east-northeast wind buffeted her superstructure leeward. The position data were deteriorating at a rate of a half nautical mile per hour.

Meanwhile, the crew was plotting the ship's position by reading off the latitude and longitude from the GPS display, assuming the position portrayed there was accurate, as always. And since the GPS unit was spitting out dead reckoning positions—completely oblivious to what the sea and wind were doing to the ship—the trail of hourly plotted fixes on the chart showed as if she was exactly on her intended track. Therefore, nothing appeared amiss—the navigation actually looked perfect. And since the time-honored rule of taking a backup fix (from another source, e.g., LORAN-C, radar, or celestial navigation), was not practiced on the *Royal Majesty*, her officers believed that all was well. But at any rate, at 5 P.M. on June 10, after 28 hours in DEAD RECKONING mode, the *Majesty* was 14 miles southwest of her intended route.

Entering the Lanes

"When do you expect to see the first buoy?"

The chief officer, answering the phone, responded quickly to the Captain: "we are about two and a half hours away from the Boston traffic lanes." Then they went on to discuss how they were going to pass a mile or so to the right of the entrance (BA) buoy, which marks the entrance into the Boston traffic separation lanes (see figure 8.6[a]). The BA buoy, about six feet wide and ten feet tall, had a boxy-looking radar reflector on it—making the small buoy detectable by radar at a range of about 10 nautical miles. But the ship was still 35 miles out, so they could not detect it yet on the radar.

"Call me when you see it," said the captain as he ended the conversation.

In preparation for entry into the Boston traffic separation lanes, the chief officer lowered the speed to 14 knots. This was good practice in coastal waters where small vessels abound and fog banks appear without notice this time of year. At 5:30 P.M., after tea, the captain was up on the bridge again. He checked the vessel's progress by briefly looking at the chart and then walked forward to examine the radar map. There he could see it all. The radar map not only showed the radar image, but overlaid on top of the radar image was a map-like

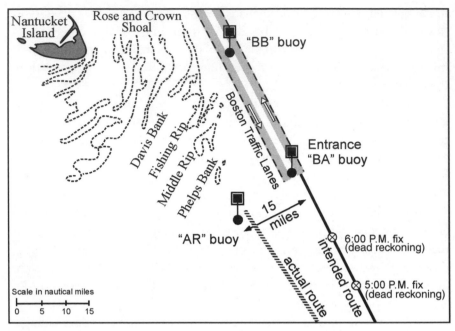

Figure 8.6 (a). Actual and intended route of the Royal Majesty.

rendering of the traffic lanes and buoys. On the perimeter of the display there was a plethora of related information such as heading, course, speed, drift, set, autopilot data, and the latitude and longitude of the vessel (see figure 8.6[b]).

At 6 P.M., the radar map showed the vessel sailing on track and heading accurately toward the right side of the traffic lanes. Yet in reality, the *Royal Majesty* was 14.5 miles west of the lanes, close to shore, and heading toward the dangerous Nantucket Shoals (figure 8.6[a]). The radar map deceived the crew—and the reason for this is important for us to understand because it exposes the kind of problems that plague automatic control systems, especially of the navigational kind. You see, the source of ship position for the radar map display was none other than the crippled GPS unit; whatever the GPS said, the radar map displayed. And since the GPS position did not reflect the ongoing and increasing position deviation, it provided information as if the ship was following its pre-programmed route. That's what the radar map displayed—which was, of course, a false description.

The only way for the officers of the *Royal Majesty* to recognize the discrepancy was to compare the information displayed on the radar map with another (independent) source of position data. One such source could have been the radar image, which is part of the radar map display. The radar does not deceive; it cannot tell you where you are, but it always tells you where other buoys and landmarks are, in relation to you. But although the radar was on, the ship was

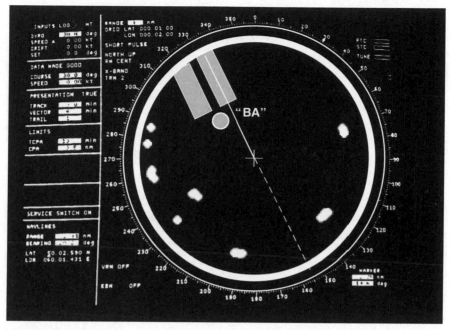

Figure 8.6 (b). The radar map, showing the radar screen and rendering of the traffic lanes and entrance buoy (BA).

more than 15 miles away from the entrance buoy to the Boston traffic lanes, which was too small to be detected. And since the radar's range was set to 24 miles, they did not see land on the radar. As a consequence, the map shift was not detected.

At 6:45 p.m., the chief officer saw a blip on the radar, approximately seven nautical miles away. The little blip was collocated with the rendering of the entrance buoy on the radar map (figure 8.6[c]). Based on the GPS position data, which indicated that the *Royal Majesty* was following its intended track, and on the fact that the blip had been detected on the radar at the time, bearing, and distance that he had anticipated detecting the BA buoy, the chief officer concluded that the radar blip was indeed the BA buoy. But most of all, it was the fact that the blip appearing on the radar coincided with the plotted position of the buoy on the radar map display, which sealed his conviction that this was the BA buoy. There it is, he thought, just where it should be. The chief officer was now certain that the ship was well protected from the dangerous Nantucket Shoals.

Mariners, to this day, are very wary of the entire area. It is considered one of the most treacherous areas to navigate in the entire U.S. coast, and piloting books warn that "the whole area should be avoided by deep draft vessels when possible." The Coast Guard had planted several large buoys to mark the area.

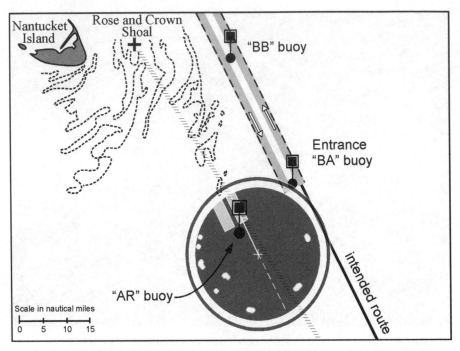

Figure 8.6 (c). Superimposition of the radar map and buoy rendering on top of an actual nautical map. Showing how the actual location of the "AR" buoy matched the expected location of the "BA" (entrance) buoy.

One of them, the Asia Rip buoy, flashing yellow light at a 2.5-second interval, was guarding the southeast tip of the shoal area. Anchored by a long and heavy chain to the wreck of an unfortunate vessel, the Asia Rip buoy was 15 miles to the west of the entrance buoy. And 15 miles is an important number here, because by a rare and unfortunate twist of fate, at that time the *Royal Majesty* was exactly 15 miles off her intended route! And therefore, on the radar map, it was actually the radar return from the Asia Rip buoy that coincided with the rendering of the entrance (BA) buoy (see figure 8.6[c]).

But on the spacious and modern bridge of the *Royal Majesty*, all of this was hidden from view. On the radar screen everything looked picture perfect; the ship was heading toward the rendering of the lanes and the little tick marks on the chart were there to prove it. A feeling of complete security, the most fatal delusion a seaman can entertain, fell upon the bridge. And it is true that the great majority of accidents at sea are not due to the violence of weather or fragility of a ship, but over-confidence as to the ship's position.

At 7:20 P.M., a large iron buoy, with a light and an awkward-looking boxy radar deflector on its flat top, was passed less than 2 miles off the port side of the ship. The chief officer had been tracking it on the radar for the past half an hour, assuming it was the genial entrance buoy. Like a disregarded messenger, the Asia Rip buoy bounced in the wake of the *Majesty*, and then quietly faded out of sight.

Several minutes later, the captain telephoned the bridge and asked the chief officer for the third time whether he had seen the BA buoy. "Yes, captain, we just passed it 10 minutes earlier." The captain then asked whether the chief officer had detected the buoy on radar. "Yes, captain, I did." The captain was now relieved of his concern. Unfortunately, the chief officer did not tell the captain that he had been unable to visually confirm the identify of the BA buoy, and the captain did not ask whether the buoy had been visually confirmed. And the good ship *Royal Majesty*, blowing white smoke out of her smokestack, continued toward the shoals—leaving a tranquil wake behind.

The 8 P.M. Shift Change

On Saturday evening, June 10, the winds were blowing on the starboard side. The temperature was falling slowly and the gray daylight was fading away. The ocean was running to the shore with increasing force. As the second officer walked to the bridge, lights started to appear on the decks, the smell of burned fuel was in the air, and the ship rolled gently with the three- to four-foot waves. New to this ship, but hardly a novice mariner, he walked briskly up the flight of stairs, opened the large wooden door, and entered the bridge.

Inside, it was already night. An array of digital lights were coming from the displays, and radio calls from the speakers. A ship is like a city that never sleeps. During the change-of-shift briefing, the chief officer informed the second officer that they were well inside the Boston traffic lane. "Course 336, speed 14 knots, wind from the east-northeast at 15 to 20 knots. I already turned on the navigation lights for you." The second officer went over to the chart table to look at the fix. Meanwhile, two lookouts took their positions on the port and starboard side.

The evening watch slowly transcended into that calm so common on a ship's bridge during the night. The second officer, now alone on the spacious bridge, was standing by the radar map. Noting several radar returns on the port side, he moved his head up and looked outside. As the night spread over the waters, navigation lights slowly replaced the masts of vessels in the vicinity. Red over white, he noticed; fisherman working along the nutrient-rich Nantucket Shoals. Noting that all the fishing vessels were far away and almost stationary, posing no danger to the *Royal Majesty,* he reduced the radar range to six miles—giving the traffic lanes his full attention.

The lookout on the port side reported a yellow light on his side. "Thank you." Several minutes later, both lookouts reported seeing red antenna lights on the port side. Then they reported seeing more red flashing lights off on the horizon. Like a set of city traffic lights, these lights announced the presence of nearby land. The port and starboard lookouts continued to report uncommon sightings to the officer. He acknowledged each with a "roger" but did not take in the significance of these sightings. With his radar scale set on 6 miles, the phosphorous image of Nantucket Island, now 25 miles away and almost dead ahead, was hiding beyond the scale.

Good Neighbors

The Portuguese fishing vessels *Rachel E* and *San Marco* were pulling heavy nets along Nantucket Shoals all day. From inside the salt-sprayed cockpit windows, Antonio Pimental, a fisherman on the *Rachel E,* was looking at the evaporating horizon and checking the sea conditions. The light of day was fading fast and the sea appeared a deep gray, turning black. And then he gasped. He refocused his eyes with amazement. In front of him, less than a mile away, looming like a disoriented leviathan with gloomy green and red eyes, was a huge ship. And there was no question about it—she was heading straight toward the shoals. As he was considering what to do, he heard the following transmission, broadcast in English, on channel 16, the maritime distress radio frequency. (The following are transcripts of marine radio transmissions, recorded on Coast Guard receivers, on the evening of June 10, 1995):

"*Fishing vessel, fishing vessel, call cruise boat.*"

There was no reply. Antonio waited a minute and then, recognizing the voice, he called his friend Toluis on the *San Marco.* The ensuing conversation was in Portuguese:

"*Hey, it's Antonio Pimental. That guy is bad where he is. Don't you think that guy is wrong in that area?*"

"*I just tried to call him,*" replied Toluis. "*He didn't answer back. He is very wrong.*"

Antonio Pimental picked up the microphone and made the following calls (in English):

"*Calling cruise ship in the position 41-02 north, 69-24 west.*" "*Calling cruise ship 41-02 north, 69-24 west.*"

No answer.

After several additional attempts, Toluis picked up the radio and commented, in Portuguese: "*Maybe nobody on the bridge is paying attention.*" "*I don't know,*" replied Antonio, "*but he is sure not going the right way. . . . *"

Machine-Machine Interfaces

By now the large and ponderous vessel was well inside the treacherous shoals, her 20-foot-deep hull passing over Phelps Bank, Middle Rip, Fishing Rip, and

Davis Bank. The two huge screws were sending commotion to the bottom of the ocean, raising yellow sand, but not touching. And with the depth meter (fathometer) not selected for display, and its alarm set at a meaningless 0 feet, no one on board knew that she was sometimes only 10 feet from running aground. Although a common practice in any ship passing through costal waters, the officers of the *Royal Majesty* failed to turn on the depth meter for display. They did not consult the depth meter nor did they set its alarm properly. Thus they robbed themselves of critical information that would have forewarned them that they were approaching shallow water. The *Royal Majesty* continued her mindless tack toward Nantucket Island, while on the bridge, high above the waterline, the second officer was observing his progress along what he thought was the Boston traffic lane. As the gray sky turned black veil, the phosphorus-lit radar map with its neat lines and digital indications seemed clearer and more inviting than the dark world outside. As part of a sophisticated integrated bridge system, the radar map had everything—from a crisp radar picture, to ship position, buoy renderings, and up to the last bit of data anyone could want—until it seemed that the entire world lived and moved, transparently, inside that little green screen. Using this compelling display, the second officer was piloting a phantom ship on an electronic lie, and nobody called the bluff.

Why?

Although both the autopilot and radar map had a variety of internal tests to check for erroneous data, all of them failed to warn the crew. To understand this matter, we need to plumb the depths of the machine-machine interfaces involved here. It turns out that there were serious design problems in the communications between the GPS unit, the autopilot, and radar map display that circumvented every attempt to warn the ship's officers about the increasing deviation between the ship and her intended track. Here is why: the GPS sends its position data to the autopilot and radar map display by spitting out a sequence of sentences. Just like the transcript of a theatrical play, each sentence starts with the name of the machine sending the information—followed by the time, the latitude and longitude, and other information. The GPS sentence normally ends with the word "valid."

The GPS sentences sent out to the autopilot and radar map while the ship was heading out of the harbor (before the cable broke) looked like this:

GPS, 12:10:34, latitude, longitude, valid.
GPS, 12:11:51, latitude, longitude, valid.
GPS, 12:13:01, latitude, longitude, valid.

When the antenna cable broke, the GPS defaulted to DEAD RECKONING mode as it was designed to do. It continued sending latitude and longitude, only this time the values were based on approximated dead reckoning computation. At the time these integrated marine electronic systems on board the *Majesty* were designed, there was no industry standard as to how the GPS unit indicated that the data it is sending was based on dead reckoning calculations (and not on the precise satellite-based GPS solution). To deal with this lack of standards, the designers of this particular GPS unit found a simple and labor-free solution: they used the last word of the sentence as a means of "flagging," or indicating, that the data was inaccurate, while retaining the entire structure of the sentence as shown above. The last word in the sentence would say *invalid*. Therefore, after the cable broke and the GPS defaulted to DEAD RECKONING mode, the sentences read:

GPS, 1:11:46, latitude, longitude, invalid.
GPS, 1:12: 09, latitude, longitude, invalid.
GPS, 1:12: 51, latitude, longitude, invalid.

The only problem was that the autopilot and radar map system were not programmed to look at the last word of the GPS data sentence. Therefore, the *valid* or the *invalid* statements were completely ignored. You see, the designers of the autopilot and radar map system (which is built by another maritime electronics vendor) never expected, even in their wildest dreams, that a device that identifies itself as a GPS would send anything other than accurate satellite-based data. (What was common among other manufacturers of GPSs was to send a null position data such as 00-00 north, 00-00 south, or to send a unique sentence saying that the GPS has failed, or to halt sending GPS sentences altogether.)

But were there other internal checks on the accuracy of the position data? Actually, yes. The autopilot was designed to constantly calculate its own independent dead reckoning position and compare it with the GPS position data. If the difference between the two positions was within tolerance, the autopilot considered the input from the GPS as "good," accepted the data, and steered the ship according to the GPS data. If the difference was too large, it signaled and warned the crew about the anomaly. That kind of a check should detect a 15-mile offset, right? Not in this particular design. Since both the GPS and the autopilot used the same gyrocompass and speed inputs to compute the dead reckoning information, the two calculations were actually the same and therefore always in agreement, which, of course, eliminated the built-in redundancy and rendered this entire check futile.

One after the other, internal checks that were in place to protect and warn against flawed and erroneous data collapsed like a line of dominos.

In the same way that there is an interface between the user and the machine, there are similar interfaces between machines. Through these interfaces machines exchange data. There was an interface between the GPS unit and the autopilot and radar map. Through this interface, the GPS unit provided multitudes of information. The autopilot and radar map, on the other side of the interface, accepted some information, and, as we now know, ignored some. The main issue here, just as in interface design, is about abstraction. Here, also, we want a machine-machine interface that is *correct* insofar as it does not *hide* critical information, nor does it *lie* about the actual state (or mode) of the machine.

The interface between the GPS unit and the autopilot/radar map *lied*, because although the GPS was sending dead reckoning data, the interface passed along the data as valid satellite-based data. The GPS was in DEAD RECKONING mode and the interface took the information as if it was in SATELLITE mode; we therefore must conclude that this machine-machine interface was *incorrect*. (Later in this book, we will be discussing verification and abstraction techniques for designing correct and succinct user-machine interfaces; it turns out that the same concepts can be extended to the analysis and design of machine-machine interfaces.)

The Grounding

By now, the rift between the radar map and the actual ship position had increased to 16 miles, and the ship was coming closer and closer to shore. With an incorrect interface, all internal checks failing, and the officers of the watch not corroborating the accuracy of the position data by any other independent source—the ship was quickly approaching its destiny on Nantucket's inner shoals as so many other ships had done in the past. With his trust in technological marvels, man tends to lose the sense that the ocean is still something to reckon with. And the cruise ship *Royal Majesty*, now fully confused and disoriented, was scudding towards disaster.

At 10 P.M. the captain came to visit the bridge. The second officer told him that they were inside the traffic lanes, but failed to mention that he actually did not spot the second (BB) buoy on the radar. Satisfied that the positions marked on the chart and the position on the radar map matched and showed the vessel on its intended route along the Boston traffic lanes, the captain left the darkened bridge and walked slowly down the lighted corridor to his office. He felt the deliberate and assuring throb of the engines, as if signaling everything was fair.

Outside, however, things were not all right and starting to deteriorate at a rapid pace: "Blue and white water dead ahead," shouted the port lookout as he noted the change of ocean color, indicating very shallow waters and breaking waves. "Thank you" replied the second officer and continued his fixation on the green radar map. Normally, the sight and sound of breaking water would raise the hackles of any mariner, as this is an unmistaken indication of shallow water, but the lookout's warning went unheeded by the second officer.

Seconds later, the entire ship was jerked to the left as if shoved by a giant hand, with all her 32,000 tons taking the unexpected side blow. The bow shook first. And then another grinding sound from the side. Vibrations raced along her keel, from compartment to compartment, all the way to the stern. Many of the passengers in the lower decks woke up in fright. Like a massive pendulum, the ship swung from side to side, heaving to the right and then to the left, but still moving forward at 14 knots. The second officer was confused and thunderstruck; the captain, working at his desk, sprang up and ran like a madman through the long white corridor back to the bridge. When he got there, the second officer was already steering the clobbered ship manually. The captain dashed to the radar map, kicked it open to 24 miles—and saw an unmistakable crescent-shaped mass of land: Nantucket Island. Looking ahead in dismay he shouted.

"Hard right, Hard right, haaard righttttt!"

But it was too late. A deep guttural quiver came from below as the ocean-going vessel plowed into the shoals. The ship was slowing down. A sense of helplessness descended on the bridge. And then, with cavernous pulsations, as the hull carved through the cruel underwater shoal, the ship came to a grinding halt. They all noticed the awkward stillness of the situation. She was not moving, but her propellers were still lumbering forward. "Engines to idle," he shouted to the engine room. The engines stopped. Only the comforting sound of the generators and the blowing of the wind broke the unfamiliar silence. The captain quietly ordered, "engines to reverse." Nothing moved. The propellers were turning backwards but the ship was not moving with them. The heavy propellers were beating the water, but to no avail. The ship was hard aground.

The captain called the engine room for an immediate hull inspection. He couldn't fathom how the ocean had betrayed him; how his elegant and highly maneuverable ship all of a sudden had become paralyzed and immobilized. With these thoughts and feelings screaming through his body he ran to the chart table to check the GPS and LORAN-C positions. It was then that he realized, for the first time, that the GPS was in error by 17 miles! When he plotted the LORAN-C position on the chart, he had to dismiss the dozens of explanations and speculations that chiseled through his head. The inescapable and humiliating realization of a gross navigational error descended upon him like a dark cloud: his proud ship was aground on the Rose and Crown Shoal, only 10 miles from Nantucket Island. The calm silence was broken with shouts, orders, curses and accusations. At 10:25 P.M., she was stranded on the

Figure 8.7. The *Royal Majesty*, grounded on Rose and Crown Shoal and surrounded by two tugs and a Coast Guard cutter (source: National Transportation Safety Board).

shoal, resting on her right side, showing her dark red bottom, like a matronly aunt caught with her pants down (figure 8.7).

End of Story

All attempts to pull back from the hold of the shoal failed. By midnight, the captain gave up on his efforts to free the ship by using the engines' forward and reverse thrust. Ironically enough, it was a passenger with a cell phone that alerted the United States Coast Guard to the *Royal Majesty's* grounding. Upon the Coast Guard's radio call, the captain confirmed his situation and requested assistance. At noon the next day, two ferryboats that were chartered by the shipping company arrived on the scene. But the plan to off-load the passengers into the ferries was canceled because the sea conditions were too hazardous. In the late afternoon, five large tugboats arrived. At 10 p.m., 24 hours after the grounding, the *Royal Majesty* was finally pulled out. Her double bottom hull saved her. She did not take on any water, did not leak fuel, and fortunately, nobody was hurt. After a thorough inspection she was permitted to travel to Boston Harbor to disembark the passengers and undergo a more thorough examination and Coast Guard inquiry. Several days later she left for a shipyard in Baltimore, Maryland, where she was dry-docked and repaired. Total structural damage was estimated at $2 million. On June 24, the vessel was declared safe and resumed regular passenger service.

But something else was definitely *not* safe—and it was not just the failure of the GPS antenna, the internal checks inside the autopilot and radar map, or the so-called navigation error and glaring "human error" that were immediately broadcast on every media channel. The real problem, of course, lay in the interaction between humans, on the one hand, and automated control and navigation systems on the other.

The fact that three officers and the captain did not notice the small DR and SOL symbols on the GPS display may suggest a breakdown in bridge-keeping practices and poor interface design on behalf of the GPS manufacturer. The symbols were very small (about 1/8 of an inch), inconspicuous, and somewhat misleading (SOL). Furthermore, the GPS alarm was not connected to the general alarm system in the bridge, so that when the cable broke and the alarm sounded, only someone who was standing by the unit would hear the one-second beep, which was barely audible to begin with for such a critical failure. Finally, as mentioned earlier, the interface between the GPS unit and the autopilot and radar map was incorrect.

It is also clear that the crew of the *Royal Majesty* over-relied on her automated system. They conducted no cross-checking of navigational data by means of other independent sources such as the LORAN-C navigation, celestial, radar, or compass bearings, to corroborate the GPS fixes. The depth meter data was not used and the depth alarm was set to 0. The second officer was so sure of his track position that he ignored any clue that suggested otherwise. Furthermore, the captain and bridge officers, like most of us, did not fully understand the internal workings of the automated systems under their command; nor did they fully understand how the GPS and autopilot reacted under deteriorating conditions. But how could they? There was no training, whatsoever, provided to the officers about the working of the GPS, autopilot, and radar map.

Mariners have always learned seamanship the old way, the hard way, and for a long time the only way—in the school of hard knocks and on-the-job training. But this ancient tradition no longer holds when it comes to automated systems such as the highly automated bridge of the *Royal Majesty*. The GPS, autopilot, and radar map are complex systems that exchange information, conduct internal checks, employ default modes and settings, and display considerable amounts of information. Understanding how these automated machines behave is a requisite nowadays for knowing "the way of a ship in the midst of the sea."

Chapter 9

''Walk-in'' Interfaces and Internet Applications

Remember the airline reading light we discussed earlier? Well, as with everything, the machines that passengers interact with on today's airplanes have grown ever more complex. In this chapter, we look at the new personal entertainment system available on many aircraft. We start by pressing the video button on the personal control unit in the armrest (figure 9.1). The blank screen awakes with a flicker, but not much else. You hit the video button again but still nothing. That's not very entertaining. You play with the brightness controls thinking maybe that is the problem—but it's not. So you try every button, hoping that something might turn up on the screen. Not today.

Maybe there just aren't any movies on this flight—but two rows down someone else's screen is vivid with action. You might consider calling the flight attendant but you decide to try it again on your own. Maybe the in-flight magazine will explain how this entertainment system works, but if it does you cannot find where. You conclude that either the screen is broken or maybe you are just inept when it comes to technology.

Actually, nothing's broken, and you're not inept. The screen is just fine and all nine movies are playing. It's only that this user-interface has a problem, or to be more precise, the people who designed it were just not sensitive enough to the subtleties of how people interact with kiosks, or "walk-in" types of devices—you know, the Automatic Teller Machines (ATM) outside banks, electronic check-in devices at the airport, and credit-card payment devices at gas stations. We are expected to use such "walk-in" machines without any prior knowledge; there is no user manual, and forget about any training.

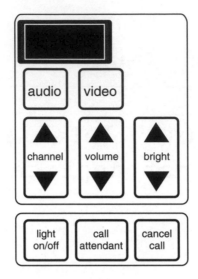

Figure 9.1. The armrest control unit.

Inside the Machine

So let's examine this in-flight entertainment system to understand the kind of problems that exist with these "walk-in" interfaces. Figure 9.2 shows you the states and modes of the system. When you are seated in your cushiony seat at the start of the flight, the system is IDLE and the screen, naturally, is BLANK. You operate the entertainment system by bringing up VIDEO, or AUDIO, or a FLIGHT MAP showing the aircraft position and some geography.

When you select VIDEO, you not only engage this mode but also trigger three of its associated reference values: the video (movie) channels, the brightness, and volume. All three reference values must be in coordination for VIDEO to become active. Here, we shall focus only on the channel reference value, which, as you can see in figure 9.3, runs concurrently to VIDEO. When you select VIDEO, the system wakes up as follows: the on screen turns silver and at the same time, the initial video channel is set to 0 (note the little arrow pointing at 0).

And this is where the problem arises. The A mode, as we know from chapter 7, carries with it a variety of reference values. For us to see the movie, we need to select VIDEO and also a video channel (because the initial channel, "0," has nothing in it). Unless you manually select an active channel (1-9), you will not see anything. You can hit the "video" button as many times as you wish, but it won't get you anywhere because you will just loop around to see more and more empty silver screens.

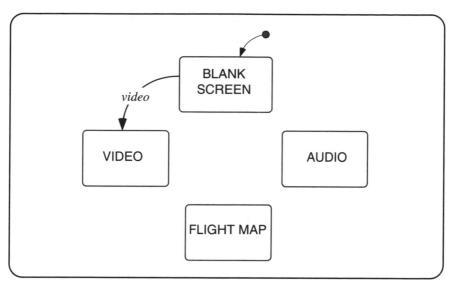

Figure 9.2. Four basic modes of the flight entertainment system.

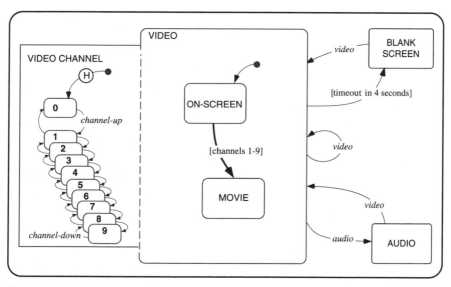

Figure 9.3. Inside the VIDEO mode.

If you are patient and hold off from hitting the "video" button endlessly, waiting for the movie to come on—this system will confuse you more, because in four seconds the screen will automatically turn blank. As shown in figure 9.3, there is a four-second timeout transition out of ON-SCREEN back to BLANK SCREEN state. The only way to see a movie is to press the channel (up or down) button, after you select VIDEO. But nobody tells you that. And it doesn't help that when VIDEO is selected, the little display on the armrest control panel turns dark. If it would display 0, then perhaps we would get a clue.

So why must you press the channel button for the device to transition to a movie? Is it reasonable to expect that people will immediately press "channel" after "video"? Maybe; the evidence, however, shows that many don't.

Initialization

The problem with the in-flight video system starts with the way this system initializes—namely, the initial mode and initial reference values. And then, of course, there are the user's expectations (such as seeing some kind of a screen image after selecting "video") that are violated. Top all this with poor feedback, and you get the kind of confusion that makes us, at times, hate technology.

The underlying concepts here—which are rather subtle, yet present in every device we use—are important to understand: we already know that every machine must start up in some initial configuration. These initial settings are pre-set and programmed by the people who designed the device. The initial configuration for this machine (after turning it on by hitting "video") is ON-SCREEN and CHANNEL 0—a mode and reference value combination—that appears on the screen as silver snow. The passenger who uses the device, however, expects to see a video image. When he doesn't see a movie after pressing "video," he gets a bit confused. After a couple of unsuccessful tries, frustration enters.

A sure way to deal with initial mode and reference value settings is to display and indicate them to the user. But, unfortunately, in this device the initial channel setting ("0") is not shown because the channel indicator on the armrest turns blank. One way to fix this design problem is to make channel 1 (or any other active channel) the initial setting. If that's not available, then perhaps a video image, which will indicate to the user that the screen is indeed functioning. A more comprehensive approach is to consider the initial status (such as the entry screen) of any device, and especially "walk-in" devices, as an "anchor." The initial screen must provide accurate feedback to the user about the working of the device. It can also be designed in a way that it provides the user with information/instructions on how to proceed (and perhaps even some troubleshooting tips).

Mode and Reference-Value Settings

Any time we change a mode of a device, the new mode carries with it one or more reference values. Sometimes these reference values are the same as the previous mode (e.g., the volume stays the same when we switch from VIDEO mode to AUDIO and back). Sometimes they change as a consequence of the mode switching. For example, when we change the mode of a microwave from COOK to POPCORN, the power setting (a reference value) changes to high, no matter what it was before. In COOK mode, the user must enter the cooking time, which is a changeable reference value; in POPCORN mode, the time is pre-set to 4 minutes and 20 seconds.

The issue here is that each mode carries with it a variety of reference values. Understanding this mode and reference-value marriage, and how this relationship changes over time, is an important aspect of user interaction with automated devices.

History Settings

Sometimes, the reference value(s) of a mode depends on the past setting. Consider the following scenario from the in-flight entertainment system: we select a video channel, say 3, and then decide to listen to some angelical Bach on audio channel 8. Afterwards, we decide to watch some earthly video. When we return to VIDEO mode, the channel setting remains at 3. That's what we all expect, right? This way of automatically re-selecting a reference value (channel 3), once we are back in VIDEO mode is called *history* and is depicted in figure 9.3 with an (H). The little arrow shows that on initialization, the reference value is channel 0—however, on any subsequent entries to VIDEO mode, the last setting (e.g., channel 3) is the one that is re-selected.

This history setting can be quite helpful to the user. It saves time and also reduces the amount of interaction with the device. If you look around you will see many examples of this. Your car's audio system, for example, has it. Here is a scenario that I'm sure is familiar: you play your favorite CD, and then decide to tune in to the news. After the newscast is over, you go back to your music. Your CD continues playing from exactly were you left it. But try that on your VCR/TV system at home and you will see that the video continued playing and when you return the movie is way beyond where you left it. On DVD players, however, when you switch modes (e.g., from DVD to TV) the DVD player stops automatically, and then resumes automatically from the spot you left off.

Both good and bad uses of the history feature are seen on the Internet. You go online to get some service and you are asked to fill in your personal information. Then for some reason you click out of the page, and then return back. Will the page contain all your previously entered data, or has it been erased, forcing you to start over again?

Default Settings

Another design problem with the in-flight entertainment system is the default transition out of ON-SCREEN. Here, when the user fails to select a channel (1–9), the screen automatically turns blank. This unfortunately gives the impression that the device has failed, when in fact the system is working just fine. To many of us, this is the last straw—from here we go immediately into the usual litany of self-criticism: "I'll never understand, I'm not good with these things, I hate machines," and on and on.

Like initializations, defaults are also pre-set actions that are programmed by the designers of the system. The term *default* comes from its legal meaning of failing some timely obligation—a loan, an appearance in court, a duty. Defaults exist in every interaction—human-human, human-machine, as well as when machines interact among themselves (such as during a secure interaction between computers). We may tell a friend "Meet me by the subway's entrance at 4 P.M., but if I'm not there in 20 minutes, go home and I'll call you later." *Go home* is the default action and 20 *minutes* is the time to wait it out.

We find defaults when we interact with businesses. Have you noticed those ads about a "month-long free trial" of a weekly magazine? You give them your credit card number and you start receiving the magazine. You can cancel the subscription anytime. However, if you don't cancel by the end of the trial period, *by default* your credit card gets (automatically) charged. Here the default action is to extend the subscription service. An example in reverse, with respect to the default action, is frequent-flier programs: when you reserve a seat with your hard-earned miles, the reservation will cancel, *by default*, if you don't book it within two weeks. Here the default action is to cancel the service (because it is a perk).

In human-machine interaction, default conditions are in place when the user fails to interact—the machine waits for some user interaction, and when it does not come, the system takes control. The classic example is when your ATM machine "eats" your card if you fail to remove it within a minute or so after the transaction is over. Defaults are also in place when the user places contradictory demands on the machine. By default, the machine reacts to one demand and ignores the other. We all understand defaults and are quite accustomed to them, yet they do pose a challenge for designers. Why? Because unlike initial settings, which you can always display and indicate what is the current mode and setting, default transitions will happen in time. It is not trivial to consistently inform the user what will be the consequences of his or her *inaction*. For machines that people own or use for work you can perhaps train and teach the user about defaults and/or describe them in the user manual. But what do you do when it comes to "walk-in" devices where training and manuals are not an option?

When we are dealing with walk-in devices, where the range of user skills in operating the device is huge, the designer has to be very careful. Any expectations on part of the designers about how the casual user "will" interact with a system are merely conjunctures. If there is an available path that the user can take—no matter how unlikely—there will be someone (or more than one) that *will* take it!

This is where models of user interactions, such as the ones described in this book, are useful. Once you have a model that details user interaction, undesirable paths can be eliminated or blocked off, while desired paths may be broadened and illuminated. It is further possible to eliminate much confusion by providing the users with "anchors," like screens that are inserted to help the user navigate his or her way along the system. A well-designed initialization screen can be used as an anchor. Defaulting back to such "anchors," whether it is the initial screen or some other screens that provide feedback on how to proceed, can also help.

To conclude this section, please note that the four concepts discussed here—*initial mode setting, initial reference-values, history settings,* and *default settings*—are all too often confused and used interchangeably. By understanding them and the subtle differences between them, it is possible to provide users with better feedback, and avoid the kind of design problems encountered in the in-flight entertainment system.

Visitors on the Internet

We are now ready to move on along our tour of "walk-in" interfaces. With only a few exceptions, all Internet interfaces are designed for immediate, no training necessary, use. Yet it is crucial for e-businesses that customers stay and not flee with frustration. It is also important that all of us—young, old, computer literate or novice—will be able to interact with Internet applications, otherwise we will be locked out of important information and resources. But how many times have you found yourself lost inside a web site? How many times have you had to retype your personal information, address and credit card number over and over, and wondering if the transaction did in fact take place or not—or if you will be charged twice? Why are so many Internet sites difficult to use?

The World Wide Web (WWW), or more accurately the idea behind what we now know as the WWW, first surfaced at CERN, the world-renowned physics research facility on the Swiss-French border, in 1989. For years, scientists and engineers have used information highways to transfer data files. The foundation was in place, but it lacked an interface that would make the interaction straightforward and usable. It was only with the first browser, Mosaic, developed by Marc Andreessen at the University of Illinois, that suddenly there was a real interface for looking at the underlying information. Mosaic, and then the Netscape browser, provided a graphic interface to the Internet in the same way that the Macintosh desktop provides a window into the working of the desktop computer.

We begin by considering an Internet application that is part of a small information system. The application allows employees of a large research and development institution to log, in advance, the visitors that they are expecting. Many people get confused while using this Internet application. On the machine's side, user confusion and resulting errors cause disorders in the databases, demanding frequent and costly maintenance to the software system. So let's go in and figure out for ourselves what the confusion is all about.

The initial page, VISITOR REQUEST MENU, has a small menu that allows you to "submit" a new visitor request and "browse" current requests, and "browse or copy" expired requests (see the top screen in figure 9.4). It looks simple. We start by clicking on the "submit" button. We land on a page labeled "visit information." Here you patiently enter dates of arrival, dates of departure, time of day, the purpose of the visit. In addition, they ask for your telephone number, your office number, an alternate person to call if you are not there, and so on. Now you're ready to "continue to the next step."

If, for whatever reason, you have tried to cut corners and skip entering the required information, the Web-warden will catch you. Conditional (c) on skipping required information, it displays an error message, a big white banner titled "Sorry, Error #185, you did not enter all the required information," telling you that you've skipped over this or that item. Otherwise, you are welcomed to the next page, which is titled "visitor submission" in figure 9.4.

On the VISITOR SUBMISSION page (at the bottom of the figure), we enter the visitor's information: first name, middle name, last name, affiliation, and so on. At the bottom of the screen there are three buttons: "save visitor," "show visitors," and "done." The "done" button seems most appropriate because we think we are. And as you can see on the right side of figure 9.4, a large page comes up. "Do you want to save visitor information before exiting"? But of course. Press the "yes" button and you are back to the visitor-request main menu, which is where we started from. And since there are eight dignitaries in this group of visitors, which means quite a lot of data entry, we need to repeat this cycle over and over: The visit information, the visitor information, and so on. Done. Back to the main menu. And then the same visit and visitor information again. Done. On and on for each visitor.

But wait. Actually, there is no need for such arduous work. As it turns out, there is a much easier and quicker way of entering visitor's information. The alternate route is shown in figure 9.5. If you click the "save visitor" button,

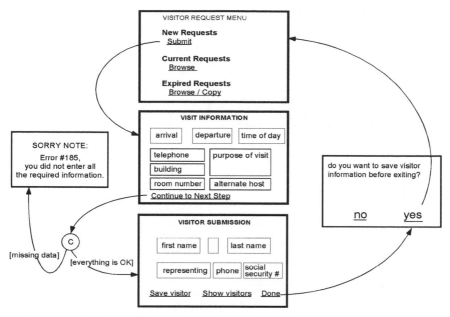

Figure 9.4. Visitor Request main menu and subsequent add/submit pages.

instead of "done," you immediately get another "visitor submission page." Fill it up with the visitor's name, click "save visitor" again, and get another VISITOR SUBMISSION. And so on and so on. This way you can do the data entry in a flash; there is no need to go back all the way to the main menu and there is no need to keep entering the repetitive information about the dates, the time, the purpose, etc., of the visit, which, of course, are common to all the visitors in the group.

Analysis

How would you even know that this labor-saving path is available? There are no cues for it on the interface. The underlying structure of this visitor request process, which can make life in the data entry lane a bit easier, is hidden. In figure 9.5, you will note that the "visit information" page is like the parent and the visitor information pages are the siblings. Once you fill in the visit information, each visitor inherits this information from the parent, and there is no need to repeat it. Part of the misunderstanding lies in the fact that there is a subtle difference between *visit* and *visitor* here: the word *visit* in this Internet application relates to the global information of the visit (the dates, purpose, host, etc.); *visitor* relates to the individual (first name, middle, last, and so on). If you click for the "Help" page and read it carefully, you might find some explanation. But let's be honest, how many times have you gone into and

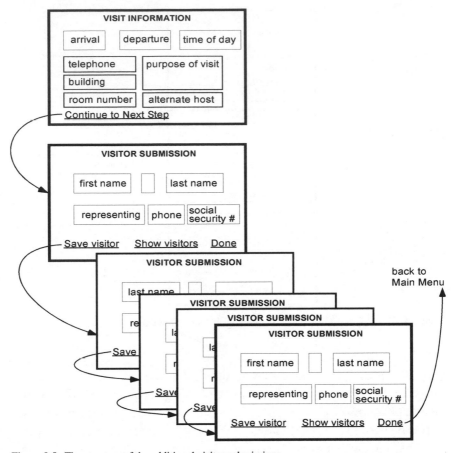

Figure 9.5. The structure of the additional visitor submissions.

carefully studied the help information? And even if you read it, you quickly find that it is not clear and straightforward given the subtle difference between visit and visitors.

Whether you do take the time to read the "Help" page or not, the heart of the problem lies with the hidden structure of this site. The hierarchical relationship between visit (parent) and visitors (siblings) is not portrayed on the page, nor is there any cue or hint of this underlying structure. The visitor information site takes you by the hand and moves you from one page to another. If you follow—and what else can you do?—you'll end up doing much more work than necessary. It is very similar to getting travel directions: "Drive three blocks, then make a right, then turn left at the fifth traffic light, and continued until you reach a dead end." The problem with such navigation, as we all know, is that when the instructions are incomplete—if you miss a turn or make a mistake in counting traffic lights—you are doomed. The only road to salvation is to find the starting point (if you are lucky) and begin again.

Navigation

What is common to all good navigators of land, sea, and air is their ability to understand and take advantage of the structure of their environment. Walking down Seventh Avenue in New York City, an urban "navigator" knows that Sixth Avenue is running parallel on the left and street numbers are descending. Similarly, a desert caravan leader knows that sand dunes run parallel and that a change in their shape may indicate the presence of an oasis. Ocean navigators in the age of discovery closely observed the color of the ocean, the smell of the air, and the shape of distant clouds as indicators of nearby land. The key to successful navigation is the ability to "read" the terrain. Once you know how to read this structure, you can fix the current location and simply proceed. Without an ability to read the structure of the environment, any navigator is lost.

We tend to think that navigation, of any kind, is all about getting there. But really it is about the process of finding one's own way within a certain context and making constant adjustments. The trick to any navigation is to have a reasonably good description and understanding of the underlying structure. Once you have that, you can deal with any contingencies. So if you're driving down Seventh Avenue and your destination is Seventh between 53rd and 54th Streets, and there's some kind of a jam on Seventh, you can slip onto Sixth, drive down a few blocks, and then return back to Seventh Avenue. We all understand that intuitively. But if you put a nomad, who may be a wonderful navigator in the desert, in New York City, there's no chance that he will be able to perform the seemingly (for us) simple maneuver.

Solutions and Exit

So what can be done to improve this visitor information site? One immediate solution is to provide, in the "Help" page, the structure of the application (depicted in figure 9.5). Better yet is to design the pages so that it will be clear to the user that there are two options: if this is a single-person visit, then just complete the "visit information" and return to the main menu; if there is more than one visitor, then there is another route to take. The good news is that when the user is made aware of the structure, things become much easier. It is possible to recover from mistakes, perhaps even finding a novel path when the unexpected occurs.

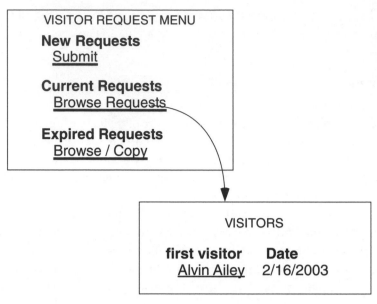

Figure 9.6. Where have all the flowers gone?

OK, so now we are finally done. Now we want to exit. But how? If we click on the "save" button, we're going to get another visitor page, which is definitely not what we want. So we hit the "done" button and get the page with the question "Do you want to save visitor information before exiting?" Yes—but what about the rest of the visitors? Will they be saved too? Well, there are no other choices here. Click "yes," and we are back in the main menu. Click on the "current requests" to make sure that all visitors are saved in there (figure 9.6), and all you see is . . . one visitor on the list!

Where are all the other visitors that you painstakingly entered? Only the first one is listed. What on earth happened to the rest?

Actually, all of them are in there and well saved—it's just that the page doesn't show it. If you click on the first visitor, you can see the "visit information." Press on the "show visitors" button on the lower left side of the screen, and the rest of the visitors show up on the screen (figure 9.7).

Here again, there's actually an underlying structure: the first visitor is like an anchor for the rest to follow. You can see this relationship in figure 9.7. The VISITORS page only contains the name of the first visitor—there is absolutely no clue to this abridged scheme on the page.

The effect of this problematic interface, which obscures the underlying structure, on user interaction is threefold. One, the user is uncomfortable with the site: you get unclear messages after clicking buttons that seem reasonable to click (e.g., the "done" button leads to more questions), and you find yourself

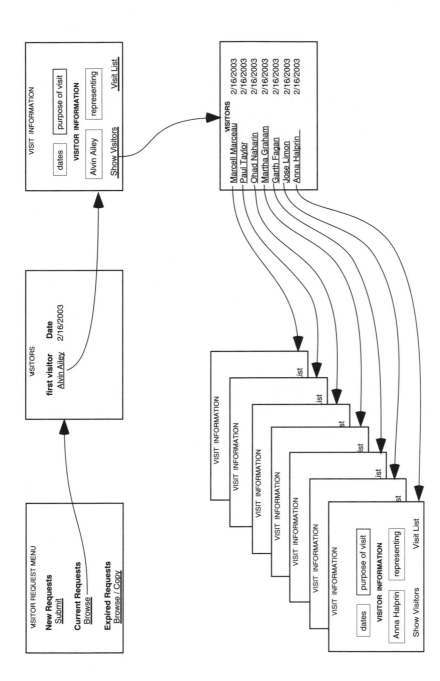

Figure 9.7. Checking the status of the visit.

baffled by what you get (how come only the first visitor has been saved?). Two, most people end up spending far more time than needed in doing this registration task. Worst of all, since nobody really cares to understand the workings of this application—most people just want to get it done, and leave— the problem is never aired. But then, of course, you end up coming back, to the same confusion and inefficiency the next time you have to use the application.

To conclude, the structure of the web site is the foundation of the interaction. The relevant structure must be made apparent to the user, especially when it comes to "walk-in" interfaces. In the beginning, ATM machines were our first encounter with such interfaces. But now we find walk-in interfaces when we buy bus tickets, pay for parking, and call up customer service ("listen carefully as our menu selections have recently changed"). But most of all, almost any web transaction is a walk-in interface. In the not-so-distant-future, we will see more of them in supermarkets, at the dry cleaners, at the post office, and believe it or not, also at the pizza parlor. We will miss the familiar face behind the counter and will have to get used to the interface. But so that all of us—the young, the old, and even the ones that define themselves as "computer illiterate"—will be able to shop, interact, and get errands and chores done, we will need much more sophisticated interfaces. And sophistication is not about fancy colors, flashing icons, and an endless amount of buttons to zing from one place to another—it's about understanding the user's tasks and expectations and building the interface to fit.

Chapter 10

Navigation and Internet Interaction

"This isn't just an ordinary up-and-down elevator!" announced Mr. Wonka proudly. "This elevator can go sideways and longways and slantways and any other way you can think of! It can visit any single room in the whole factory, no matter where it is! You simply press the button and zing! you're off!"

—Roald Dahl, *Charlie and the Chocolate Factory*, Puffin Books, 1998

We are now deep in the world of the wide web. And here is another example of a misleading navigational structure, albeit this one done on purpose. It is about the banner ads that you find on many web pages. You know the ones; you finally come to an informative page about current events in the news, and at the top you see a window-like box with a flashing yellow sign, scrolling words, or one of those flicking "click here to win $100." You click on the little "x" button on the upper right corner of the window to close it and remove the irritating image from your sight (see figure 10.1).

And then you find yourself staring at that ad page full screen. Now you're being presented with more than you ever wanted to know about a porn site, web gambling scheme, new credit card application, mortgage and car loans, or

Figure 10.1. The flashing ad window.

whatever is being advertised. That's not what you wanted! All you wanted to do was to close the window and its flashing yellow triangle. What happened?

Tricky Ads

Well, remember our discussion about population stereotypes from chapters 6 and 7? It's the same thing here. The format and style of interaction with Microsoft Windows is a very strong stereotype of our modern times. We all recognize and intimately understand the function of the three little buttons on the upper-right corner of the window: minimize, maximize, and close. And therefore, without even thinking about it, we intuitively assume that when we click on the "close" button, the window will close. And why shouldn't we?— the graphical window on this web page has every feature of Microsoft Windows style and looks. But it doesn't act the same!

The actual working of this web page and the banner ad is very different from the population stereotype that we are all accustomed to. The three little buttons don't work. They never did. *They are fake.* The entire window is just a big button in disguise because regardless of where you click in the window, you will immediately find yourself transported into the advertiser's world— learning more than you wanted to know about a 100 percent silk camisole for only $27.99. The window is just a big landmine.

The confusion arises because of the contrived divergence between what the user expects and how the machine (or interface software) behaves. Figure 10.2(a) is our user model based on our long experience with Microsoft Windows: upon clicking on the "x" button, the banner-ad window should close, while we stay put on the original page. The machine, however, transitions immediately into the advertising page (figure 10.2[b]). Look at the two models. Both start at the same initial page, but the same event (*click on the* x) takes you to different pages. Does this sound familiar?

To understand the underlying structure of this advertisement trick, we slide the user model on top of the machine model. In figure 10.2(c), the user model and the machine model now form the composite model. In this composite, the initial page is the same for both the machine and user model. The event *click on the* x is also the same in both models. But the end-pages are different: in the user model we end up in the original page while the machine model takes us to the ad page.

The composite model is non-deterministic and the resulting end state, which is a composite of the original page and the ad page, is called an *error state*. In general, error states should be avoided in the design of interfaces

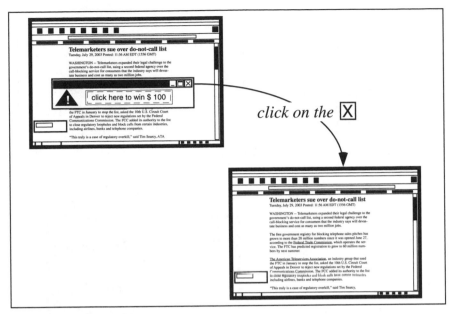

Figure 10.2 (a). User model.

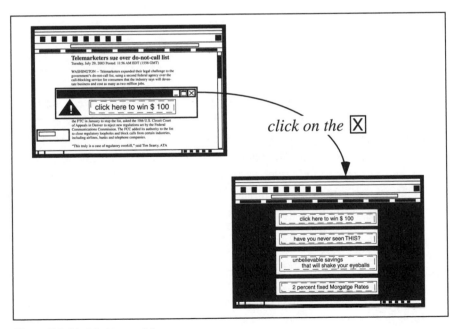

Figure 10.2 (b). Machine model.

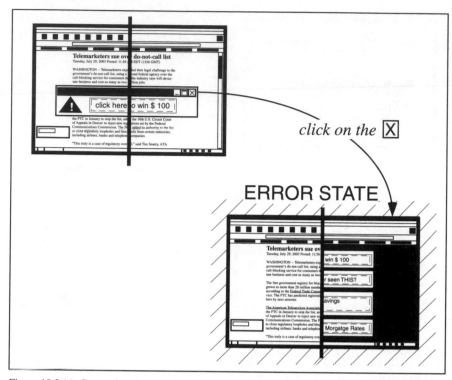

Figure 10.2 (c). Composite of user and machine model.

because they breed false expectations, confusion, and error. (In later chapters we will further explore error states and show that they are indicative of an incorrect interface.) But before we leave the ad example, it is important to emphasize that the error state here is created on purpose. The designer of the window-like ad deceitfully exploits the well-entrenched population stereotypes to induce an error state—which, of course, works well for the owner of the web site that we end up in.

Magic Cards

There is fascinating and strong resemblance between magic tricks and interface design. A successful magic trick is all about cloaking the underlying structure that produces the trick (and *creating* surprise); good interface design is all about revealing the underlying structure to the user (and *averting* surprise). Understanding the principles of magic tricks can be very illuminating and help us design better user interfaces.

Figure 10.3. Choose one card and stare at it.

Here is an example: six playing cards are laid in front of you (see figure 10.3). Please select one with your eyes. Focus on it for 5-10 seconds and remember it.

Now the magician quickly shuffles the cards, spreads the cards in front of you, and quickly places one card on its back (figure 10.4). You look at the five cards in front of you and your card is not among them.

How did he know which card you selected?

Again, the trick is achieved by manipulating the underlying structure: When the magician shuffled the cards, he replaced *all* six cards with different ones. Now, when he proudly places one card on its back, you scan the remaining cards and fail to recognize any of them. You therefore erroneously conclude that "your" card is the one facing down. But, of course, every card from the original deck is missing and, therefore, your card, too.

The tricky ad example is an illusion in the same way the card trick is an illusion. In both cases, the underlying structure was changed. There was no correspondence, whatsoever, between the first deck and the second deck—every card in the second deck is an error state. In both cases, banner ads and magic cards, the error states are contrived. The only difference, of course, is that we come to a magic show to be purposefully misled—but that's not our intention when we are on the Internet.

Now the worst of the lot—which is really a dirty trick—is when an Internet ad window pops up with the following question (figure 10.5): Would you like a free vacation in the Bahamas? Below the statement are three innocent looking buttons: "yes," "cancel," and "no." You hit the "no," and? . . . You guessed right. Since the *entire* window is a button to the Bahamas, it doesn't matter if you click the "no" button or the "cancel" button. Like it or not, you're already in the travel agency's web site.

This trick is very annoying. Granted, some people perhaps understand it and know how to avoid it—but many of us don't. Most of us don't get what has just been done to us, how we were snatched, and how we ended up in the travel agency's web page. Thinking it may be a legitimate part of what this

Figure 10.4. Your card is the one that is face-down.

Internet site is all about, or that we just did something incorrect, we blame and shame ourselves, and continue, never really understanding what actually took place. In some web pages you'll find a sequence of these landmines, sucking you from one page to the other. After surfing these pages with no apparent control you wipe out, exhausted and frustrated.

Souvenirs

Advertisers will always find novel ways to push their products on us. One scheme that is used on the Internet is to give you a (concealed) going-away present. You visited a page that sells a consumer product. Regardless of whether you bought something or not, upon leaving the site, a "souvenir" ad page is automatically launched into the background behind your current browser. You will not immediately notice it because it is hiding behind your browser's window, but eventually you will face it (and some of them still flash and dance even though you are off the Internet).

So what's the problem? Simply close those unwanted ad pages and go on with your net-life. True. Sounds reasonable, doesn't it?

But wait a minute. Would you accept that after you finally close the door on an annoying salesperson, he or she would glue an advertising leaflet on your door? Would you accept that the telemarketer, who bothered you during dinnertime, would also, unbeknownst to you, plant a hidden message on your answering machine?

And then there is the nasty one where you got lured into some kind of "special" 0 percent credit card, a "free" prize, or some other "unbelievable" deal. You click and a new browser is launched for you. After you check out the deal you decide to escape immediately by quitting the browser altogether. You press the little "x" button to close, but then the site automatically launches a new browser on you with a "Leaving so soon?" inscription and additional options. Note what is happening here: after you consciously decide to quit, the site launched a new browser from your own computer!

Figure 10.5. The dirty trick.

Both advertising schemes are done without your consent. Both are intrusive—you may think that you are done and have moved on or quit altogether, but in fact, it is another opportunity to catch your attention and hopefully get you to click-in with your wallet. Advertisers have always taken advantage of those of us who don't understand new rules and new systems—and the Internet is no exception.

Night Dancer

Tired of bouncing ads? Appalled with porn traps and unreal deals? Sick of getting things that you don't want shoved into your face with advertising zeal? Let us now leave the world of virtual advertising and turn to a real performance by a famous dance company. You go to a performance and are wowed by the choreography. Back at home you decide to learn more about the company. Why not go to their web site and check it out? Information is what the web excels in. You surf to learn more about the history and repertoire of this troop. Their web page is marvelous—light blue, purple, and ever-flowing dance images. As soon as it loads, another browser is immediately launched. You can see this automatic transition as a broken line in figure 10.6.

Slowly, images and menus come by. The new browser that got launched for you has no toolbar. It is a window-like browser devoid of all the control buttons that are associated with a regular browser (the "file," "edit," "view" and "go" are gone, and forget about "backward," "forward," and "stop" buttons). On it you learn about the company, the staff, the school, upcoming tours and more. Browse the photo gallery if you like. One item on the new window's menu is the company's e-boutique. Consummate shoppers at heart, let's see what's in there to buy. You put your cursor on it, and click.

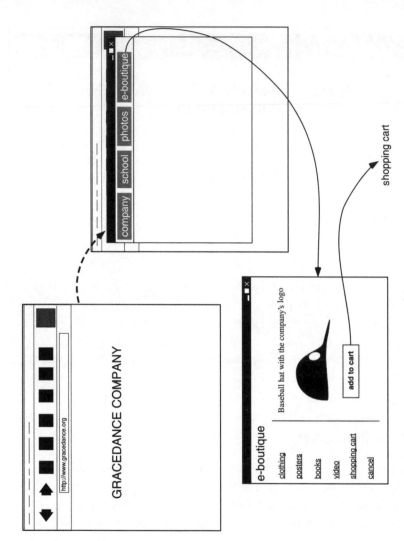

GRACEDANCE COMPANY

http://www.gracedance.org

company | school | photos | e-boutique

e-boutique

clothing
posters
books
video
shopping cart
cancel

Baseball hat with the company's logo

add to cart

shopping cart

Figure 10.6. From homepage to baseball hat.

E-boutique: clothing, posters, books, videos, and more. Who can resist a cool black baseball cap with the company's logo? Add it to your shopping cart.

Proceed to Checkout

Figure 10.7 shows us the items in our shopping cart: one baseball hat. It is affordable enough so let's proceed to checkout. Here is the form-filling page that we are so familiar with: first name, middle initials, last name, street address, town, state, zip, country, phone, e-mail, company. You start entering all the information in with some irritation, but the hat is worth it. Press "continue" and the bill of sale is ready. The last thing to do is enter payment information.

Ready with your credit card in one hand, mouse in the other, you click the "enter payment information" button. But instead of a new page for credit card information, a warning page informs you that your e-mail is invalid—attention to detail is not a common user trait when entering data on the web. You go back to correct your error by clicking the button on the error page labeled "<< back" and it puts you . . . . back into the bill of sale page!

You scratch your head. There are no e-mails to modify on this bill of sale page, so you carefully look around for some clue how to get back to the form filling page.

How? Is there a way out?

Let's consider the options here: you can try clicking the "backward" button to get to the infosheet. But no. Recall that the "backward" and "forward" navigation buttons were removed from this browser. Try the "go" pull-down menu to retrace your steps, but that doesn't work for the same reason.

So what should you do?

Well, you can always click the "x" button and quit the window. But you know what that means—if you still want the baseball cap you will have to go through the same thing again. You can also "cancel," which is one option on the side menu of the bill of sale (see figure 10.7). But wouldn't that just wipe out the whole transaction? Who wants even to try? With no patience for exploring workarounds and side tricks, most people will vote out with their fingers, taking with them their wallets, clicking away to better sites.

Was all this frustration necessary? Why wouldn't the form-filling page check for the correctness of the information, including the e-mail, after you click out of it? Why wait until after the bill of sale to do this? And finally, why doesn't the "<<BACK" button take us back to the form-filling page? Isn't that where we need to go and where we expect it to take us?

Designing user interaction with a web site is a complicated and difficult task. The designer must anticipate the user's clicks and moves and provide escape maneuvers. The problem is compounded in e-business, such as online

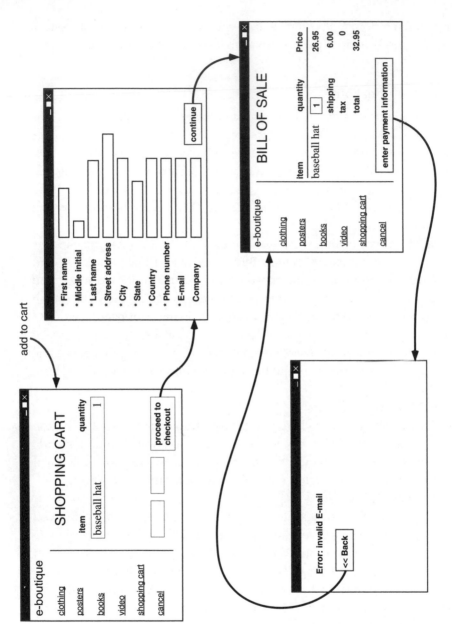

e-boutique

clothing

posters

books

video

shopping cart

cancel

SHOPPING CART

item	quantity
baseball hat	1

proceed to checkout

add to cart

* First name
* Middle initial
* Last name
* Street address
* City
* State
* Country
* Phone number
* E-mail
Company

continue

e-boutique

clothing

posters

books

video

shopping cart

cancel

BILL OF SALE

item	quantity	Price
baseball hat	1	26.95
	shipping	6.00
	tax	0
	total	32.95

enter payment information

Error: invalid E-mail

<< Back

Figure 10.7. The dance lock.

banking and stock brokerage, because of the delicate sequence of transactions that must be maintained. Sequential blunders here can create havoc in the underlying databases. And that's why common navigation tools such as "backward" and "forward" are usually disabled or removed altogether.

Conclusions

In virtual shopping, shoppers no longer care about what a store looks like, the inviting shelves, or the well-dressed and cute salesperson. E-commerce relies heavily on user interface design. The key for success on the Internet is all about face: the interface between users and the web page. Making users interact safely and comfortably with a web site, so they will stay, shop, and visit again, is the ultimate goal. And the mantle of this responsibility lies on the designer. In this context, it is important to remember that if the structure of the user interaction is fundamentally flawed, no fancy colors or jumping icons will suffice to cover up the intrinsic problems.

Although the ad example takes the issue of hiding the underlying structure to its extreme, many times designers of web sites do the same thing unintentionally with the same resulting confusion on the part of the user. An interesting question is why we all accept this. Would you buy a pair of shoes that are too small? Would you even look at a shirt with missing buttons? Would you accept a defective sofa? So why do we all accept defective interfaces?

The only plausible explanation here is that we do not really understand what is being done to us. Unlike a piece of furniture where you can visibly see the defect, the kind of user interface deficiencies we seen in this, as well as in previous chapters, are hidden from view. And because we can't explicitly see these defects, we all tend to assume that the fault lies with our own lack of aptitude, experience, and understanding. Worst of all, we sometimes tend to blame and shame ourselves for being "computer illiterate" and dumb.

Part Three

User Interaction with Automated Control Systems

Chapter 11

On a Winter's Night

The elementary unit of information—
is a difference which make a difference.

—Gregory Bateson, *Steps to an Ecology of Mind*, Part V, chapter 5

I n part two of this book we have examined a variety of everyday devices to see the many characteristics of machines—such as concurrency, hierarchy, relationship between mode and reference values, default setting—that directly affect user interaction. We also considered several important principles of human interaction such as population stereotypes, generalizations, and consistency. In part three of this book, we will see how these characteristics of machines and principles of user interaction play out in much more complicated systems such as cruise controls of a car, medical equipment, and aircraft automation.

Cruise Control

One automated convenience that has made long distance driving more pleasant and endurable for drivers is cruise control. Out on an open highway, it is quite a luxury to switch on the cruise control and give your feet a rest. However, this automation takes a large portion of the control of a powerful and potentially dangerous vehicle out of the driver's hands. It is therefore important to examine how this control system and interface work and what potential confusions may result.

In many modern cars, the buttons for the cruise-control system are mounted on the steering wheel (figure 11.1[a]). On the left side of the wheel is

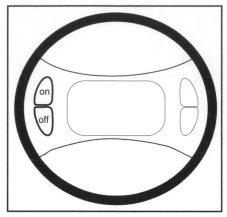

Figure 11.1 (a). The on/off rocker buttons on the wheel.

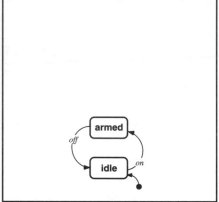

Figure 11.1 (b). *On* and *off*, from IDLE to ARMED.

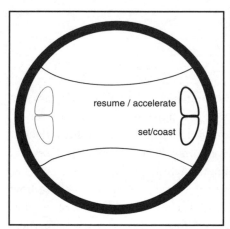

Figure 11.2 (a). Set, coast, resume and accelerate.

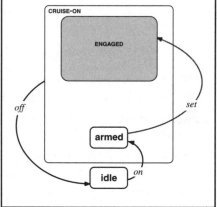

Figure 11.2 (b). From armed to engaged and back to idle.

a rocker switch, which you can use to ARM the cruise control (figure 11.1[b]). The rocker button on the other side of the wheel, like its companion on the left, has two positions. The one on the bottom is called "set/coast" and the one on the top is "resume/accelerate" (figure 11.2[a]). Each button has two functions, indicated by the slash: SET and COAST, RESUME and ACCELERATE. You squeeze the "set/coast" button and the cruise-control system engages with a slight surge; you can sense the smooth suction in the pedal. Now a pneumatic device is holding the pedal at the set speed of 40 miles per hour, allowing you to drop your aching foot and relax and let the trusty machine take over.

The commuter traffic is slugging along at 40 miles per hour and the highway is long. So we can now turn our attention to the underlying cruise-control system. In figure 11.2(b), we note that the *set* event immediately took us from ARMED to ENGAGED. The car drives on its own and the speedometer tells you that your speed is held precisely at 40 miles per hour. However, there is no visual indication to let you know that the cruise control is engaged and alive—but your resting leg knows.

If you wish, you can play with the cruise speed: if you press the "set/coast" button again, the system takes it to mean COAST. And sure enough, the car decelerates. Every push reduces the speed by 1 mile per hour. If you hold it down steadily, the car's speed will drop down much faster. Hearing the honk from the driver behind you, the "resume/accelerate" button is now the savior; if you press on it, the car accelerates slowly to whatever speed you desire.

If at any moment you press the "off" button, the whole thing disengages. The car slows down, and you must use your foot again to maintain the speed. The *off* event triggers a global transition (see figure 11.2[b]). Regardless of where we are, ENGAGED or simply ARMED, pressing the "off" button will always send us down to IDLE.

Cautious Driver

The cruise control is also sensitive to braking. If you tap or step on the brakes, it transitions from cruise control ENGAGED to a configuration called CANCELED (figure 11.3). In CANCELED, as in ARMED, you must control the speed with the gas pedal, because the cruise-control system is no longer engaged. The good thing is that you can always re-engage the system by pressing the "set" button as figure 11.3 shows. The cruise control will engage and maintain your current speed.

It is possible to re-engage and also restore the speed to that at which the traffic is actually moving by pressing the "resume/accelerate" button. This re-engages the cruise control, and the car slowly resumes its previously set (40 miles per hour) speed. Now the cruise control is active and the car, like every other car on this slow-moving highway, is maintaining 40 miles per hour.

Time passes by and the traffic picks up. You gently press on the gas pedal to move ahead. The car accelerates smoothly and the engine is pulling along at a legal 65 miles per hour.

Your foot is hard on the pedal, and your body enjoys the speed. You are now in manual OVERRIDE, which means that the system is not actively controlling speed, yet the cruise control is engaged in the background and ready. If you let go of the gas pedal, the car will slow back to 40 miles per hour, and the cruise control will become active again. In figure 11.4 this automatic transition is guarded by the condition (when speed = 40). Nothing new here, as every driver knows.

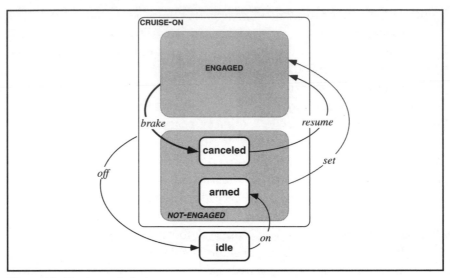

Figure 11.3. Tapping on the brake.

Into the Intersection

You have reached your exit, and now it's time to leave the highway. As you pull onto the exit ramp, you let go of the gas pedal. You have done this a million times. The speedometer falls gently back as the car slows down the long exit ramp, which ends in an intersection on top of a small hill. The car's speed drops to 50 miles per hour and lower as it encounters the hill; 45 miles per hour and you can see the red traffic light, but you know that there is no need to brake yet because the uphill grade will slow you down.

And then a jolt!

The car jumps forward and before you can slam on the brakes, you lurch into the intersection out of control, just barely missing a car crossing in front of you.

Under the Hood

How did this happen? How did a dull commute nearly turn deadly?

We know for sure that prior to the jolt the cruise control was on and engaged. As for the jolt itself, it happened as the cruise control transitioned from OVERRIDE to ACTIVE while the heavy car was climbing up the hill and decelerating. This transition, as shown in figure 11.4, is quite unique: while all other transitions in this system are initiated by the driver, this one takes place on its own, automatically (note the broken transition line).

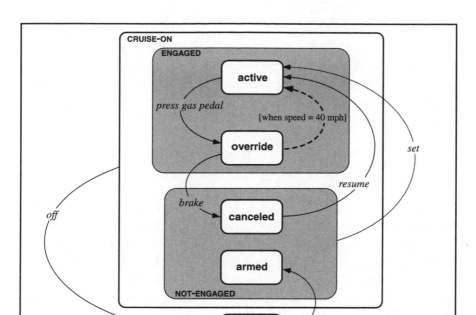

Figure 11.4. The whole cruise story.

So why was this change from OVERRIDE to ACTIVE unexpected? It seems like everyone would know very well how to work cruise control. Evidently, our driver was unaware of this impending re-engagement. The driver expected the car to slow down to a halt, but contrary to expectations, the cruise control re-engaged at 40 miles per hour.

When it comes to investigating such confusions while using automated devices, it is useful to apply the lessons learned in previous chapters. We know that in order to anticipate the behavior of a system, the user must be able to:

1. *Identify the current* mode of the machine;
2. *Know the behavior* of the machine—how and when it moves from one mode to another;
3. *Sense the events* that trigger the transitions between modes.

As in any investigative approach, we first rule out the impossibilities: We know that our driver did in fact release the gas pedal and was fully conscious that the foot was off the pedal. So scratch off item (3). Our driver was also aware that OVERRIDE returns to the set speed after releasing the gas pedal. Scratch item (2). That leaves item (1); namely, did the driver know the current mode of the car?

Let us look at the options here. We know that while driving on the highway, before entering the exit, the driver was actively pressing the gas pedal (at 65 miles per hour). So we may assume that the driver knew that the cruise control was *not* in ACTIVE mode. Figure 11.4 shows that there are several modes in which the cruise control is ON and the driver is actively pushing on the gas pedal:

1. OVERRIDE;
2. CANCELED;
3. ARMED.

How can we tell which is the current mode of the cruise control? Usually there are interfaces and displays to help us figure it out, but in the case of this specific cruise-control system, like others, there is no display whatsoever to tell you if the system is in OVERRIDE, CANCELED, or simply in ARMED mode.

All right, then, so you don't know which one of these three modes is the current mode—but does it matter? What are the consequences of not knowing? Let's find out.

For each of the three modes, consider the state of affairs, after releasing the gas pedal:

1. If we are in OVERRIDE and lift our foot off the pedal, the speed will go down until we end up at 40 miles per hour and return to ACTIVE mode.
2. If we are in CANCELED and release the gas pedal, the speed will go down to 0, just as if we were driving manually.
3. In ARMED the speed will also go down to 0.

You can see all of this in figure 11.5(a). This is the story from the machine's side.

Now let us consider the consequence of the event *release gas pedal* from the driver's perspective. We have already established that the driver cannot differentiate between the three cruise-control modes in which the driver's foot is on the pedal. For the driver, the three modes—OVERRIDE, CANCELED, and ARMED—are indistinguishable (see figure 11.5[b]). Likewise, we already know that when the event *release gas pedal* occurs, there are two possible outcomes: if in OVERRIDE mode, the cruise control will transition to ACTIVE and cruise at 40 miles per hour; if in CANCELED or ARMED, the cruise control will stay at that mode and the car will gradually slow down to 0 miles per hour. But since the driver cannot distinguish between these three modes, he or she will not be able to tell which outcome (ACTIVE and cruise at 40 miles per hour, or gradually slowing down to stop) will take place. The driver can look at the buttons, stare at the dashboard, or do whatever—there is simply no available feedback. And since the driver cannot tell which is the current mode, he or she will not be able to anticipate the next mode.

Sounds familiar?

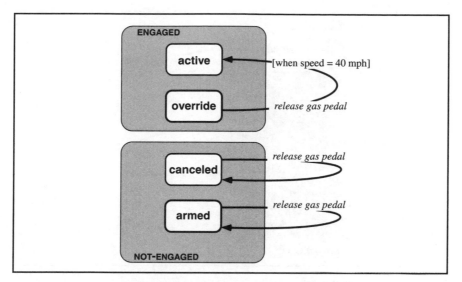

Figure 11.5 (a). Machine model (release gas pedal).

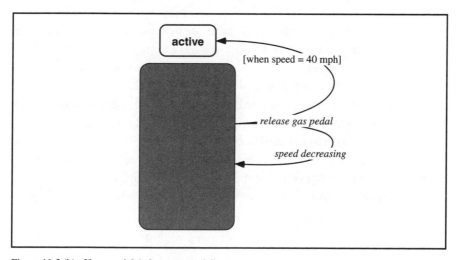

Figure 11.5 (b). User model (release gas pedal).

Non-determinism, Again

The cruise-control system, from the driver's point of view, is non-deterministic. By looking at *all* available feedback, the driver cannot resolve where the system will go after releasing the gas pedal. Moreover, it wouldn't help one bit even if the driver had a complete understanding of how this system works. There is just simply not enough information here.

We can now penetrate deeper and pinpoint the underlying structure that creates this non-deterministic interface. There are two factors at play: the event

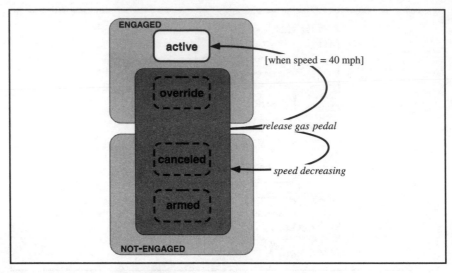

Figure 11.5 (c). Composite model.

release gas pedal and the resulting behaviors (ACTIVE and cruise at 40 miles per hour, or gradually slowing down to 0). It is important that the driver will be able to predict in which of these two behaviors the system will end up. Nevertheless, after the driver lets go of the gas pedal (event *release gas pedal*), he or she cannot predict if the system will end up in ACTIVE or NOT-ENGAGED (which contains both CANCELED and ARMED). You can see this in figure 11.5(c), which is a composite model of the machine and user models: when the driver has his or her foot on the pedal and then *releases* the *gas pedal*, sometimes the cruise control will end up in ACTIVE and at times in NOT-ENGAGED. Clearly, the interface is non-deterministic and we therefore must conclude that the interface for this cruise control system is incorrect.

So What Happened at the Intersection?

The driver completely forgot about the cruise control. The driver assumed, probably even without thinking about it, that since the foot was on the gas pedal, letting go would slow the car. This, however, was not the case on the machine's side. But as we understand now, there was no indication available to the driver about what was really going on; the cruise control was silent. And it did not help that after letting go of the gas pedal, the car did in fact reduce it speed from 65 down to 40, lulling the driver into thinking that all was well. That is why the driver roused with a jolt as the car re-engaged, quite abruptly, to maintain 40 miles per hour in an upward grade, and ran the intersection.

Evaluation and Redesign of the Interface

Now let us turn a more general topic of evaluating and designing interfaces. In our pursuit of understanding how humans interact with automated system, we are not only interested in what happened in this cruise control, but also in what we can learn from it. How would we evaluate a similar device and design it so that non-deterministic problems do not occur? Since we already have the cruise control in mind, let us continue with it, but the focus is now on the general principles for evaluating and designing interfaces that are "correct."

We begin by looking into which modes the user needs to differentiate between. The general idea is that modes that need *not* be differentiated can be safely and correctly lumped, or abstracted, into a single indication, thereby simplifying the interface. In the cruise control example, we already know that we must be able to differentiate between the engaged modes (ACTIVE, OVERRIDE) and the not-engaged (CANCELED, ARMED). Otherwise, as we discussed earlier, the event *release gas pedal* makes the system non-deterministic from the user's perspective.

The next step is to examine the two engaged modes: ACTIVE and OVERRIDE. Do we need to differentiate between the two? Sure. In this design we can distinguish between the modes by noting whether the foot is on the pedal or off. Fair enough, it's a reasonable distinction.

There are two modes in which the system is not engaged: CANCELED and ARMED. Do we need to differentiate between them via some interface indication? Well, it's not as clear-cut as before. How do we go about finding out if they need to be set apart? We do this by considering all the events that take place in our system. They are:

1. *off*
2. *on*
3. *release gas pedal*
4. *press gas pedal*
5. *brake*
6. *set*
7. *resume*

We simply go down the list and check the consequences of these events on both CANCELED and ARMED: *off* (1) works the same for CANCELED and ARMED; it will turn off the cruise control completely, and *on* (2) does not do anything (because the cruise is already on). So with respect to these two events, we don't need to distinguish between CANCELED and ARMED.

Let's go down the list and check the rest of the events: The *release gas pedal* (3) *press gas pedal*, (4) and *brake* (5) events have the same effect whether you are in ARMED or CANCELED. Therefore, according to these three events, we do not need to split the modes. Event (6) is *set*. This event will take the system to ACTIVE, no matter if we are in CANCELED or in ARMED. Great, maybe there is a chance to combine the CANCELED and ARMED modes into a single indication.

But look what happens with resume (7): if we are in CANCELED, the event *resume* will transition us to ACTIVE; in ARMED it will do nothing. You can see this in a graphic depiction in figure 11.6: if we were to lump CANCELED and ARMED together (the shaded area), the system becomes non-deterministic, because the same event takes us to two different end states. We therefore cannot abstract CANCELED and ARMED as we might have hoped; we need to display them separately on the interface.

We have examined the engaged, the not-engaged, and all their modes. What's next? Well, the last possible grouping is between CRUISE-ON and IDLE. In this cruise control there is no indication to tell the driver whether the system is on. (This is because the "on" button is spring-loaded and bounces back to a neutral position. Therefore, it cannot serve as an on/off indicator.) What are the consequences? We proceed along the same lines as before: Here is the list of all the system's events again:

1. *set*
2. *resume*
3. *off*
4. *on*
5. *release gas pedal*
6. *press gas pedal*
7. *brake*

Our goal is to try to lump together and abstract away as many states as possible because we want a simple as possible (succinct) interface. So we try to lump CRUISE-ON (which includes all the modes of the system) and IDLE together. This is the shaded area in figure 11.7. The first event on our list is *set*. Looking at the figure 11.7, we immediately recognize that such an abstraction creates a non-deterministic interface: if we press "set" when the cruise is ON, we transition to ACTIVE; if we press it when in IDLE, it does not do anything (and we transition back to IDLE). Therefore, if we abstract the CRUISE-ON and IDLE into a single indication we end up with a non-deterministic interface—in other words, an incorrect interface.

A New (and Correct) Interface

We have a machine model of how this cruise-control system works, and after going through the above evaluation, we know which modes we need to present distinctly so that the driver can tell them apart. We therefore can

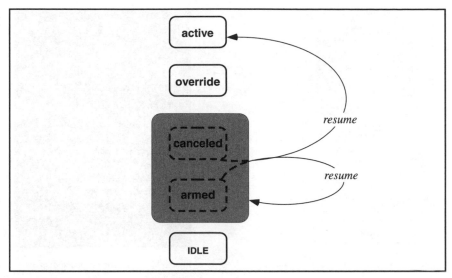

Figure 11.6. Resuming.

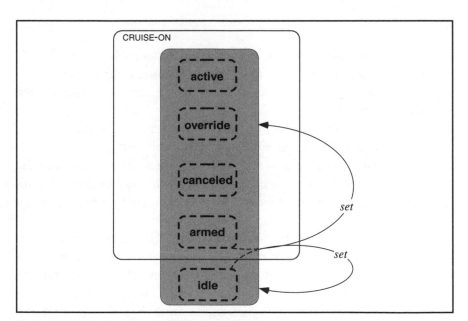

Figure 11.7. Differentiate between IDLE and CRUISE-ON?

prudently proceed to redesign the cruise control interface so that it will be correct. Figure 11.8(a) is a model of the cruise-control system that portrays the modes and the two regions of behavior. We show regions of behavior as levels: at the first level there is the IDLE and CRUISE-ON. ARMED and CANCELED belong to the NOT-ENGAGED level. Above them is ENGAGED.

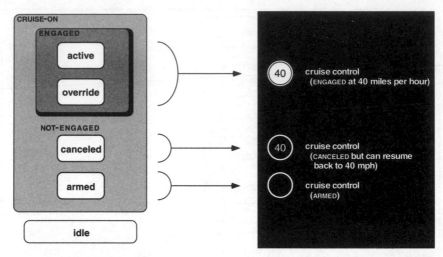

Figure 11.8 (a). The structure of the interface.

Figure 11.8 (b). The proposed display indications.

There is a clear hierarchical structure to this system, where every level is necessary for the other (e.g., you must be in CRUISE-ON in order to transition to ENGAGE). We are going to take advantage of this structure in designing the interface indications.

There are probably dozens of graphical arrangements and display icons that one can use to present this structure to the driver. Here is one suggestion: when the system is turned on, an open circle, along with the word "cruise control," appears on the dashboard (this way we can differentiate between IDLE and ARMED). You can see this open circle at the bottom of figure 11.8(b).

After the driver presses the "set" button and the system transitions to ACTIVE mode, the inside circle is filled and the speed appears in the middle (differentiating between NOT-ENGAGED and ENGAGED). If you press the gas pedal and OVERRIDE the cruise control, the speed value and the filled circle will be there to remind you that the cruise control is still engaged in the background (at 40 miles per hour), even though you are manually pressing on the gas pedal and driving at a higher speed. When you brake and thereby transition to CANCELED, the filling disappears (but the set speed, 40 miles per hour, is still there to jog your memory). And we are done. With the structure that we have identified and the interface suggested here, we have taken care of all the necessary indications in this cruise-control system. Now the interface is correct.

Final Remarks

The source of the user interaction problem stems from the unequivocal fact that the interface between the user and the cruise control was incorrect. Given all the information provided to the user and all available feedback, it was impossible to tell if the system was in OVERRIDE, CANCEL, or ARM. Consequently, the driver could *not* discern if, after letting go of the gas pedal, the car would gradually stop or the cruise control reactivate. Approaching the intersection, the driver assumed the car would stop, but instead the cruise control reactivated and the car ran the intersection at 40 miles per hour.

The transition from OVERRIDE to ACTIVE did not occur instantaneously. At first the car slowed down gradually as the driver expected, and then the automatic transition took place (unexpectedly). In any given system, it is important to pay special attention to such transitions, especially those that are guarded by conditions that are not visible to the user. Such transitions must be well announced, and sometimes it is prudent to alert the user before they are about to occur. Another important point to remember is the role of the interface as a memory jog—a log book that keeps track of the current configuration (including mode, state, parameters) of the machine. If the driver kept a log of his prior actions (engaging the cruise control before taking manual override of the speed), then it would have been possible to anticipate the reactivation. But we cannot expect users to keep track, in their minds, of previous modes and events in the system. That's an unreasonable expectation that leads to an unreasonable design.

But most of all, the purpose of this example was to illustrate the process of evaluating an interface. We understood the problem with the existing cruise-control interface and made sure that the new interface is correct for the task. This is our foremost concern. Beyond that, another important requirement is that the interface should be as simple as possible. And by simple I mean with as few modes and indications as possible. There is a clear advantage for simpler interfaces—they usually are cheaper to produce, they make user manuals smaller and less confusing, and place less of a burden on the user. In most cases, simpler interfaces are probably more "user-friendly." While the cruise control is an example with few modes, in a more complex system with hundreds of states and modes, there is a high premium for such simpler, or succinct, interfaces.

In the course of evaluating the cruise control, we went through a systematic, step-by-step process that can be applied for designing correct interfaces. This process, which was presented here in an intuitive and simplified way, is the foundation of a formal methodology for constructing correct and succinct interfaces. The goal of this process is to determine which modes should be presented on the interface and which modes can be grouped with others (and therefore abstracted away from the interface). The resulting abstracted interface is one that provides the most elementary amount of information (e.g., modes) necessary to control the system—that is, it displays only differences that actually make a difference.

The above process of designing a correct and succinct interface identifies the building blocks necessary for laying out the graphical user interface (GUI). It is important to note that there are many human factors and ergonomic considerations that must come into play for our suggested interface to be not just correct and succinct, but also suitable, efficient, and practical. We must work hard to make sure that the color of the interface indicators is appropriate (will it work well with other indications on the dashboard? Can it be read by people who are color-blind?). We must also consider the size of the indicators and identify the most appropriate location for them on the dashboard. We can perform simulator experiments and conduct subjective assessments to explore whether it will be helpful to add a dedicated indicator for the OVERRIDE mode. This will be redundant information, but it may prove to be safer. There are many ways to further improve the user interface and to design it in a pleasing and aesthetic way, but the foundation of any design is the information content. And with the above analysis, we have taken all the steps to make sure that the information content provided to the driver is both correct and succinct.

Chapter 12

Incorrect Blood Pressure

Limbo: *In some Christian theologies, the eternal abode or state, neither heaven nor hell.*
   1. *any intermediate, indeterminate state or condition.*
   2. *a place of confinement, neglect, or oblivion.*

—*Webster's New World Dictionary,* Third College Edition, 1988

With fast highways and gloomy intersections behind us, we enter into the world of hospitals and operating rooms. The story here is about a medical device, a blood-pressure machine. These machines are very reliable and are used in wards and operating rooms. Nevertheless, the overall reliability of a system depends not just on whether the internal components work well. Reliability, as a design and evaluation criterion, should be extended to include correct interfaces and suitable user interaction.

The Blood-Pressure Machine

The device, as the name implies, automates the blood-pressure measurement sequence. Placing a cuff on the patient's arm is done manually by the nurse or physician, but the sequence of pumping the cuff, finding the peak pressure in the arteries, deflating the cuff, and measuring the lowest pressure—is all automatic. The device we will examine here is in the operating room. An operation will start in a short time, and while the patient is being prepared in an adjacent room, an anesthesiologist and his resident are checking the array of machinery scattered around. The boxy-looking blood-pressure machine sits on a metal cart by the surgery table (figure 12.1[a]).

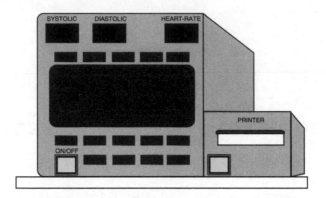

Figure 12.1 (a). The blood-pressure machine.

The machine comes to life by pressing the "on/off" button at the lower right of the interface (figure 12.1[a]). But then it takes about two minutes for the blood-pressure machine to go through an internal examination called a built-in test, which checks that all interior circuitry and software routines are intact. Such built-in tests are quite common in safety-critical systems. Their role is to check critical hardware and software functions, making sure the machine has no hidden failures, and then displaying the results of these internal checks in a short and binary message, either "BUILT-IN TEST—OK" or "BUILT-IN TEST—FAIL."

Today, at least for now, all is well. The built-in test is OK and the rectangular screen lights up with a glowing phosphorus green. Figure 12.1(b) shows that once the test is OK, the machine transitions automatically to OPERATIONAL.

Operational

Anesthesiologists and operating-room staff spend a considerable amount of time preparing and checking equipment to discover potential malfunctions, making sure that the equipment is operational and safe. In helping the surgeon and staff perform the operation safely, anesthesiologists have two goals: one is to induce and then maintain anesthesia throughout the surgery; the other is to detect, and promptly correct, deviations from the planned sequence of actions. Anesthesia, contrary to common belief, is not about "putting the patient to sleep." Rather it is a complex process that begins by gradually disarming the nervous system with extremely potent drugs; keeping the patient anesthetized, but alive, while he or she is subjected to excruciatingly painful surgical interventions; and then safely returning them to consciousness. Throughout this process, recording and assessing bodily signs such as blood pressure is vital.

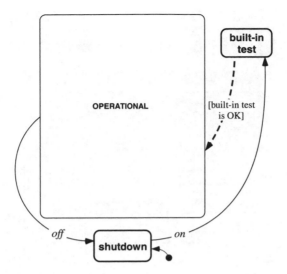

Figure 12.1 (b). From SHUTDOWN to OPERATIONAL mode (via the built-in test).

Now the patient is brought into the operating room and is quickly hooked up to the machines; the gray cuff is placed over his left arm, close to the heart.

Manual Measurement

The anesthesiology resident makes an initial measurement. He presses the "start" button (see figure 12.2[a]). The machine awakes and transitions into a SINGLE MEASUREMENT cycle (figure 12.2[b]). The cuff, which is now tightly wrapped around the patient's left arm, starts to inflate, producing a tightening pressure. Several seconds later, the machine has reached the maximum cuff pressure and stops. The cuff is fully inflated, and the brachial artery, under the triceps muscle, is blocked.

A hissing sound follows as the cuff deflates. The pressure on the arm is slowly relieved. In the old and familiar apparatus where the doctor pumps up the cuff with a black rubber bulb, the doctor listens carefully with the stethoscope, patiently waiting to detect the blood's return into the arm, which sounds like a turbulent stream as it passes by and overcomes the surrounding cuff's pressure. The heart, contracting and expanding, pumps blood into the arteries. It flows in a wave-like, undulating rhythm. The blood-pressure measures the *peak* and *trough* of that wave. In this blood pressure machine, a (Doppler) flow sensor is used to detect the blood's return. The cuff's pressure, at the point in which the blood starts returning, is equivalent to the *peak* pressure at which the heart pumps blood into the body. This peak pressure is the systolic pressure.

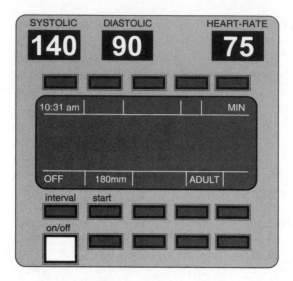

Figure 12.2 (a). Interface (in MANUAL mode).

As the pressure in the cuff is relieved, the brachial artery gradually opens up, and more and more blood gushes through. Seconds later, the cuff no longer restricts blood flow. At this point, the cuff's pressure is equivalent to the blood pressure inside the artery. This is the *resting*, or diastolic, pressure. The machine deflates a bit more, but the measurement is over. With the peak and rest values recorded and stored inside, the cycle is complete, and the machine transitions to OFF (figure 12.2[b]).

The systolic and diastolic pressures that correspond to the just completed measurement cycle are recorded and stored inside the machine and displayed on the top of the interface. In figure 12.2(a), the systolic blood pressure is at 140 millimeters of mercury and the diastolic is 90. The large rectangular screen, just below, provides information about the modes and setting of the machine: the current time, as you can see on the top-left corner, is 10:31 A.M., the patient is an ADULT, and the cuff is set to inflate all the way to 180 millimeters of mercury before it automatically deflates.

For the anesthesiologist, this is a wonderful labor-saving device, alleviating the time-consuming and attention-demanding measurement and recording with the old stethoscope and squeezer cuff. With the automated machine, all you do is press "start," and from there on everything is automatic. Furthermore, everyone in the operating room can look at the interface and see the patient's blood-pressure values.

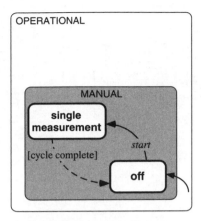

Figure 12.2 (b). Measurement cycle in MANUAL mode.

Automatic Measurement

Now the anesthesiologists are ready to sedate the patient with powerful drugs, bringing him slowly out of consciousness. During this descent, frequent measurements are required. So, short of pressing the "start" button every minute or two, the sequencing of measurements can also be done automatically, like an automatic timer. This way, the machine can run independently—initiating the measurement cycle, inflating and deflating—and then recording and presenting the blood-pressure measurement for display.

The anesthesiologist can choose among several automatic configurations: the machine can be set for CONTINUOUS operation, in which case the device is continuously measuring the patient's blood pressure (i.e., one cycle after another, no intervals). This setting is used if the patient's condition is severe and the surgeon and anesthesiologist need constant measurements. Another option is timer-interval, in which you set up the interval, or the hiatus, between successive measurements, and choose between 1, 2.5, 5, 10, 15, 20, 30, 45, 60, or 120 minutes. Note that these intervals, as well as CONTINUOUS, are all reference values for the automatic mode.

Here is how the anesthesiologist actually sets up the machine for automatic mode to work: he first presses the "interval" button (lower left corner in figure 12.2[a]). This triggers an internal event, called "interval," which takes the machine from OFF to CONTINUOUS (see figure 12.3[a]). Another button press takes it to the one-minute interval setting. Press "interval" again and the machine is set for 2.5 minutes, another press and it's 5 minutes. The interval setting appears on the screen, just above the interval button (see figure 12.2[a]). During the course of the operation, the timer-interval is usually set for 5-, 10-, or 15-minute intervals, depending on the patient's condition. A

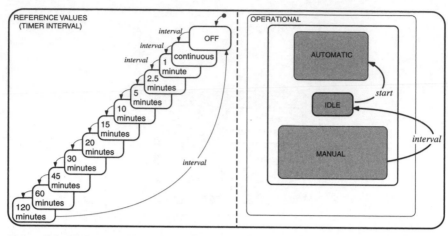

Figure 12.3. Modes and reference values.

common timer interval during the start of anesthesia is 5 minutes, and this was indeed the setting during the operation. (The longer interval settings, such as 60 minutes and 2 hours, are commonly used while the patient is in the recovery room.)

The event *interval*, in addition to setting the timer-interval value, also takes us out of MANUAL mode. As you can see in figure 12.3, the transition takes us initially into IDLE. And there the machine waits until the anesthesiologist presses the "start" button to begin the automatic measurement cycle.

At the completion of the first measurement cycle in automatic mode, the machine starts a time counter. (In figure 12.4[a], on the upper right corner of the screen, you can see the time counter—now showing that 2 minutes have elapsed.) Meanwhile, the anesthesiologist is carefully monitoring the patient's response to the drugs that he is providing through the intravenous tubes. By now the counter has reached five whole minutes. The blood-pressure machine wakes up and the cuff inflates with a sigh. At the end of this measurement cycle, everything looks normal—blood pressure is 135 over 83 and heart rate is at 70 (figure 12.4[a])—and now the counter starts again dutifully, counting up to the next measurement cycle.

Mode and Reference Values

The main task of the anesthesiologists is to assure the well-being of the patient through constant monitoring of the patient's vital signs. The process of anesthesia involves inducing sleep, bringing the patient down to a state of no pain (analgesia) and paralysis, and relaxation. Anesthesia is a very fragile place, kind of an in-between state. During anesthesia the intensity of pain is delicately

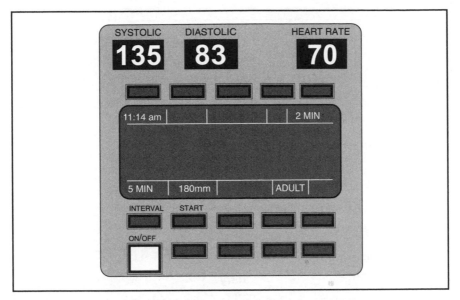

Figure 12.4 (a). In AUTOMATIC mode (2 minutes after start of measurement cycle).

countered by depth of sedation: if the patient is too awake he will experience excruciating pain, if he is too sedated he may be sliding down a dangerous and potentially lethal slope. Therefore, there is not a fixed amount of drugs that are used in anesthesia; rather, the drug quantities are titrated by their effects on the patient. Because of the close link between drug administration and the functioning of the heart, control of blood pressure is critical. And while the anesthesiologist and his resident are closely monitoring the patient's descent, and the machine cycles properly every 5 minutes, we can turn our attention back to the internal characteristics of the machine's behavior.

In considering user interaction with this device, we focus straight-away on the relationship between modes and reference values, which, as we have seen in previous chapters, is a source of many user-interaction problems. The event interval scrolls the machine between numerous interval reference values; now look at the event *interval* in figure 12.4(b). It appears there three times: it takes us out of manual mode into idle (*interval=continuous,* or 1 minute up to 120 minutes); it also transitions us from AUTOMATIC to IDLE, and finally, when we are in AUTOMATIC mode and would like to change back into manual engagement, we do this by pressing interval and setting it to off (*interval=off*).

The point is that the event *interval*—although appearing as an event that only changes (the timer interval) reference values—also switches modes. This multifunctionality of *interval* is a subtlety that will become crucial later. But for now, it is important to remember that quite frequently, during the operation, while the machine is in AUTOMATIC mode, the anesthesiologists may need to

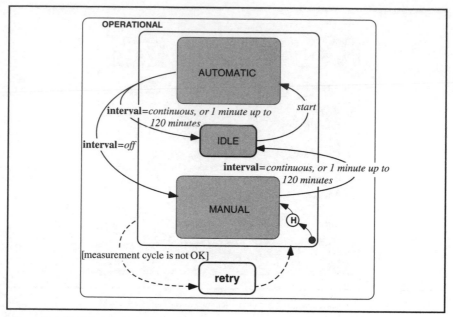

Figure 12.4 (b). Modes and envelope protection.

change the timer interval to either a higher or lower setting depending on the surgical needs and the status of the patient. To do that, they first need to select the desired timer-interval setting (say to 1 minute), and then press the "start" button to reactivate the machine in automatic mode (figure 12.4[b]).

Automatic Protection

Before we complete the description of the machine, it is worthwhile to mention that in addition to the presence of a watchful anesthesiologist, there is a built-in protection mechanism making sure that each measurement is performed properly and accurately. In figure 12.4(b), the model of the blood-pressure machine is expanded to include a protection mode. The RETRY mode is like a security blanket that covers both the manual and the automatic mode. While the machine is performing the cycle, there is a sensor and a software routine that constantly evaluate the quality of the measurement. Think of it as an ever-present quality-control team. If the measurement cycle is bad, it takes over and makes the machine repeat the cycle.

Such protection systems—that constantly make sure that the system does not wander beyond its zone of safe operation—are called *envelope protection systems*. You find them in machines that operate in safety-critical domains such as medicine, aviation, space flight, and nuclear power, making sure that these automated machines will work within the safe operating envelope and not produce bad outcomes. (In chapter 15, we will talk more about such envelope protection systems, in the context of aircraft and autopilots.)

Coming back to this blood-pressure machine, once the measurement triggered by the RETRY mode is okay, the machine automatically returns to normal operation. Figure 12.4(b) shows that the machine will return to the previously held mode. But how does the machine know which mode was previously active? The machine actually keeps track of this historical information (and we denote this ability to remember the previous mode with a history [H] transition that we have already discussed in detail in chapter 9). However, if the measurement triggered by the RETRY mode is still bad, the machine will try again. After three consecutive bad tries, the machine will abort automatically, and an "UNABLE TO MEASURE" message will appear on the screen.

A Critical Incident

Surgery can last for hours, while the anesthesiologists sit there and monitor the journey undertaken by the patient. But there is always a potential danger lurking just below the surface. Among bridge officers, anesthesiologists, pilots, and technicians who supervise critical automated systems, there is a colloquialism that the job is characterized by "hours of boredom—moments of terror." When those moments come, the anesthesiologist is usually faced with a complex situation that stems from the fact that many events and bodily functions are highly interconnected.

And things have gone *very wrong* with the surgery we have chosen to observe. The following is an excerpt from the investigation into a mishap that occurred with this specific blood-pressure machine:

"The patient's blood pressure read high ("hypertension") and the surgeon complained of profuse bleeding in the surgical field. The anesthesiologists set the device to read more frequently and became quite busy trying to lower the blood pressure to help the surgeon control the bleeding. Periodically the anesthesiologists glanced back at the blood-pressure device, only to see that it still showed an *elevated* blood pressure. They did not realize that it showed the same blood pressure as before and had never made another measurement. So, for 45 minutes they aggressively treated the hypertension with powerful drugs that dilated blood vessels. The bleeding continued to be a problem until the surgeon got control of the bleeding sites. Finally it was discovered that there had *not* been a recent measurement of blood-pressure, which was in fact found to be very *low*." (A low blood pressure is especially dangerous as vital organs may suffer irreversible damage if they are not sufficiently perfused by oxygen-rich blood). In the end, the anesthesiologist was able to restore normal blood pressure and fortunately there was no lasting harm to the patient.

Analysis

Let's start by considering what happened here in the context of what we already know about this blood-pressure machine. At first, everything was normal and the machine was in a fully automatic mode at five-minute intervals. Later, the incident erupted: the blood pressure increased, and abdominal bleeding was out of control. The surgeon asked the anesthesiologists to reduce the patient's blood pressure, and they began administering drugs. But first, to better monitor the intervention, they set the machine to one-minute intervals.

What changed this episode from a routine intervention to an almost fatal complication was one (missing) button press. After resetting the timer interval from five minutes to one, the anesthesiologists probably forgot that a subsequent press on the "start" button was necessary to restart the measurement cycle.

At this point, we could place the blame on the anesthesiologists and classify it as "anesthesiologist error." But what assurance is there, for potential patients, that a similar error will not happen again? We need to go beyond the blame and try to find out what prompted this medical error, what was the sequence of events, and what can be learned from this incident that will help us better understand user interaction with automated systems.

Modes

From the user's perspective, the machine can operate in two distinct modes: (1) MANUAL, in which the user initiates each measurement cycle by pressing the "start" button, and (2) AUTOMATIC, in which the user pre-selects the interval and the machine initiates each measurement cycle automatically.

However, if you look carefully at figure 12.4(b), you will note that between MANUAL and AUTOMATIC there is another configuration. And this configuration is IDLE. When the machine is in MANUAL mode, selecting any interval setting sends the machine to this IDLE; a subsequent press on the "start" button is required to engage and activate the AUTOMATIC mode. And when the machine is already in AUTOMATIC, changing the interval (for example from five minutes to one), also sends the machine to IDLE, where it will stay, unless the user presses the "start" button. Therefore, for all practical purposes, we really have three different distinct modes here—MANUAL, AUTOMATIC, and IDLE. The user, however, is not fully aware of this intermediary IDLE mode. This, as you will see soon, played a significant role in the confusion that brought about this medical incident.

Feedback

Now let us consider the feedback provided to the user about the modes of the blood-pressure machine. When we are in MANUAL mode, the interval indication at the lower-left corner of the screen says OFF (see figure 12.2[a]); and when we are in AUTOMATIC, the same indicator shows the interval setting (five minutes in figure 12.4[a]). It appears straightforward and clear.

But consider what happens when we change the interval from five minutes to one. We press on the interval button, and now "1 MIN" is displayed on the screen. But has the machine transitioned to AUTOMATIC? Not really. Will the system sequence a future cycle? No; it will be stuck in IDLE (awaiting subsequent "start"). Nevertheless, the screen shows us the same indication *as if* we were in AUTOMATIC mode. The user assumes the machine is in the fully AUTOMATIC mode, when in fact the machine is stuck in a limbo mode—idling indefinitely in this in-between mode.

To see the problem, look at figure 12.5, where the indications (OFF, 1 MIN) are superimposed on top of the machine model. The resulting composite

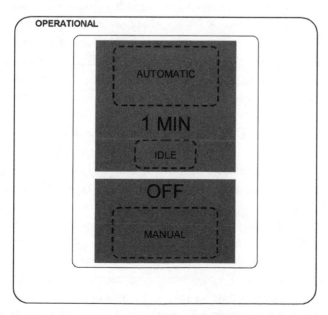

Figure 12.5. Superimposition of the interface indication on top of the machine; showing that the "1 MIN" indication covers both in AUTOMATIC and IDLE modes.

shows that the "1 MIN" indication can occur in both the AUTOMATIC and IDLE modes. But as we know well, in IDLE mode the machine will not make any measurement whatsoever. The interface provides the anesthesiologist with misleading information. Specifically, the interface is incorrect because two very distinct modes (AUTOMATIC and IDLE) are abstracted into a single indication ("1 MIN").

Population Stereotypes

Beyond the fact that the interface here is incorrect, it is also worthwhile to try to find out why the anesthesiologist or his resident forgot to press the "start" button immediately after selecting the one-minute interval. If these two user actions (select "interval" and press "start") had been conducted sequentially, the system would have gone immediately into AUTOMATIC mode. (This, of course, would not eliminate the incorrect interface problem—but it would certainly bypass it.)

Recall that all the anesthesiologists wanted to do was simply to change the timer-interval setting from five minutes to one. They were not after any mode changes. The kind of change they sought is not very different from changing the temperature setting in your toaster oven. For example, say that the salmon is not cooking fast enough and you want to increase the temperature. What do you do? On almost every oven, you simply grab the temperature knob and rotate it from 300 Fahrenheit to 450, and that's it. You are not expected to tell the system that you want it to stay in OVEN mode—you know it will.

As for light switches and TV/VCRs, there are strong population stereotypes when it comes to operating automated devices. But remember that there is nothing inherent in our brains about this or that population stereotype; the way we interact is simply based on our accumulated experience with automated devices. In particular, the relationship between a mode and its reference values, such that a simple change in reference value does not affect the mode, is one such prevailing stereotype. The anesthesiologists, like all of us, under stress and hurry, probably have fallen back into this stereotype. They assumed that changing the timer interval reference value is just that, and that the machine would stay in AUTOMATIC mode and measure at a faster rate. It didn't.

You see, the underlying design problem in this blood-pressure machine is the fact that the mode and reference-value settings are combined. The reference value is used both for setting the timer interval (such as five minutes and one) and for switching modes (MANUAL, AUTOMATIC, and OFF). Such designs usually result in hidden behaviors that many times even the designers of the system are unaware of. As a consequence, such designs are a breeding ground for confusion, frustration, and error. (In chapter 16 we will see a somewhat similar mode and reference value problem in autopilots of modern commercial aircraft that results in hidden behaviors.)

Recovery

There is a two-pronged approach to dealing with such mode and reference-value problems: one is to improve the design in order to stave off potential confusion. The other is to provide feedback that will allow for early detection and ways for a quick and safe recovery. So the next thing we will try to understand is why it took so long for the staff in the operating room to detect that the systolic/diastolic measurements were never updated, because with every minute that elapsed and drug dosages that were added, the patient's condition was getting worse. Just like the machine, he was also in limbo, holding tight, but slowly losing grip on that line between life and death.

Almost everyone in the room could see the blood-pressure values. The displays for the systolic and diastolic values, located above the screen, were rather large (about one inch high). However, there was no direct cue or salient feature anywhere on the interface to indicate the "age" of the values. Therefore, it is very likely that both the anesthesiologists and the rest of the surgical team believed that the blood-pressure measurement values were being updated, when in fact they were not.

The only indication that could cue the anesthesiologist that the device was stuck in idle mode was the small "time elapsed" indicator on the upper right corner of the screen (see figure 12.4[a]). This indicator displays the counter that starts ticking at the beginning of a new interval. Nevertheless, this was only an indirect cue that required that the anesthesiologist compare the elapsed time to the interval time. Apparently, it just wasn't powerful enough to compete with the prominent and compellingly misleading systolic and diastolic values on the interface.

In Closing

Several design features contributed to this critical incident: The root of the problem is the way modes and reference values were combined in this machine. As a result, it was necessary to design the interaction in such a way that required the user to always press the "start" button after changing the reference value, which, in turn, is a violation of a strong population stereotype. Finally, the lack of indication about the measurement's (systolic and diastolic) "age" hindered the surgical team's ability to recover from the error.

There are many potential solutions to the design problem in the blood-pressure machine. It may be possible to modify the design such that after changing the reference value, there is no need to press the "start" button. A more viable approach, one that both corresponds to the prevailing population stereotypes and provides salient mode feedback, would be to completely split the modes and reference values, such that the modes are clearly defined and displayed separately from the reference values. Finally, the indications of the systolic and diastolic values must include a feature that indicates the measurement's age, so that the nurses and doctors know when they are looking at old readings. It is noteworthy that the problems identified in this blood-pressure machine cannot be labeled as a manufacturing defect in the narrow sense of the term—because nothing broke and nothing worked different than advertised. Nevertheless, a design deficiency does exist in this machine. Furthermore, the resulting user interaction problem lies dormant, and when triggered, can easily turn a routine situation into a critical and harmful one.

There are numerous ways to improve the user interaction and the interface of the blood-pressure machine. Coming up with a "good" design is still an art form. But in terms of making sure that the design is good, we first want a correct interface. That is, it is essential that all the necessary information is there and that the interface does not tell you that the machine is in one mode when in fact it is in another. Second, we want to make sure that there are no violations of population stereotypes and there are no situations when, considering human interaction and human limitations, the interface "appears" non-deterministic and therefore incorrect to the user. Third, we need to make sure that there is consistency in the design of menus and that the labels on the buttons and interface are meaningful for the task. Finally, it is most important to conduct extensive usability testing with actual users and to evaluate the design in context. All of the above steps are important for consumer devices and are essential for safety-critical systems such as medical equipment.

Chapter 13

Procedures, Synchronization, and Automation

It is only with the heart one can see rightly;
what is essential is invisible to the eye.

—Antoine Saint-Exupéry, *The Little Prince*, chapter xxi

Around midnight on a stormy night, a medium-sized commercial jet was making its way into a midwestern airport. For the crew, this was the last flight in a four-day trip, with numerous takeoffs and landings, many hours of flying, and too little sleep.

"Turn left 150 degrees."

"Roger, heading 150," replied the captain to the air traffic controller's instruction. The copilot was flying the aircraft. On this flight, it was his turn. A young man in his early 30s, the copilot was sitting close to the control wheel, holding it tight with his hands, trying to keep the wings level in the turbulent air.

They were gradually descending from 5,000 to 2,000 feet, and the captain was helping with the radios and procedures. He was also closely monitoring his copilot's actions. The captain placed his hand on a small handle above the side window and pulled himself closer to the instrument panel. Passing 3,000 feet, he checked the instruments, compared the altitude in his altimeter to the altitude depicted on his copilot's altimeter, and said, "One thousand feet to go." He then rose up in his seat and peered outside. They were deep inside thick and heavy rain clouds, completely engulfed by darkness.

The plane was an older model aircraft, built in the late 1960s. There were just two pilots in the small cockpit. An aluminum-covered door was positioned behind them, and a large instrument panel—a myriad of dials, switches, and

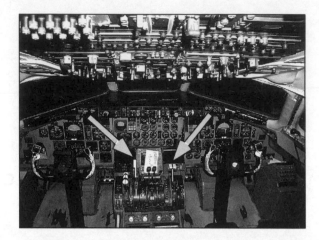

Figure 13.1. The cockpit (showing the two control wheels and the instrument panels). The throttles are located in the center of the pedestal between the two control wheels. The white arrow on the right points to the flap handle and the arrow on the left points to the spoiler level. Photo courtesy of Bevin Shively.

levers—in front (figure 13.1). On the right, the copilot was leaning forward toward the instrument panel, adjusting a switch. They were in a turn, descending toward an unseen runway that lay below.

As the aircraft continued its turn, the copilot looked at the captain and said, "Flaps extend, flaps five, please." Between them stretched a long pedestal filled with levers, handles, and two small, sporty-looking throttles. The flap mechanism—which looks similar to a gear handle in an automatic car—was located by the copilot's leg. The captain grabbed the flap handle, and pulled it back.

The sharp metallic sound of the flap handle moving toward the five-degree notch merged with the crackling voice of the air traffic controller that came through the speakers: "Turn left heading zero seven zero, maintain one thousand eight hundred feet until established on the approach, cleared for the approach."

The controller was directing them toward the long runway that was hiding beyond the thick clouds. "Roger," replied the captain. The copilot dialed 070 on his compass card and began the turn. Descending through the cloud deck, in the turn, they were blind to the outside—they could not see the horizon, nor could they see the runway. Only their instruments provided a clue as to their whereabouts. As the altimeter passed 2,000 feet, the copilot began pulling back on the control wheel, gradually leveling the aircraft at their assigned altitude of 1,800 feet. It was going to be yet another instrument approach to yet another wet and slippery runway.

"Flaps fifteen," called the copilot.

The captain hesitated for a second, and then reached to grasp the flap handle with his right hand. A sudden gust from the left pushed the aircraft to the side. His hand hit the green metal pedestal between the two pilots and bounced back. He quickly retracted his hand and then came at the handle from above. Supporting his body on the glare shield with his left hand, he pushed the weight of his arm on the handle and firmly drove the flap handle down to the 15-degrees notch.

On the flap gauge, located just in front of the flap handle, a little needle plummeted down and settled close to the 15-degree mark. But the captain's eyes did not stay there for long. He quickly withdrew his gaze and moved it up the instrument panel toward his own instruments. Settling on a large display that showed the artificial horizon and a symbol of the aircraft wings (see figure 13.2), he straightway realized what had attracted his attention: on the left side of the display, a small diamond-like symbol was pulsating continuously. He coughed to clear his throat and then said: "Localizer alive."

With the airport somewhere beneath the dark accumulation of clouds, the pilots relied on the instrument-landing system to help them fly toward the runway. At first, the horizontal (localizer) diamond was telling them that the runway was all the way to the left. But soon after, the little white diamond moved slowly toward the center of the display, indicating that the aircraft was progressively aligning itself with the center of the runway, 8 miles away and 1,800 feet below. The captain bent forward, glanced at the copilot for a second,

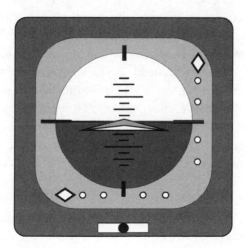

Figure 13.2. The artificial horizon display, showing the localizer diamond on the bottom (left) and the glide-slope diamond on the right side (top). Now the aircraft is below the glide-slope and the right of the runway.

took a quick glance at the instrument panel, and then started reciting the litany ingrained in him, one that had served him for hundreds of flights in this aircraft.

"We were cleared for the approach and the flight director is armed." "Roger," responded the copilot as he made a slight turn to further align the aircraft with the localizer diamond.

"Hydraulic pumps and pressure," continued the captain, making sure that all pumps were on and their respective needles were in the green range. "Good," he remarked quietly to himself as his eyes moved to inspect the other gauges on the instrument panel: the hydraulic-fluid quantity gauge indicated that enough pressure was stored to extend the gear and push out the flaps when necessary. Braking pressure was also in the green range. Everything here looked just fine, he thought to himself as the aircraft took another unexpected jolt. Seven miles out—and there was no end to the gusts.

Another little diamond appeared at the top-right side of the artificial-horizon display and began to shake. The captain welcomed it with a firm tone: "Glide-slope alive."

"Gear down, Before Landing checklist," came the immediate response from the copilot. The captain reached forward for the gear handle, pulled it out and then down in a quick, mechanized way. He kept his hand on the handle for a second or two, holding it in place, as if waiting for some response from the landing gear itself.

He did not have to wait long, because within two seconds a rumble sounded through the entire body of the aircraft—the large and heavy metallic doors housing the landing gear inside the aircraft opened into the icy air and disrupted the smooth flow of air under the aircraft. This was a normal response of the aircraft, expected, just like in any of the hundred landings that he had made in this aircraft, and the captain continued with the checklist by raising his arm to a large panel over his head. There, among a multitude of switches, buttons, and indicators, his hand reached for the engine ignition switch. He moved it from OFF to OVERRIDE, making sure that if the engine decided to quit during the descent to the runway, there would be plenty of ignition power to make a quick re-start.

The captain turned to his left and pulled a long, narrow piece of paper from his side—the checklist (see figure 13.3). Worn from use, its creases showing the signs of hundreds of not-so-clean fingers, the checklist looked like a restaurant menu. The captain gave it a quick look, and moved his gaze back to the instrument panel and his artificial-horizon display. The vertical (glide slope) diamond was sinking slowly toward the center of the display, the aircraft's wings were level, and the copilot was also keeping the horizontal (localizer) diamond in the center. Assured that the aircraft was properly aligned with the runway, the captain lifted the checklist card closer to his face,

```
┌─────────────────────────────────────────────┐
│  ┌───────────────────────────────────────┐   │
│  │          APPROACH CHECKLIST           │   │
│  └───────────────────────────────────────┘   │
│  ┌───────────────────────────────────────┐   │
│  │                                       │   │
│  │                                       │   │
│  │                                       │   │
│  └───────────────────────────────────────┘   │
│  ┌───────────────────────────────────────┐   │
│  │            BEFORE LANDING             │   │
│  └───────────────────────────────────────┘   │
│     IGNITION..........................OVERRIDE │
│     LANDING GEAR...........DOWN, THREE GREEN   │
│     SPOILERS.............................ARMED │
│     FLAPS.................EXTENDED, 40 DEGREES │
│     ANNUNCIATOR PANEL.................CHECKED  │
│  ┌───────────────────────────────────────┐   │
│  │             AFTER LANDING             │   │
│  └───────────────────────────────────────┘   │
│  ┌───────────────────────────────────────┐   │
│  │                                       │   │
│  │                                       │   │
│  │                                       │   │
│  └───────────────────────────────────────┘   │
│  ┌───────────────────────────────────────┐   │
│  │          SHUTDOWN CHECKLIST           │   │
│  └───────────────────────────────────────┘   │
│  ┌───────────────────────────────────────┐   │
│  │                                       │   │
│  │                                       │   │
│  │                                       │   │
│  └───────────────────────────────────────┘   │
└─────────────────────────────────────────────┘
```

Figure 13.3. The aircraft checklist (showing only the BEFORE LANDING items).

adjusted the overhead light, focused his eyes on the small black letters that were bouncing with the never-ending turbulence, and began reading.

"Before Landing checklist:"

"Ignition," the captain paused for a minute, looked up at the switch, and then announced—"Set to override."

"Landing gear?"

He could hear the unmistakable noise and the constant vibrations indicating that the gear doors were wide open and the landing gear was sliding out of the aircraft's belly. He looked down toward the triangle of three indicator lights just above the gear handle. The two outermost lights were shining bright green. Yet the center light was dark, indicating that the nose wheel was still coming down. Two seconds later the copilot called, "Gear down, three green." The captain looked down again and confirmed that all three lights were shining green; the left, right, and front (nose) wheels were now fully extended and locked into place underneath the aircraft. Nevertheless, the landing gear was still not ready to touch the ground, because the gear doors that had opened to let the wheels drop out were slow to retract and fold back into the aircraft body.

The next item on the checklist was to arm the spoilers. The spoilers are large panels that pop up from the aircraft's wing after touchdown and help stop the airplane on the runway. The captain reached his hand forward to grab the spoiler lever and arm it—but he had to wait. There was a condition in the checklist telling him that he could not arm the spoilers until the gear doors were closed. And a small light above the gear handle indicated in bright amber that the doors were still open. He kept on looking at the amber light, thinking that within several seconds the doors would close, the amber light would turn dark, and he would continue with the checklist and arm the spoilers.

"Flaps 25," said the copilot in a sharp voice.

The captain had to suspend the checklist to comply with his copilot's request. Without losing a second, he reached forward and brought the flap handle down to the 25-degree notch. Meanwhile, the copilot was making small corrections with the wheel, focusing on the two diamonds that are part of the instrument-landing system. In conditions of poor visibility, like this night, the instrument-landing system, which sends a radio beacon from the runway up into the sky, was their trusted friend. The antennas in the aircraft received the radio signal, and the localizer and glide-slope diamond symbols indicated to the pilots where the beacon was. The copilot was keeping the localizer diamond in the center and waiting for the glide-slope diamond to come down toward the center of the display.

As the flaps were coming out, the aircraft's speed slowed down to 175 knots, and the vertical glide-slope diamond was almost centered; a second later, it reached the center of the display. Now they were perfectly aligned with the beacon and the copilot did not hesitate for a second: "Glide-slope capture, flaps 40." The captain responded immediately and moved the flap handle to 40 degrees. Sometime during these moments, the gear doors finally closed up and the small amber "gear door open" indicator went dark. But the captain was busy setting the flaps in response to the copilot's requests; he was no longer looking at the gear lights. At the time, he couldn't know just how many times later he would return to these seconds in his mind, wondering what went wrong and why he forgot to arm the spoilers.

With more flaps coming out and enlarging the surface area of the wing, the aircraft began to pitch down. The copilot pushed the steering wheel slightly forward, and the aircraft began to descend. The flaps were slowing the aircraft, and the copilot added some power on the engines. Flaps 40 was their final flap setting and also an item on the checklist (see figure 13.3). The captain gave the flap handle a jiggle, making sure it was secured in place. With his right hand still holding the flaps handle, he brought the checklist card up, refocused his eyes again on the black letters, looked at the copilot, and declared, "Flaps extend 40, flaps set, annunciator panel—to go."

The small airspeed indicator read 160 knots. They were descending, riding down the invisible radio beacon toward the runway. The copilot stayed focused, constantly aligning the aircraft so that the two diamonds always stayed centered. As long as he kept them centered in the display, they would continue to ride the beacon. And at the end of the beacon they would eventually find the runway. With the effects of wind and gusts, the aircraft might deviate from the beacon, but the display and the two diamonds were always there to tell the pilot how much to correct, either vertically or horizontally, in order to re-align with the radio beacon. The copilot corrected vertical deviations by adding or reducing power on the engines. As for horizontal deviations, although he could use the rudder pedals to make quick corrections, the copilot used the control wheel to turn the aircraft back toward the beacon.

The outside air was rushing toward them at 150 knots, but the wind was coming from the right. To keep the aircraft from deviating from its flight path, they were listing slightly on their right side, in kind of a constant right turn, sustained by the spread of the wings and the thickness of cold air. The altitude indicator showed 1,500 feet.

A loud chime sounded inside the cockpit!

Both pilots immediately looked up in surprise. "Disregard," said the captain as he identified the source of the chime; the copilot's body sank back in his seat and his eyes returned to the artificial-horizon display. The altitude alerter chime was simply warning them that they had deviated from their previously assigned altitude of 1,800 feet; they had forgotten to reset the altitude alerter when they began their descent toward the runway. The captain reached forward and reset the alerter to 200 feet. Finally, the gusts were subsiding. "One thousand feet above minimums," said the captain. At that moment, a little light flickered in his peripheral vision. Above the dark window, on the overhead annunciator panel, the "rudder unrestricted" light was now shining blue, indicating that the rudder was available for use.

"Annunciator panel checked," he affirmed, and then, lifting the checklist card, he announced: "Before landing check complete."

They were coming down fast, but there was no sign of the runway; they were completely surrounded by a veil of clouds. The captain closely watched the altimeter and glanced at the engine gauges. He roused himself and said: "Seven hundred feet above minimums." Since they could not see the ground, every-thing related to their altitude was referenced to "minimums." The minimum altitude for this approach was 200 feet, which meant that if they could not see the runway when the aircraft altitude was at 200 feet, they would have to break off the approach, go around, and try again, perhaps at another airport.

One hundred and forty knots. The two diamonds were centered, which meant that they were tracking straight toward the runway and coming down the glide-slope perfectly—yet the captain looked around the cockpit with some apprehension. Something felt wrong. Something in this liturgy of actions, requests, checklists, and communications with air traffic control was missing, but he just couldn't figure out what. Frustrated, he looked outside in anticipation of signs of terrain through the clouds. Nothing. Then he suddenly realized: the tower-approach control did not switch them to the tower frequency! They needed the tower's permission to land the aircraft.

He quickly keyed the microphone. "Approach control." "Approach," he said in a louder voice, "this is Flight 847."

"Approach, do you want us to go to tower?" The aircraft's speed was 137 knots. At least the copilot was doing a good job of keeping the speed on target and the aircraft on the glide-slope.

They were going to break out of the thick overcast soon, and he certainly did not want to meet the runway without a landing clearance. The captain grew a bit nervous as he called, "Five hundred feet above minimums," to the copilot, and continued with his efforts to hail the controller. "Approach, are you there?"

Still no radio contact.

The captain tried to figure out if the microphone's jack was out of the socket or maybe something else was malfunctioning. As he was about to push the microphone switch for another attempt, a soft and sleepy voice entered the cockpit: "Sorry, contact tower on frequency 128.2." He hastily ratcheted the new frequency into the radio and immediately pushed the microphone's talk button: "Tower, this is Flight 847, we are with you on the approach, ILS runway zero four."

"Cleared to land," came the immediate response.

The captain, relieved, made a quick scan of the cockpit instruments: 400 feet above minimums, 1.7 miles out, 137 knots. The instrument landing system (ILS) was working just fine and they were tracking it in—now they were a bit above the glide-slope and the copilot was making a correction to close in. He quickly moved his eyes above the instrument panel and into the gloom beyond the window, and caught a glow of light. "Three hundred above minimums." Ahead and below, through the dark gray clouds, he could now see a misty pink light bouncing forward. Then the pink became red. And a string of bouncing red lights lay ahead.

He focused on the red lights and his eyes followed them as if pulled forward by some invisible string. The clouds opened up and he saw the runway's lights.

"Runway in sight."

"Going visual," came the quick response from the copilot as he raised his head. The runway scene was coming alive—the set of red lights shining ahead marked the beginning of the runway, with white lights bordering the runway on both sides, and soft blue taxiway lights fading beyond. The runway's center line greeted them with its dotted white lights. The copilot focused on the centerline lights, aligning his body and spine with them. A little bank to the right, a quick push on the rudder pedal, and now the aircraft was in line with the runway.

"In the groove," said the copilot in a satisfied tone. The captain made a final check inside the cockpit to make sure that everything was set properly for landing. All the landing gear lights were shining in green, "good"; the flaps handle was in the 40-degree notch, "check"; and engine thrust was "okay." But something was out of place; he could almost feel it in his body. Noting the copilot's hands tightening on the throttles, knowing the plane would be landing in a few seconds, the captain made another scan up and down the pedestal for evidence of any problem. Then he saw it. There, by his right leg, the spoiler lever was down; the spoilers were not armed and ready for automatic deployment.

"Missed it on the checklist," he muttered to himself with a sharp resentfulness. He hurriedly reached his hand forward to grab the lever and arm the spoilers. And then he hesitated.

Years of experience told him that this was not the right thing to do. His muscles and sinews held his hand back from arming the spoilers at this altitude. They were only 300 feet above the runway. Changing the aircraft's configuration at this low altitude was dangerous; any fault in the automatic spoiler system could result in automatic deployment of the spoilers in mid-air and cause the aircraft to sink precipitously. Arming the automatic spoilers now was not a good idea. No way around it, he thought to himself; he would have to manually deploy the spoilers once they touched the runway.

"Two hundred feet to touch-down," called the captain, and then he silently said to himself, "And when we touch down, I reach for the lever, pull it up, back, and then up again." This was his mental preparation for manually deploying the spoilers once they touched the runway.

"One hundred feet."

The copilot pitched up the aircraft slightly for landing. There was a slight updraft at the same moment, the aircraft wallowed upward, and the copilot's left hand slowly moved the throttles backward. A second later the engines were whining down. The airplane, about to land, was in a vulnerable state—pitching up, slow, and very close to the ground. A gust from the right lifted the wing, and the copilot quickly corrected with a quick turn of the wheel to the right. They were slow, almost hanging in the air. The wings leveled and their eyes were glued to the runway-center lights. The captain's hand moved to the spoiler lever. The black runway widened and his hand closed around the lever. As the runway lights rushed up toward them, his right hand clutched the spoiler lever hard. An updraft from the left jolted them and the entire aircraft vibrated in response. The copilot corrected to the left and pushed the throttles to the very end—"clack," came the metallic sound of the throttles hitting the hard backstop. "Reach, pull it up, back, and then up again," the captain said to himself as his hand pulled back on the spoiler lever.

The bottom suddenly dropped out from beneath them.

The aircraft was no longer flying. It sank like a brick. The copilot instinctively pulled back on the control wheel to stop the aircraft from sinking further. But this corrective action was futile, because within a second the aircraft's tail hit the ground with a thud.

"Damn," said the captain as the entire hull shuddered. The left wheels took much of the impact, then the right wheels shortly thereafter. The sound of aching metal echoed from the tail of the aircraft through the cabin and then up to the cockpit. They felt the violent and jarring shock in their tailbones, up their spines, through their ribs, chests, and necks, all the way to their heads and eyes. The landing gear produced hollow guttural sounds as the hull's frame and bulkheads responded with pain. It seemed that every piece of metal wobbled and every rivet in the aircraft shrieked. Then the nose wheel came down hard on the black pavement, bouncing them in their seats.

The copilot pushed anxiously on the brake pedals to stop the aircraft. The captain grabbed two small levers on the throttles and pulled them back; the engines reversed and slowed down the aircraft.

"80 knots," the captain's voice rang out.

"I have control," said the captain as he took the aircraft in his hands; and with those three simple words he took upon himself the emotional weight of the impact and relieved much of the copilot's anxiety.

They slowed to a walking pace midway down the runway, and from there throttled the aircraft into the next taxiway. Now everything was silent, but their bodies were still reverberating with the shock. The captain understood what had happened and was quick to realize the consequences of this incident for him. He apologized to the copilot, and with a heavy heart picked up the checklist card from his side and began:

"After landing checklist;

"Lights—nose lights dim."

"Flaps—15."

"Antiskid—off."

Later the next morning, the mechanics checked the aircraft thoroughly. There was serious damage to the tail, which hit the runway before the wheels. The impact registered almost 1.5 times the weight of the aircraft. Support structures and metal frames in the tail section sustained imploding damage. Some rivets had popped out. The skin of the aircraft around the tail was wrinkled and scratched. One of the supporting belts that connects the landing gear to the airplane was broken. The landing gear, which absorbed most of the impact, was bent in agony. The center fuel tank, lying close above the tarmac, had sprung a slow and weeping leak.

What Went Wrong?

The captain's premature deployment of the spoilers while the aircraft was about 20 feet above the ground caused the hard landing. It looks like another one of those "human error" stories we read about so often in the newspapers. The captain was the definite culprit: first, he forgot to arm the automatic spoilers. Second, he failed to follow the checklist sequence and skipped over the "spoiler arm" item. Third, when he finally realized the omission, he manually deployed the spoilers at the wrong time.

"Pilot error." Case closed!

But as we have seen in the preceding chapters, human-machine interactions, especially in highly dynamic environments, are far from simple. To truly understand what happened, we first need to look carefully at the automatic spoiler system, the procedure to operate it, and the checklist. Only then can we truly judge what really happened.

From Air to Ground

Landing an aircraft is a delicate and complicated maneuver. The aircraft, flying at speeds in excess of 150 miles per hour, must be guided from the wide-open sky to a narrow strip of runway and then brought to a stop. In flight, the aircraft's weight is supported by the air; once on the runway, the landing gear takes the load. The transition from flying to rolling down the runway is a gradual process, because, even after the wheels touch the ground, there is still air flowing beneath the wings, supporting the weight of the aircraft. Prolonging this gradual process is something to avoid, or, more precisely, to minimize. Why? Because the longer this transfer of weight takes, the longer the aircraft will roll on the runway before finally stopping. Runways are limited in length. And when there is rain, snow, and slush, they effectively become even shorter.

And this is where the spoilers come into the picture: they pop up at touchdown to help the aircraft settle firmly on the ground. The spoilers, as the name suggests, spoil and disrupt the flow of air over the wings and thereby reduce lift. You may have seen them on the aircraft's wing during landing: big wide panels, several feet long, which rise up from the wing's surface immediately after touchdown and stay up until the aircraft slows down.

Once the aircraft is firmly on the ground, the pilots stop the aircraft by pressing on the brake pedals, just as you do in your car. When the airplane's weight is mostly on the landing gear, the brakes are more effective. This allows for a short and quick landing. Efficient braking is even more critical while landing on a wet and slippery runway, because, just like in your car, if the asphalt is wet it will take longer for braking action to be effective. It is therefore critical to make sure the spoilers are deployed to help the aircraft settle firmly on the runway. Aerodynamic calculations show that if the spoilers are not deployed after touchdown, the wings support about 70 percent of the aircraft's weight, and only 30 percent is carried on the landing gear. However, if the spoilers are properly deployed, 80 percent of the weight is quickly transferred to the landing gear. When most of the aircraft's weight is on the landing gear, the aircraft will stop faster. Spoilers, as you can see, make a huge difference in efficient braking. Forgetting to arm the spoilers and/or improper deployment of spoilers has caused many aircraft to slide on the runway, including several accidents in which the aircraft overran the end of the runway (see the notes for this chapter).

The Automatic Spoiler System

Having considered the function of ground spoilers, we can now focus on how they work and how the pilots operate this system. The spoilers can be deployed manually or automatically. As soon as the aircraft touches down, the pilot can manually pull the spoiler lever and deploy the spoilers. Just like a handbrake in a car, the more the pilot pulls the spoiler lever toward him or her, the more spoilers come out of the wing to disrupt the airflow. That's the manual option.

The other option is to have the spoilers pop up automatically during touchdown. There are two advantages to automating spoiler deployment: one, having the spoilers pop out just at the right time, not too early and not too late, provides for maximum braking efficiency (and perfect timing is something that a machine is well-suited to do); two, automatic deployment reduces the pilot's workload, which, as we have seen, is quite high and challenging during landing. So this is not a bad idea, but how does the aircraft know when to deploy the spoilers? We want to be assured that the spoilers will come out only when the aircraft is on the ground (and not in the air). Inadvertent deployment of spoilers in mid-air will cause the aircraft to sink rapidly—a situation that can easily deteriorate to a stall condition in which the aircraft's wing can no longer produce lift and the aircraft falls out of the sky.

Since it's important that the spoilers deploy as soon as the aircraft's wheels touch the ground, there is a sensor on the nose wheel for this. When the sensor is activated, an electric pulse is sent to the aircraft's logic circuits signaling contact with the ground. This triggers the spoiler mechanism to automatically deploy the spoilers. Because the spoilers are so critical for landing, there is a built-in redundancy so that another event can also trigger the spoiler mechanism (in case the nose wheel sensor fails). The sensor for this event is located on the main wheels; it sends an electric pulse when the wheels are spinning (as opposed to skidding). Specifically, the sensor will send the pulse when the wheels are spinning at an RPM that's equivalent to 80 knots. Therefore, either one of these two conditions,

nose wheel sensor = GROUND
or
main wheels' speed = 80 knots

will automatically deploy the spoilers.

Figure 13.4 is a model of the automatic spoiler system. It describes the system and the pilot's interaction with it. The system starts in state *A* with the landing gear up. Then the spoiler is armed (state *B*) and the landing gear comes out (state *C*). It takes some time for the landing gear to come out of the aircraft's belly, unfold, and lock into place. Once the gear is down and locked in place, there is a transition to state *G*—the gear is down and the spoilers are armed. When the aircraft touches down, the spoilers pop out automatically (state *H*), and everything is well.

But things do not always go as planned. It turns out that the above sequence (*A-B-C-G-H*) has some potential hazards. Years of operational experience with this aircraft has shown that state *C* is dangerous because of the mechanics of the landing gear and the sensor that signals "GROUND." This sensor is located inside the nose-wheel strut, which is similar to the shock absorbers on cars and on mountain bikes. When the strut compresses to less than three inches, the sensor signals that the aircraft has touched the ground. But there is a real possibility that the strut will also compress to less than three inches when the nose wheel slides out of the aircraft's belly and is greeted by extensive pressure from the rush of the air around it. When this happens, the strut may compress to a point that the GROUND sensor activates, and the spoilers will deploy while the aircraft is in mid-air. The consequences are potentially serious: the aircraft will sink fast and may even stall (see endnotes).

A Disturbance Event

Having the ground spoilers pop out unexpectedly in mid-air is not a welcome prospect, because it can lead to a catastrophe. State *D* in figure 13.4 denotes an unexpected spoiler deployment in mid-air. Reaching this state, which is always a possibility, is *unsafe* and we certainly want to prevent it from happening. But how do we do this?

Let's first try to consider how we may inadvertently find ourselves in this unsafe state where the spoilers pop out in mid-air. We get to this unsafe state only when the *errant* GROUND *signal* event takes place in *C*. Note that this type of event is different from all the ones we encountered in previous chapters: it is not *manual*—the user did not trigger it—and it is not *automatic*, because it is not a timed event or an internally computed event (like ones we have dealt with in the clock radio and the autopilot of KAL 007). So what is it? The dotted transition in figure 13.4 is a *disturbance*, an event that is triggered by the environment or by some fault in the system.

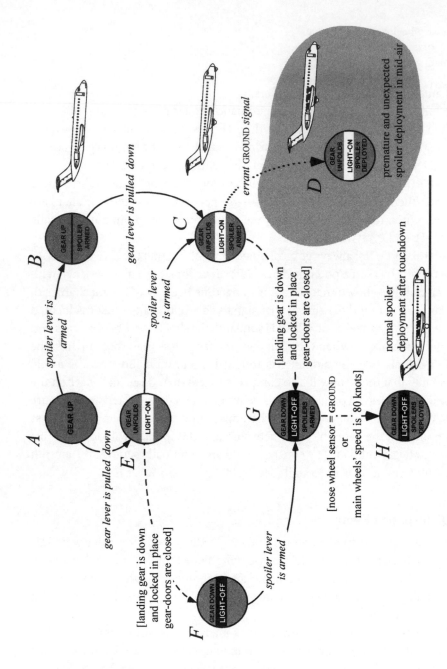

Figure 13.4. Machine model of the automatic spoiler system.

The occurrence of a disturbance is always unpredictable. We do not know when it will happen (and we do not really care, for the purpose of this analysis, if it will happen once in every 10 flights or once in every 1,000 flights). All we know is that it *may* happen and it is dangerous. We therefore want to block it from happening, and the sure way to do this is to avoid reaching configuration C. That is, we should avoid altogether arming the spoilers before lowering the gear. But note what we just did: we are restricting entry into configuration C, which by itself is not dangerous, only because it harbors the *potential* for a disturbance event. For all practical purposes, state C is also *unsafe* for us.

So how can we avoid state C and still lower the gear and arm the spoilers? Let us look again at figure 13.4. One way to avoid these two unsafe states (C and D) is to use another path. Consider the following sequence: Lower the gear (and transition to state E), wait for the landing gear to fully come down and lock into place (state F). But how do we know when the gear is down and locked in place, and that it is therefore safe to arm the spoilers?

Recall that there is an indicator light in the cockpit to announce when the gear extension sequence has been completed. It takes the landing gear system somewhere between 17 and 39 seconds to come out, unfold, lock into place, and then for the gear doors to close. During this period, a little amber light, "gear door open," which is located by the gear handle, illuminates. When the landing gear locks in place, ready for landing, and the gear doors finally close, the light goes off. Only then should the pilot arm the spoilers (and transition to state G).

The new sequence, A-E-F-G, guides the pilot through lowering the gear and arming the spoilers and guarantees that we will never enter the unsafe state D. In other words, to avoid inadvertent spoiler extension in mid-air, the pilots must follow a specified sequence of actions.

Such a sequence is called a procedure.

Procedures

A procedure specifies a unique sequence of actions; it is a recipe of sorts. It shows us the necessary steps to perform a task, such as arming the spoilers for landing. Figure 13.5 is a description of the spoiler arming procedures. The procedure also tells us when to begin and end the sequence of action. We begin the procedure by lowering the landing gear, and we terminate it after the spoilers are safely armed, the gear is down, and the gear doors are locked in place. Procedures, however, are not limited to aircraft operations—you find them almost everywhere you see a human interact with a device, and especially if it is done in a risky environment.

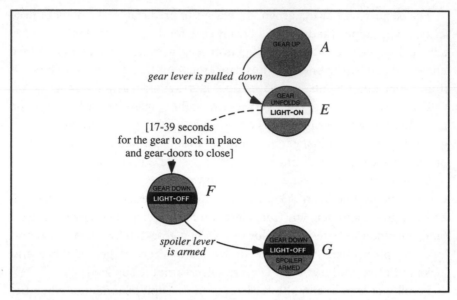

Figure 13.5. Spoiler arming procedure.

Sometimes the sole reason for a procedure is an inherent problem in the design of the system. This is the case with the spoiler deployment procedure here. In the case of the blood-pressure machine (chapter 12), it is possible to specify a mandatory sequence of actions for changing the timer interval. Using procedures and instructions to bypass an unsafe state may appear, on the surface, to be a perfectly cheap and easy solution. Simply write a procedure, print it on a placard, place it by the device—problem solved. But mandating a procedure in lieu of making a design or interface change is really only putting a Band-Aid on the problem. Why? Because the underlying assumption is that the user will *always* follow the procedure. Yet humans, like it or not, are not machines. Either advertently or inadvertently, at one time or another, a procedure may go unused. And if you think it is always possible to make sure that a procedure is followed, just think about all the safety procedures and instructions that you have violated while using power tools and installing electrical devices at home. There is never a full assurance that procedures will be followed, no matter how many are specified.

Nevertheless, not all procedures are Band-Aids for design deficiencies. Some procedures are in place to provide guidance to the user on how to work the machine. In order to operate a complex system successfully, the user must be well supported by procedures. In high-risk systems, such as aircraft operations, space flight, nuclear power, chemical production, and high-technology medical practices, it is essential that such support be flawless—as the price of errors can be high. Therefore, first and foremost, a procedure must be correct in the sense that it must not lead the user into unsafe states as well as situations from which the user cannot recover.

Instrument-landing Procedure

Flying a commercial aircraft involves many complex and critical tasks. To make sure that these tasks are performed safely and to maintain a high level of safety on every flight, airlines mandate procedures. The intent is to provide guidance to the pilots and to ensure logical, efficient, and predictable means of carrying out tasks. Procedures are also in place so that each crew member will know exactly what each other crew member is doing, and what action he or she will take next.

An instrument approach for landing is a highly complicated task, and indeed all airlines provide their pilots with a specific procedure for performing it. The procedure tells the pilot what to do and provides cues for when to do it. Figure 13.6 is a description of the sequence of actions that were part of the instrument landing procedure used during the accident flight. There are three steps that we focus on:

1. The gear should be lowered as the aircraft starts to receive the glide-slope signal;
2. Flaps should be set to 25 degrees as the aircraft nears the glide-slope.
3. Flaps should be set to 40 degrees as the aircraft captures the glide-slope.

The timely execution of this procedure depends on the aircraft's speed. It is possible to calculate the time it takes the aircraft to fly from one point to the other. Below the instrument approach profile in figure 13.6 is a timeline: For an approach speed of 175 knots, it takes 10 seconds to fly from point 1 (gear down) to point 2 (flaps 25). It takes the aircraft an additional 6 seconds to get from point 2 to point 3 (flaps 40). Altogether, it takes 16 seconds to fly from point 1 to point 3.

Synchronization of Procedures

Hardly any aircraft procedure is conducted in isolation. Many procedures are executed concurrently. In our case, both the spoiler extension procedure and the instrument landing procedure run concurrently. The question now is whether they are well synchronized, so that one procedure sequence does not block or disrupts the other's sequence (and vice versa).

Naturally, the difficulty of evaluating such concurrent procedures increases when there are three, four, and perhaps more procedures running at the same time. Multiple (and concurrent) procedures are the reality in aircraft opera-

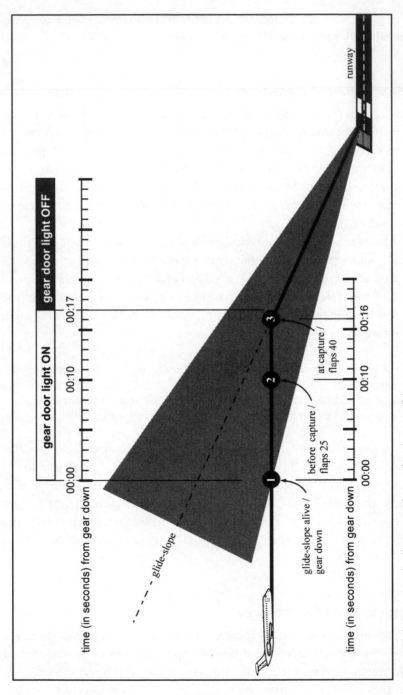

Figure 13.6. Instrument-landing approach procedure (not to scale).

tions and other safety-critical systems. Moreover, having well-synchronized procedures becomes even more critical during high-tempo phases of a flight, such as during an instrument approach for landing.

Synchronization is all about timing. Recall that both the spoiler extension procedure and the instrument-landing procedure start with lowering the landing gear. That is the initial event for both procedures. The spoiler extension procedure tells the pilot that he must wait for the gear door light to turn off before arming the spoilers (and we know that from gear-down it takes at least 17 seconds for the light to extinguish). The instrument-landing procedure tells the pilot to set flaps 25 and then flaps 40 (and we already know that from gear-down it takes 16 seconds to reach flaps 40). At the top of figure 13.6 you can see the timeline for arming the spoilers; at the bottom of the figure is the timeline for flying the approach and setting the flaps. Looking at figure 13.6, we note that flaps 40 (point 3) will always occur *before* the gear light turns off. Therefore, spoiler arming can be initiated *only* after flaps 40. Does this limitation have any relevance here?

Checklist

Let's turn to the checklist to find out. The pilot's tasks in preparing and checking that the aircraft is ready for landing are listed in chronological order on the BEFORE LANDING checklist (which, of course, is yet another procedure). The checklist is in place so that the pilot won't forget or skip items and fail to prepare the aircraft for landing. The pilot must follow the checklist item by item, making sure that all actions have been accomplished. Figure 13.7(a) is the BEFORE LANDING checklist sequence. It tells the pilot to:

1. lower the gear, then
2. arm the spoilers, and only then
3. check that the flaps are set at 40 degrees.

But when we take into account the time it takes to accomplish items (2) and (3) we find a sequential problem. It takes at least 17 seconds (from gear-down) until the pilot can arm the spoilers; yet it takes 16 seconds (from gear-down) to reach the flaps 40 point. Therefore, flaps 40 (item 3) will occur before the pilot can arm the spoilers (item 2). See the problem? The checklist's mandated sequence 1-2-3 can be performed only as 1-3-2 given the actual dynamics of the aircraft.

Thus, in the accident described above, the captain had to wait for the gear door light to turn off before he could arm the spoilers. But in the meantime, the airplane was moving and the copilot instructed him to set flaps 25, and, shortly after, flaps 40. Flaps 40, as shown in figure 13.7(a), is a checklist item. So what

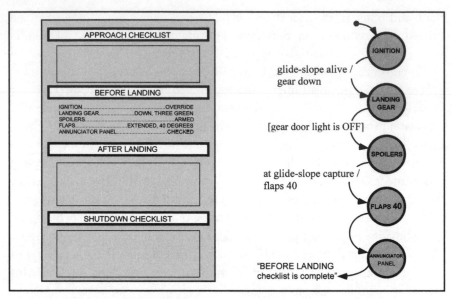

Figure 13.7 (a). The BEFORE LANDING checklist card and sequence.

does he do after moving the flap lever to 40? He calls "Flaps 40 set," picks up the checklist and continues on—completely bypassing the spoiler item!

Notice what unfolded here: the specified and mandated sequence of "gear–spoilers-flaps 40" got switched with the truncated sequence of "gear-flaps 40." You can see this visually in figure 13.7(b). As a consequence, the spoiler check was out of the picture, and the pilot forgot to arm them.

Possible Solutions

How do we go about solving this procedure synchronization problem? One possible solution is to modify the BEFORE LANDING procedure sequence to account for the fact that "flaps 40" can come before "arming spoilers." This seems to be a simple fix. However, we must be very careful here, because such fixes must first be evaluated in the context of all procedures.

A better (and much safer) solution is to redesign and modify the landing-gear mechanism and/or the spoiler deployment logic with the intent of eliminating the possibility of inadvertent spoiler deployment. This way, the spoiler can be armed anytime and there is no need for the doors to close before the pilot can arm the spoilers. Nevertheless, such modifications can be rather costly, especially when applied to dozens of aircraft. So it is not surprising that although the manufacturer of this aircraft sells a modification kit for the spoiler/landing gear mechanism, no U.S. airline has yet bought it.

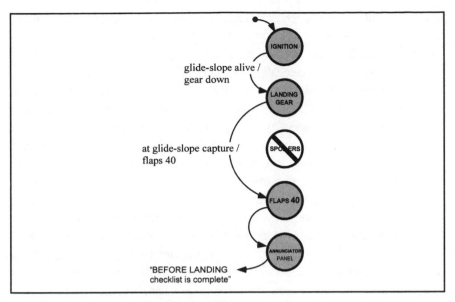

Figure 13.7 (b). The actual sequence during the flight.

Regardless of which of the above solutions is adopted, there always exists a possibility that the pilot will forget to arm the spoilers for landing. In the current design, there is very little feedback to the pilot that the spoilers are not armed and ready for landing. And since forgetting to arm the spoilers is a known problem that has contributed to many incidents and accidents, it may be also beneficial to install an indicator (or another design feature) that can sense when the spoilers are not armed and warn the pilot.

The Effect of Priming

The above discussion showed the synchronization problem and we now appreciate why the arming of the spoilers was skipped over and missed on the checklist. We have also seen that checklists, being a temporal list of actions and checks, are extremely vulnerable to omissions and timing constraints. Once an item is skipped over, there is often no return because the user assumes that all is well and moves on. In this accident, reaching the end of the checklist gave the captain a false assurance that the list was complete. But the captain noticed, just before the landing, that he forgot to arm the spoilers; he then pulled the spoilers manually, but at the wrong time.

Why?

This question is difficult to address because now we are in the not-so-exact realm of human performance. We know, however, that before the captain manually pulled the spoilers, he rehearsed the action sequence verbally to himself: "And when we touch down, I reach for the lever, pull it up, back, and then up again." In a way, he primed himself to pull the spoilers. Such priming is something that we have all encountered and seen before. We see it sometimes in the Olympics, when swimmers are ready to jump off the starting blocks at the sound of the gun. Someone in the audience claps, or some noise is heard—and then one swimmer jumps off to a false start.

Competitive swimmers prime themselves so that the sound of the gun triggers a jump. A similar sound or a related cue causes them to jump. Considering that the captain primed himself to manually deploy the spoilers, something must have triggered his action. What was the event that triggered him to assume that the aircraft was on the ground and then pull the spoilers? Did he mistake a movement of the aircraft, due to a downdraft, for a landing? Did the metallic clack of the throttle, as it was pulled back, provide an auditory trigger? Was fatigue an additional factor? The sad truth is that it is very difficult to answer these questions with any sense of precision and confidence.

In Conclusion

In this chapter we evaluated the correctness of a single procedure (arming spoilers), then added a concurrent procedure (instrument landing), the checklist, and finally added time as a factor. We saw how synchronization among sequences of action and timing constraints play an important role in the design of procedures. The objective of this chapter was to show how subtle timing inconsistencies can cause a critical human-machine interaction to fail and then to demonstrate a general approach for evaluating and designing procedures that are correct.

Correctness of procedures is an important aspect of human-machine interaction and becomes even more important in human interaction with automated systems. The reason is not only that the human is a player in executing procedure steps, but that the machine can trigger events automatically. Furthermore, as we have seen in the spoiler-arming example, the environment can also play a role and create disturbances that may drive the system into an unsafe state—sometimes with detrimental results.

Beyond procedures, we also came to realize that an accident like the one described in this chapter is a complex sequence of events that come together in some unexpected way. It is impossible to completely eliminate accidents. However, it is possible to reduce their likelihood by removing much of the "fuel from the fire." Forgetting to arm the spoilers during an approach is hardly a new problem; it has happened to many pilots. Sometimes the pilots catch the omission; sometimes they do not. Interestingly enough, many pilots who have encountered this particular timing problem and skipped over arming the spoilers believe that it occurred because they were not attentive enough, failed to follow the checklist, or they attribute the omission to a similar "guilt trip." The reality, as we now know, is more complex.

It is important to note that we all have a tendency to fixate on one major flaw as the explanation for an accident: In this case, some may blame the pilot for pulling the spoiler at the wrong time, others will blame the design of the spoiler mechanism, or blame the way the procedure is written. But the naked truth is that most accidents lie in the interaction of many factors. These factors are technically complex, involve human performance issues that are not well understood and predicted, and are intricate in the sense that there are several concurrent processes going on at the same time. What makes it especially difficult to describe and fully understand many problems associated with the correctness of procedures is the delicate synchronization requirements and timing constraints that are invisible to the eye. Nevertheless, these requirements and constraints are essential for developing correct procedures for automated control systems.

Chapter 14

Automation in Modern Aircraft

If you really want to learn, you must mount a machine and become acquainted with its tricks by actual trials.

—Wilbur Wright, quoted in *Test Pilots: The Frontiersmen of Flight* by Richard Hallion, Doubleday, 1981

I n the previous chapter, the aircraft was flown manually. That is, the copilot was steering the aircraft by using a large yoke-shaped control wheel. By turning the wheel left or right, he steered the aircraft in the same way we steer a car. In addition, by pulling back on the control wheel, the pilot made the aircraft climb; by pushing the control wheel forward and down, he made the aircraft descend. Likewise, the engine throttles were operated manually: the pilot pushed them forward to increase thrust on the engines and pulled them back to reduce thrust. Many of the other flying tasks, such as navigating and flying the glide-slope down to the runway, were also done manually.

This brief chapter will introduce you to automated aircraft in which most of the flying work and navigation is relegated to the computer. The next chapter applies this knowledge to analyze an accident involving pilot interaction with a highly automated aircraft.

Traditionally, the aviation industry has always been in the forefront of technological progress. Many of the automated systems that we find in maritime technology, automotive, and other transportation systems were first introduced and tried out in aviation. (The first automatic pilot, for example, was developed by Lawrence Sperry and demonstrated in flight in the summer of 1914; it became available as a commercial product for aircraft in the early

1920s. Autopilots can be found today in boats, ships, and even in some tractors and combines.) It is therefore worthwhile to consider how automation is used in aviation, and what kind of problems emerge.

Cockpit Automation

Automation existed in pre-1980 airline cockpits in the form of automatic spoilers, autopilots, and inertial navigation systems. We have discussed some of these systems in chapter 4 (Korean Air Lines Flight 007) and in chapter 13. In the early 1980s, more advanced automation systems were introduced as more efficient computers and improved information technology became available. The Boeing 757 and 767 aircraft, and later the Airbus 320 aircraft, were at the forefront of this new design approach. The main goal of the new designs was to reduce operating costs by optimizing the climb to altitude, the route of flight, and the descent to the runway—thereby saving fuel and time. A secondary goal was to reduce overall cockpit workload and thereby reduce crew size from three pilots to two pilots. Other advantages were reduction in aircraft weight and increased reliability.

On the human side, things moved a bit more slowly. The new automated systems required a paradigm shift in the way pilots were trained to fly and operate aircraft. Pilots needed to learn practical computer skills and have a good conceptual understanding of how automation works. And then came the topic of when to engage the automation, when to disengage it, and how to monitor, supervise, and interact with automated systems. It took time to develop safe and efficient work practices, procedures, and policies—but in the meantime, the aviation industry paid a heavy price. During the 1990s, an alarming series of incidents and several accidents involving pilot-automation interaction occurred. Almost all the reports of these accidents concluded that the pilots had lacked an adequate understanding of the machine's behavior. In some accidents there was also a problem with the way pilots had interacted with interfaces and displays. As we know very well by now, both factors limit the user's ability to reliably anticipate the current and next mode of the machine, leading to false expectations, confusion, and error.

It turned out that the primary limitation on efficient and safe use of these automated systems was not the technology itself, but rather pilot interaction with it. Therefore, in the following sections we will look at pilot interaction with automated control systems, and examine one in-flight incident in detail.

Automatic Flight-Control Systems

The automatic flight-control system found in every modern aircraft is a suite comprised of several sub-systems and components. The primary component

Figure 14.1. Automated cockpit. (The photo has been enhanced to show the two flight management computers, throttles, and flight control panel. © copyright Boeing Company, used under license)

is the autopilot, which, as the name suggests, can automatically steer the aircraft as well as make it climb and descend. Then there is the automatic throttle system, which is similar in principle to a cruise-control system in a car. The automatic throttles maintain aircraft speed and engine thrust throughout the flight. Figure 14.1 is a photo of a typical automated cockpit. Note the large displays and the computers, and in particular the flight control panel, located just below the front windows. It is through the flight control panel that the pilots interact with the system by engaging the autopilot and automatic throttles, as well as by switching their modes. The most sophisticated automated flight components, located on the pedestal between the two pilots, are the flight management computers. These two computers combine navigation information (such as flight routes and approach profiles), aircraft performance information (the most economical climb profile), with information from onboard sensors (altitude, speed, distances to waypoints, and so on) to compute an optimum and fuel-efficient flight route. The flight management computers fly the aircraft with unprecedented accuracy.

To take advantage of all the smarts and sophistication of the flight management computers, the pilot engages a mode called VERTICAL NAVIGATION. Left on its own, VERTICAL NAVIGATION mode can automatically pilot the aircraft from after takeoff to just before landing. This was the mode used by the crew of a modern airliner during a climb-out from a busy metropolitan airport on the East Coast of the United States. The climb and subsequent pilot actions resulted in a speeding incident, which came about because of pilot confusion about the behavior of the VERTICAL NAVIGATION mode.

Speeding Ticket

As the aircraft was climbing after takeoff, the air traffic controller assigned to monitor the flight instructed the flight crew to climb and level off at 11,000 feet. At about 3,000 feet, the copilot, who was flying the aircraft manually, engaged the autopilot and selected the fully automatic VERTICAL NAVIGATION mode. In all other modes, the pilot has to enter the speed he or she wants the aircraft to maintain, but in this fully automatic mode, the speed (reference value) is calculated by the flight management computers to provide the most economical speed. As the aircraft was passing beyond 10,000 feet, the economy speed was 300 knots.

"Reduce speed to 240 knots for in-trail separation," came the instruction from the air traffic controller.

As soon as the copilot heard the air traffic controller's instruction to slow the aircraft, he reached forward and dialed the new 240-knot speed limit into the flight control panel. Such speed limits, or "restrictions" in air traffic control terminology, are a common occurrence in congested airspace around airports. Apparently, there was another aircraft straight ahead, and the controller wanted to make sure that this flight would not come too close to the aircraft ahead (violating the FAA's minimum allowable horizontal separation).

In response to the copilot's data entry, the throttles automatically eased back to accommodate the new speed setting. There was no mode change. The aircraft was still engaged in VERTICAL NAVIGATION, but the speed reference value was no longer coming from the flight management computer. None of this was out of the ordinary; the pilots had done what they were trained to do—and probably had done hundreds of times before. The climb profile is depicted in figure 14.2.

As the aircraft approached 11,000 feet, which was the altitude the air traffic controller instructed them to maintain, the autopilot began to ease the rate of climb. It nosed down, throttled back, and began a gradual maneuver to capture 11,000 feet. All of this, of course, was done automatically—there was no direct pilot involvement. Once the aircraft reached and "captured" the new 11,000 feet altitude, the autopilot automatically switched to ALTITUDE HOLD mode. Now they were flying straight and level at 11,000 feet, both pilots monitoring what

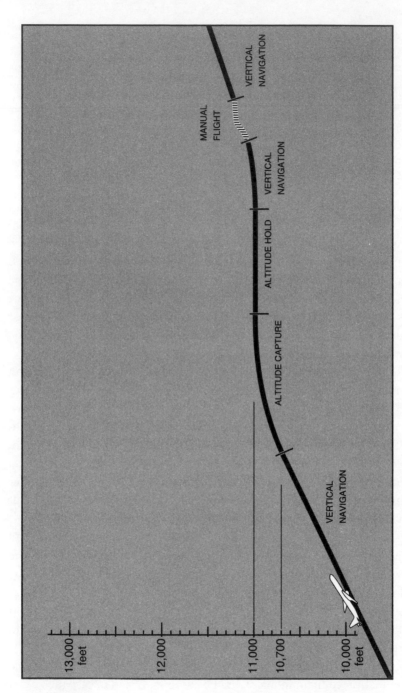

Figure 14.2. Climb profile with altitudes and mode (not to scale).

the autopilot was doing and checking on their flight route and weather at destination, which was less than an hour away. A few minutes later, the controller's voice entered the cockpit:

"Climb and maintain 14,000 feet."

The copilot reached up to the flight control panel and re-engaged the VERTICAL NAVIGATION mode in order to initiate a new climb to 14,000 feet. The automatic throttles moved forward, the engines began to whine with increased thrust, and the aircraft accelerated. A second later, the nose pitched up slightly and the aircraft began a climb to 14,000 feet. Everything was normal, or at least that's what the pilots thought.

What the crew did not anticipate was that the aircraft's speed was beginning to accelerate beyond 240 knots. (Recall that there was an aircraft ahead, and the air traffic controller told the crew to maintain 240 knots so as not to come close to that aircraft.) Unobserved by the crew, the aircraft speed increased to 250 knots, 260, 270 and then to 280 knots. Seconds later it was 50 knots above the limit they had been given, and was about to reach 300 knots.

The copilot grew uneasy. Maybe it was the noise of the engines or maybe he felt the subtle acceleration in his body. He became aware that something was wrong. But he wasn't alone in his uneasiness. "Click-clack" came the sound from his left side. It was the captain, who quickly disengaged the autopilot. A clinking of metal followed as he pulled the two large throttles backwards, banging against the metal stop. "Autopilot disengaged, autothrottles disengaged," muttered the captain. Now the aircraft was stripped of all its sophisticated attire.

"What's happened?" asked the copilot with surprise.

The captain answered by pulling back on the control wheel. The aircraft responded immediately by pitching up. With the nose above the horizon, and throttles at idle, the aircraft's speed began to drop: 280 knots, then 275, and within seconds it reached 260, then 250 knots. The captain's left hand was on the control wheel and his right hand gripped the large throttles. He shoved the throttles halfway forward, and the engines responded slowly with increased thrust; he relaxed the pressure on the control wheel slightly, and the nose of the aircraft came down. By the time the engines settled on a constant thrust, the aircraft was climbing to 14,000 feet at a comfortable rate. Speed 240 knots. In 15 seconds it was all over.

The Captain flew the aircraft manually for a while. Relaxing back into his seat, he gently pushed a button and engaged the autothrottles. The autopilot was engaged soon after. Most of the passengers on this 6 A.M. flight were fast asleep, and those that were awake hardly noticed the aircraft pitching up and the thrust coming off and then on again, thinking it was a bump from unsettled air or some random act of turbulence. The rest of the flight was uneventful.

Automation Surprises

In the above incident, the aircraft accelerated beyond the speed limit instructed by air traffic control. The copilot, who had dialed that speed (240 knots) into the machine, assumed the aircraft would maintain this speed during the new climb to 14,000 feet. It didn't. And indeed the copilot and captain were very surprised to see the aircraft reaching 300 knots, 60 knots faster than instructed. "Automation surprise" is a term that is sometimes used to describe this kind of confusion. But just labeling the above incident an automation surprise is not enough. We need to understand why it came about. And to do this we must consider the sequence of events, the machine's behavior, and the human–automation interaction that brought about this incident.

Mode Transitions

Figure 14.3 shows the sequences of events and subsequent autopilot mode transitions that occurred during this in-flight incident. We begin in VERTICAL NAVIGATION mode and in economy speed (which is progressively calculated by the computer). This is configuration A, and from there we transition to configuration B, in which speed is set to 240 knots in response to the air traffic controller's instruction.

Nearing 11,000 feet, a delicate sequence of mode transitions takes place, and as the broken lines in figure 14.3 indicate, it is completely automatic. First is the automatic transition to ALTITUDE CAPTURE mode, which guided the aircraft during the curved maneuver from climb to level flight (configuration C). Once the altitude was reached and "captured," and the aircraft was at 11,000 feet, the autopilot transitioned automatically to ALTITUDE HOLD mode, which, as the name suggests, simply held the aircraft at this altitude (configuration D) and maintained level flight.

The "automation surprise" occurred several minutes later: air traffic control instructed the aircraft to climb to 14,000 feet, and the crew responded by engaging the VERTICAL NAVIGATION mode (state E). Soon after, the aircraft accelerated well beyond the speed limit of 240 knots to almost 300 knots. Why?

Machine Model

To understand why the aircraft "ran away" from the pilots, we start by looking at the machine model of figure 14.4. The fully automatic VERTICAL NAVIGATION was engaged throughout the climb to 11,000 feet. When the aircraft was close to this altitude (10,700 feet), a transition to the ALTITUDE CAPTURE mode took place. This transition was triggered by a logical guard ("near level-off alti-

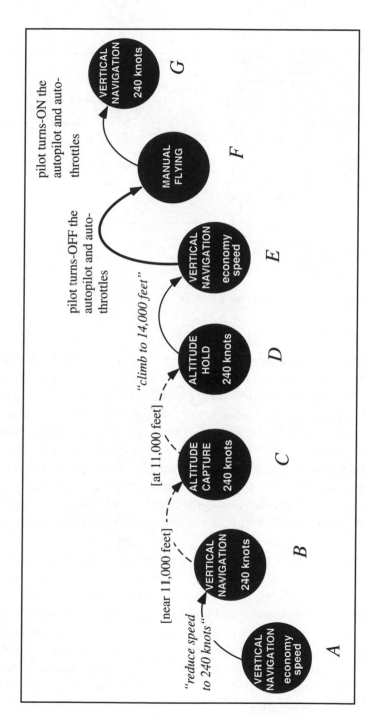

Figure 14.3. Sequence of modes and speed reference values.

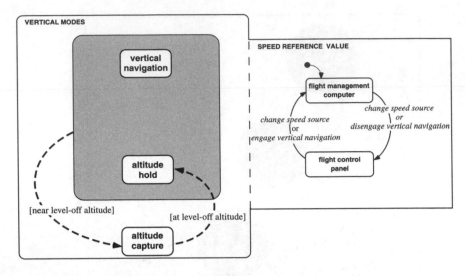

Figure 14.4. Autopilot modes and speed reference values.

tude"), which is an algebraic equation involving accelerations and the aircraft's rate of climb. The transition from ALTITUDE CAPTURE to ALTITUDE HOLD is also automatic, and is triggered by a condition ("at level-off altitude") that becomes true when the aircraft is just several feet below the reference altitude (11,000 feet).

Before, during, and after the altitude capture maneuver, although modes were changing automatically, the speed was kept at 240 knots. This contributed to the copilot's expectations that the speed he dialed in manually (240 knots) would be maintained by the automation. But then, as soon as the copilot re-engaged VERTICAL NAVIGATION, the aircraft accelerated unexpectedly to 300 knots. The reason for this runaway situation lies in the switching that takes place within the speed reference values in figure 14.4.

While in VERTICAL NAVIGATION, the speed input initially comes from the flight management computer (economy speed). Note that it is possible to override this speed by manually dialing in the desired speed in the flight control panel. In this configuration, the mode is still VERTICAL NAVIGATION, but the speed input is from the flight control panel. Once the aircraft begins the capture maneuver, the VERTICAL NAVIGATION mode automatically disengages and the autopilot switches to ALTITUDE CAPTURE mode—yet the speed reference value still comes from the flight control panel. In ALTITUDE HOLD mode the speed input also comes from flight control panel, and all is well.

However, look what happened after the air traffic controller instructed the pilots to climb to 14,000 feet. The copilot re-engaged VERTICAL NAVIGATION (assuming the aircraft would still maintain 240 knots). But as you can see in figure 14.4, when the VERTICAL NAVIGATION mode engages, the speed reference

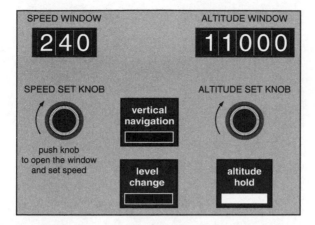

Figure 14.5 (a). Flight control panel. The speed reference value (240 knots) was manually set into the speed window by the copilot.

value switches to the flight management computer as the source (where the economy speed was computed to be 300 knots). Therefore, the aircraft immediately began to accelerate to 300 knots. The pilots, however, were caught by surprise, because they assumed that the aircraft would maintain the 240-knot speed limit still in effect. But the system can do only what it is designed to do. It took the pilots some time before they realized that the aircraft was accelerating some 60 knots beyond the speed restriction the controller had given them. The consequence was a speed violation, which probably also triggered some stress responses in the air traffic control center.

It is evident that the copilot did not recall, at the moment of re-engaging VERTICAL NAVIGATION, that this mode automatically switches its speed source to the flight management computer. He apparently forgot this piece of information, and since there was nothing technically wrong with the automation—no malfunction or failures—inquiry might stop here. But just knowing what had happened is not enough. We must try to understand why it happened, and what can be done to prevent it from happening again, because in the seeds of such an incident lie the potential for a harmful accident.

Interface

The pilots were well trained and knew how the speed reference value works and how it switches depending on the mode of the automatic flight control system. So why did the copilot fail to exercise his knowledge? Let's begin by looking more closely at the flight control panel interface, which serves as the interface between the pilots and the automation. Figure 14.5(a) is a line

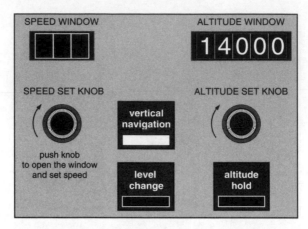

Figure 14.5 (b). Flight control panel. The speed reference value is computed by the flight management computer (and therefore the speed window is blank).

drawing showing the flight control panel, which contains the speed set knob, the speed window, the altitude set knob and window, and three mode buttons. In the context of this in-flight incident, figure 14.5(a) shows what the pilots saw on the interface while the aircraft was flying straight and level at 11,000 feet: the speed was set to 240 knots, the ALTITIUDE HOLD mode was engaged, and the copilot had already entered the new altitude (14,000 feet) into the flight control panel.

Figure 14.5(b) shows the interface after the copilot engaged VERTICAL NAVIGATION mode and the aircraft began to climb. As you can see, the speed window is blank (to indicate that the source of speed reference value is no longer from the flight control panel). So the pilots' main cue that a new reference value has replaced the old 240 knots speed is a blank display— certainly not a positive indication by any standard. This may have contributed to the copilot's inability to quickly identify the change in speeds and recover from his error. (There are two other cues that indicate the change in reference value: one is the header of the climb page in the flight management computer, and the other is the speed marker [bug] on the pilot's air speed display. Nevertheless, both cues are peripheral to the flight control panel where the interaction takes place.)

The Underlying Structure of the Interface

To understand why the pilots were surprised, it is necessary to look deeper into the interface. It is important to consider the underlying structure of the interface, because this kind of incident has happened many times, even for those who have already experienced this trap.

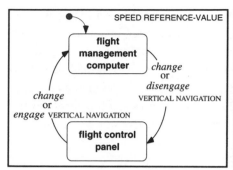

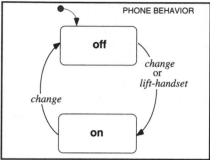

Figure 14.6 (a). Switching between speed reference-values in the autopilot of a modern aircraft.

Figure 14.6 (b). Switching in the cordless phone example (from chapter 6).

When the aircraft is in VERTICAL NAVIGATION, the speed input comes from the flight management computer; this is how this mode is initialized. The pilot can manually override this speed by pressing on the speed knob, which opens the speed window for entry—and then rotate the knob to manually enter a new speed value. Press the knob again, and the speed window blanks, which means that the source of the speed input has reverted to the flight management computer. The pilot can press on the knob and change the speed back and forth between manual and automatic as many times as he or she likes. If this was the only way to change the speed source, the system would probably not be confusing. But as figure 14.6(a) shows, there are two additional events that switch us between the two sources of speed. These two events—engaging VERTICAL NAVIGATION or disengaging it—are not directly related to the speed source change. They tend to interfere with the pilot's "muscle memory" of which speed source is indeed active.

Does this sound familiar?

Recall the cordless phone in chapter 6. In the phone example, the user can switch manually back and forth between phone ON and OFF by pressing the "talk" button. If this was the only way to turn the phone ON and OFF it would not be confusing. But there is an additional event—lifting the handset—that also switches the phone from OFF to ON. The event *lift-handset*, which is not directly related to ON and OFF, interferes with the user's muscle memory about the state of the phone. And then, of course, there is the well-rehearsed sequence of actions—lifting the phone and then immediately pressing the "talk" button— that seals the caller's fate.

The reason why events such as *lift-handset* in the phone and *engaging/ disengaging* VERTICAL NAVIGATION in the autopilot interfere with the user's memory is because they "appear" automatic—coming out of the blue, causing surprises, confusion, and error. These events are side effects: they are events that take place in one component of the system (*engaging/disengaging* VERTICAL NAVIGATION) yet trigger a side effect event in another component (speed source). I have mentioned these side effects in chapter 3 and showed in chapter 5 how they work. Because side effects "appear" as automatic events to the user, it is important to pay extra attention to such events when it comes to designing interfaces. In many ways they should be treated, in terms of the emphasis in display design and user training, with the same care as automatic events.

In Conclusion

In this and in other chapters, we have seen the subtle interaction between a mode and its reference value. This interaction always exists in automated control systems, and both users and the designers of such systems should pay close attention to the potential confusion that can result when reference values change automatically. Users have a bias toward assuming that if they *manually* enter a reference value, the machine will honor it. Nevertheless, machines are not like humans; computers and automated machines are not sensitive to situation and context. They can only do what they were pre-programmed to do. When they switch modes, they may switch to a source of new reference values—and if no reference values are available, the machine will default to some predetermined value. This value may be very different from what the user expects.

There is no simple, catch-all solution to the mode and reference value problem discussed above (although there are some novel interface solutions where speed, altitude, and other reference values are better integrated). What can be improved, however, is the feedback the machine gives the user about what it is doing: the unique interaction between VERTICAL NAVIGATION and the speed value can be better displayed so that the pilot will be aware of the consequences of his or her actions. Better feedback can be helpful in capturing the pilot's attention and allowing him or her to identify the error and quickly recover.

Finally, it is interesting to note that the kind of user interaction problem that we find in an everyday consumer product has its parallel in complex automated control systems. In the airplane as well as with the cordless phone, users know how the system works and have been using it for quite some time; some have even encountered the problem before. Yet everyone at one time or another falls into the trap. The same underlying interface structure that confuses users of a $50 cordless phone creates similar confusion in a $50 million airliner.

Fortunately, the incident described above did not end in an accident. However, there is a consensus among aviation safety professionals that incidents contain the precursor to accidents. It is therefore extremely important to understand incidents. And, indeed, the investigations of many accidents have revealed a trail of related incidents that were warning signs, yet were not heeded.

Chapter 15

Automation, Protections, and Tribulations

"On bare wings I brought them unto me."

—from a plaque in an airmen's memorial on the River Thames, London; an adaptation from Exodus 19:4.

T he long black runway sweated with haze as the elegant airliner made a swift landing. Within seconds, the wide-body aircraft reached the end of the runway and stopped. The regular stir of passengers—anxious to get out of their seats, eager to collect their belongings and disembark—was missing. In the large cabin, designed to carry more than 300 passengers, sat only three men. Two of them were seated side-by-side in the middle of the aircraft. In the forward section, behind a large console full of computer screens, panels with switches and buttons, rolls of printouts, and a mass of multi-colored wires, sat another man. He was busy flipping charts and moving switches while reading aloud into a microphone: "takeoff in flaps configuration 2, engage the autopilot at 157 knots and 6.5 seconds after takeoff, then reduce power on the right engine, shut hydraulic . . ."

This, apparently, was no ordinary flight.

Ahead of him was an empty galley, beyond it a lavatory, and then an aluminum door. Behind the door, a blue carpet led into a wide cockpit that looked like the flight deck on a starship. It was slick and clean, full of large displays and dozens of dazzling and colorful indicator lights. Above the displays was the long flight control panel, full of switches and knobs for engaging the autopilot and the autothrottles, as well as for entering speed, altitude, and other reference values into the autopilot. Below the flight control panel, in the center of the cockpit, extended a large, flat pedestal with three computers, two stout-looking throttles, and an array of radios, switches, and levers. To the right of the pedestal sat the copilot, and to the left, the captain. They were each busy in their seats, preparing for the next flight. Behind them, on blue seats fixed to the cockpit wall, sat two observers—also pilots. The captain straightened his headset and began talking over the intercom, dictating the aircraft's configurations, modes, states, and performance numbers.

This was a test flight.

The aircraft was new; its maiden flight only eight months before. It was undergoing the rigor of repeated flight tests. Every component of the aircraft, from the wheels to the engines, all the way to the avionics suite, was tested in a multitude of conditions. By now 360 flight hours had been accumulated by various test pilots, not counting the hundreds of hours spent in the aircraft's flight simulator.

The captain, who was also the chief test pilot, grasped a small wheel-like crank in his left hand, looked back through his side window, and began making a tight U-turn. As the 200-foot-long aircraft finished the turn and moved slightly forward, the white, crosswalk-looking stripes, indicating the beginning of the runway, stretched on either side of the plane. As the aircraft came to a full stop with its main wheels resting heavily on the warm runway, the two pilots sank into their seats, resting from the just-completed test flight.

But within moments, they began preparing the aircraft for its second flight. They first entered the runway information and flight profile into the flight management computer. Meanwhile they were also monitoring their systems displays and checking the temperature inside the tires, waiting for them to cool from the just-completed landing, and reviewing the status of the hydraulic and electrical systems. As they were busy entering data into the computers and making sure the engines' temperature and other parameters were back to normal, the sun was already marching toward the southwest. It was after 5 P.M., and the workday was coming to an end.

The Last Flight

This last test flight was the culmination of a long and tiring day, especially for the captain: In the morning he met with representatives from a client airline and flew with them in another aircraft for an hour and a half, demonstrating the qualities of that aircraft. Afterwards he had jumped into the simulator of yet another new aircraft, testing and evaluating the aircraft's performance. Then he had a long lunch with the clients from the morning, followed by an interview with a Japanese TV crew. By mid-afternoon he was back on the flight line, where he met the copilot and flight-test engineer in preparation for these two test flights, which began soon after 3 P.M. It was a busy and intense day, but just part of that unmistakable buzz that no visitor to a flight-test facility can ignore. The exhilaration associated with creating something new and a certain presence of danger is always in the air. Flying new aircraft and subjecting new designs to tests and maneuvers is a thrill, not unlike the excitement sustained on a movie set, in the sports arena, on a dance floor, or in any other highly creative human endeavor. Being part of it, participating in its tempo, is an engulfing lore that makes one go beyond the limits.

The flight-test engineer, from his console in the back of the aircraft, finished all his checks. He made sure that the autopilot was configured properly for the impending test. This flight was part of a required test to certify the autopilot and the engines. The test had been performed several times both in the simulator and also in the previous flight. This flight was going to be a routine flight—a repeat of what was already done. They were confident and relaxed. But the captain was tired from the long and eventful day, and as he looked to his right, he suggested that the copilot make this takeoff and fly the aircraft. The captain quickly briefed the impromptu change through the intercom. All of the test activities, from engaging the autopilot to shutting down systems, were done in coordination with the flight-test engineer who was sitting in the empty cabin. It was important for the captain to keep the flight-test engineer "in the loop" with respect to all activities that were taking place in the cockpit.

The control tower announced that wind was from the northeast (040 degrees), at 3-8 knots. The copilot mentally rehearsed the takeoff and the test sequence. They would be rolling fast down the runway, and then he would pitch the nose up and make a quick and aggressive takeoff. Seconds later, the captain would engage the autopilot—and then simulate an engine failure by reducing power on the left engine to imitate a condition in which the engine has quit on takeoff. Immediately afterwards, the captain was to shut off the left hydraulic system.

The aircraft was designed to perform adequately and climb safely with just one engine. And this capability had already been tested and demonstrated over and over during the preceding eight months. This time around, the engineers had suggested a small modification to the control logic of the autopilot. And part of this test flight was to evaluate the autopilot modification while the aircraft was flying on one engine only.

Testing and retesting the consequences of disturbance events such as engine failure on takeoff is one of the fundamental purposes of test flights. We never know when an engine will fail during a normal flight, with many passengers onboard—all we know, based on historical data, is that it might. Test flights are designed to ensure the airplane will survive the most extreme conditions it is likely to encounter. And therefore it is important to test and retest an aircraft in all kinds of flight conditions, well before the aircraft is fully operational, in order to understand how the aircraft reacts under abnormal situations.

Test pilots are well trained for such meticulous testing. The captain was a graduate of a prestigious test-flight school, with considerable engineering experience and almost 8,000 hours of flying and test flights to his credit. He had flown 123 hours on this specific aircraft to date. The flight-test engineer, supervising all activities from his console in the back of the aircraft, had logged more than 6,000 hours as an engineer on such test flights. The copilot had more than 9,500 flight hours, and now he was ready to fly the aircraft and repeat the same test that they had just completed. During the previous flight, the autopilot modifications worked so well that they were too subtle to be even noticed. This flight was supposed to be no different.

The copilot was shifting in his seat, positioning his torso for the takeoff and placing his right hand on a small stick on his side. Instead of the traditional yoke-looking control wheel, this aircraft was equipped with an elegant side-stick controller. The copilot was going to be making a manual takeoff, pulling on the side stick to pitch the aircraft into the air. Shortly after becoming airborne, the captain was going to engage the autopilot—and from there onward it was going to be an automatic flight. The autopilot they were using was part of a sophisticated flight control system that included the autothrottles, flight management computers, and a variety of flight protection modes.

Envelope Protection

Figure 15.1 is a partial description of the autopilot and speed modes of this specific aircraft. In this autopilot, the ALTITUDE CAPTURE and ALTITUDE HOLD modes for transitioning from climb to level flight are the same as we have seen earlier in chapter 14. However, note that there are additional modes in the periphery. These are protection modes that engage automatically when the aircraft is in trouble, namely when the flight profile is unsafe and the aircraft is nearing a stall. Here's how the protection modes work: An onboard system senses and monitors the flight profile, constantly checking whether the aircraft is about to transgress beyond the normal and safe operating envelope. If the aircraft is coming close to a stall, a transition to the protection mode takes place automatically.

There are two protection modes in figure 15.1. The first one, ALPHA PROTECTION MODE, commands the aircraft to lower the nose to avoid a stall. However, just lowering the nose and thereby reducing the pitch attitude of the aircraft is necessary but not always sufficient to avoid a stall. When an aircraft is about to stall, it is usually flying at a low airspeed. Therefore, an additional protection mode, THRUST PROTECTION, is there to command full power from the engines, so as to increase speed and avoid a stall.

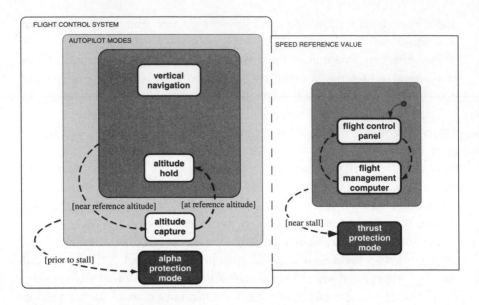

Figure 15.1. Autopilot modes and envelope protection.

As you can imagine, such *envelope protection* systems and their modes are extremely important in automated systems. We now find them in safety-critical systems such as medical equipment and aircraft, and they are also slowly migrating to automobiles. In any event, it is important to make a distinction between two categories of envelope protection systems: one, in which the envelope protection system protects against malfunctions or poor performance strictly on the part of the machine (such as in the case of the RETRY mode in the blood-pressure machine); the second category includes systems that override the user's actions in order to prevent an adverse effect. One such example is the antilock braking system (ABS), found in many new cars, that overrides the driver's braking pressure and prevents the wheels from skidding. Such systems improve on human abilities (even professional race drivers can't stop as quickly as an average driver with an ABS-equipped car) and prevent user errors and miscalculations during panic stops.

Envelope protection systems such as the antilock braking system—which is concerned with a single aspect of driving, namely braking and the prevention of skidding—have a very specific protection requirement. However, more complex envelope protection systems found onboard aircraft have to contend with multiple aspects of the flight such as preventing stall, ensuring that the pilot does not perform maneuvers that may cause structural damage to the aircraft, and preventing the aircraft from hitting the ground or other objects. Since there are multiple protection requirements (some of which interact in non-trivial ways), and since there may be more than one way to protect the aircraft, taking the optimal protection action is a complex matter. Furthermore, making sure that the automatic actions that the envelop protection takes are *always* correct, given the context of the situation, is a major challenge facing the designers of such systems, as well as their users. The subject of automatic envelope protection systems that take authority away from the user has been a topic of hot debate, some of it quite emotional (especially in the aviation industry), and we shall talk more about this issue later.

Takeoff

From his observation platform towering high above the airport, the air traffic controller assigned to monitor and supervise the test flight had a magnificent view of the airport, adjacent farmlands, and the nearby city. A long black runway stretched below his feet. At the far end of the runway, the white aircraft was poised for takeoff. Turning his gaze back into his air traffic console, he made all necessary preparations to clear the aircraft for takeoff. The flight profile was going to be just like the previous one—a short takeoff run and a gradual climb to 10,000 feet, several slow maneuvers at altitude, and then a wide descending turn back to landing. The whole affair should take less than an hour. The air traffic controller checked the flight information of other aircraft and made a quick scan on the radar, making sure that there were no other aircraft in the way. He glanced at his copy of the flight-test order and visualized in his mind the takeoff and the straight climb to 10,000 feet. He looked up toward the white aircraft and quickly shifted his gaze toward the end of the runway and up to the open skies. His right hand clutched a little black box that he held near his belt and his finger pressed on its little "push to talk" button: "Cleared for takeoff."

As the air traffic controller's instructions rang in their ears, the copilot requested a confirmation from the captain as to which takeoff power setting to use. The captain responded, and the copilot repeated to himself: "takeoff thrust." The copilot gently moved the two throttle levers forward, and the engines reacted with a long and high-octave whining. Within two seconds, the two big engines pulled the aircraft out of its parked spot. The wheels started rolling as the two large turbofans pulled all 325,605 pounds of metal, fuel, avionics, and crew forward. The aircraft was moving, its entire weight on the wheels, shaking slightly as it encountered small bumps on the runway; the wings drooped with heaviness and the tips were fluttering up and down in response. Within seconds, the speed was up to 50 knots, and increasing. And now the engines were pulling the aircraft and thrusting the long body forward.

The aircraft accelerated with every passing second and the ride became smoother as some of the aircraft's weight was gradually transferred to the wings. "80 knots," called the captain. As the speed increased further, the long wings curved upward, and the wingtips rose in anticipation. At 100 knots the aircraft was ready to soar, but the copilot was pushing his side stick forward and holding the nose down; 120 knots, and he had to flex his hand muscle to prevent the aircraft from escaping free of the ground.

"Rotate," commanded the captain at 132 knots, and the copilot pulled full aft on his side stick controller. The nose of the aircraft rebounded into the sky, and only the main wheels were still in contact with the ground. Within a second the nose was passing 6 degrees above the horizon and pitching up further. At that moment, the wings gave so much lift that the heavy aircraft had no choice but to let go its grip of the runway. The aircraft was airborne. Free of the ground, it began pitching up to 12 degrees, 14 degrees, and then 18 degrees. They were elevating into the sky at a high rate of climb.

As the long white aircraft soared to the sky, invisible contrails of data poured into the ground control center. Every word that was uttered in the cockpit was recorded. Every sensor in the aircraft—from those detecting airspeed, pitch attitude, and altitude, to those measuring engine thrust and hydraulic pressure—was recording information. The autopilot and autothrottles were not just recording and sending information about their active mode, but also about their internal states and parameters, and every reference value entered by the crew. Every two seconds, this telemetry information was batched and transmitted through a high-speed communication link to the ground control center. Most of the data was stored for further analysis; some of the data was displayed on computer screens inside the ground control center.

Four seconds after takeoff, the aircraft's pitch attitude was already 20 degrees above the horizon, which is higher than the 15-17 degrees that we all commonly experience as passengers. The airspeed reached 155 knots, but then the high pitch attitude caused the aircraft's speed to slowly decay.

But the aircraft was still full of thrust and uplifting energy when the captain called "Positive rate of climb," and the copilot responded, "Gear up." The captain instinctively leaned toward the instrument panel. The long day had exerted a toll on him, but everything here was familiar and he proceeded with business-like efficiency. He reached forward with his right hand and raised the little gear handle. "Clack," came the metallic response as the handle hit the stopper, and then the landing gear began its familiar cycle, retracting into the belly of the aircraft. The captain moved his hand upward toward the long flight control panel. There, he went straight for the autopilot engage button and pushed it in. Meanwhile the aircraft's nose was coming to a high 23-degree pitch attitude, and the copilot, noticing the unusual attitude and the gradual loss of speed, began manually pushing down on his side stick and lowering the nose.

But the autopilot did not engage.

The captain pushed on the button again. No engagement. After several successive tries at the button, each time harder, the autopilot finally engaged and he could see the proof of this on his display. Relieved, the captain's mind raced on and was already preoccupied with the next task in the flight test order. What he did not see was that the aircraft's speed was slowly coming down to 150 knots, as the engines, although in full forward thrust, could not keep up with the high pitch attitude that had already attained 24 degrees and was still going strong.

The autopilot, for its part, was still resisting engagement in a passive way. Although engaged, it did not activate any mode, nor did it take control of the aircraft. The captain, however, did not notice all of this as he was already preparing for the next task—reducing thrust to idle on the left engine to simulate an engine failure. The pitch attitude continued to inch up as the captain's hand flew toward the throttle quadrant. He grabbed the left throttle with his hand and gently brought it back; making sure it was firmly against the idle stop. He then quickly went on to his third task: shutting down the left hydraulic system.

As the captain's eyes shifted to the overhead panel and his hands reached up for the hydraulic system controls, the autopilot mode indications went blank. This was by design. The systems logic was that if the aircraft's attitude was beyond 25 degrees, certainly an abnormal situation, the pilot should not be bothered with superfluous mode information. The display now showed only the airspeed, heading, altitude, vertical speed, and the pitch attitude, which had already reached 28 degrees nose up when the left engine ceased to produce thrust. The aircraft was now flying on one engine, 500 feet above the ground, speed decreasing below 145 knots, and its nose pointing 29 degrees up into the sky. The aircraft was heading towards an undesirable, abnormal, and potentially dire situation.

Autopilot Taking Control

At that moment, the autopilot activated and assumed control and then immediately transitioned to ALTITUDE CAPTURE mode. But the displays were omitting this information, and the pilots had no way of knowing which mode the autopilot was in. A second later, the ALTITUDE CAPTURE mode began guiding the aircraft along a parabolic profile to attain 2,000 feet, which was the reference altitude setting on the flight control panel.

Initially, the autopilot gave nose-down commands, and the pitch attitude came down to 28, then 27, and then 25 degrees. It seemed as if the autopilot was finally taking command and assuredly returning the aircraft to a normal and safer climb profile. But a second later, the autopilot reversed its commands; instead of continuing to lower the nose, it gave nose-up commands.

The aircraft pitched upward, and began racing into the sky again—its nose at 26 degrees, 27, 29—as if the autopilot was desperate to climb even faster, in spite of the fact that the aircraft's thrust had been cut in half, and its left wing had begun to drop.

"Pump fault," said the captain in a casual voice, confirming the action he had taken to disable the hydraulic system in keeping with the test-flight order. With his eyes and attention focused on the hydraulic panel up in the cockpit's ceiling, he was probably oblivious to what was going on below, because at that moment the cockpit was pointing 30 degrees up in the air and the autopilot was progressively commanding higher and higher pitch attitudes.

But the laws of physics could not be fooled for long. As the aircraft began to lose its momentum and the pitch attitude only increased with fervor, the aircraft's speed began to spiral downward to 129 knots, then 120 knots. The aircraft was now losing speed at an alarming rate, passing through 118 knots, which is the lowest speed that the aircraft can still be fully maneuvered. In two seconds it was down to 113. The situation, like the speed, was deteriorating rapidly. Yet the autopilot kept coercing the aircraft on a fanatical and unattainable march into the sky. When the captain finished with the hydraulic system and his eyes and hand came down from the overhead panel, the blue horizon line was hidden below the nose, which was suspended 30 degrees up in mid-air.

"What happened?" he said in surprise.

The Captain's Takeover

It took the captain three seconds to look at the instruments and realize the severity of the situation. He immediately disconnected the autopilot and assumed manual control. He now had a crippled aircraft in his hands and he felt through his body and feet the vulnerability of the aircraft. The nose was pointing 31 degrees into the sky, speed was at a meager 100 knots, the left engine at idle and the right engine producing maximum thrust. The asymmetric thrust between the left and right engines was beginning to exert a corkscrew effect, pushing and rolling the aircraft to the left. But worst of all, the aircraft was just 1,400 feet above the ground. "Speed, Speed, Speed" chirped a metallic voice from the envelope protections system as the left wing dropped down further. The captain was fully cognizant that the aircraft was no longer fully maneuverable, and he knew he had to act fast in order to recover. But he wasn't the only one: the automatic envelope protection system also sensed that something was wrong, and immediately transitioned to ALPHA PROTECTION mode, which, in turn, commanded a nose-down attitude.

But the captain was faster.

He was already pushing the nose down, all the way, in an effort to regain speed and maneuverability. He also initiated a right turn, trying to hoist up the left wing. Sensing the heaviness of the wing, he stepped hard on the right rudder to prevent the aircraft from entering a left turn. Two seconds later, realizing that nothing was going to help—he pulled back on the right throttle.

The right engine was the only engine producing full thrust and keeping the aircraft flying forward. But as startling as it may sound (because now the aircraft became a glider), this was the only choice available to the captain, and he acted decisively. He had to bring the left wing back, and reducing thrust on the good engine was the only way to even out the corkscrewing effect that was pushing the aircraft to the left and threatening a downward spiral. But it was too late, because the left wing already began to buffet from the disruption of the airflow.

A second later another protection system woke up: the AUTOMATIC THRUST PROTECTION mode engaged, and commanded both throttles to full power, in an attempt to pull the aircraft away from the stall. But since the captain already pulled back on the right engine and both engines were at idle stop, the activation of this automatic protection mode was canceled. The aircraft was already in dire straits, and there were no more protections in store.

The pitch attitude was so high, and the speed so low, that the long and broad wing could no longer sustain itself and began sinking rapidly as the speed passed through 77 knots. The aircraft's body was shaking from turbulent confusion and the left wing dropped violently, throwing the aircraft into a brutal left turn, while a cacophony of computer-generated warnings continued sounding in the cockpit. And although the pitch attitude was steadily coming down to 0 as the captain was pushing the nose down, the aircraft wings reached 90 degrees vertical and continued to a 110-degree roll, leaving the crew hanging upside down, held to their seats only with their shoulder harnesses. But the captain kept on pushing the nose down, and the long aircraft began responding rapidly; the pitch attitude dropped to 20 degrees down as the view of the black runway and surrounding countryside rushed to fill the cockpit window. The speed began to increase: 90 knots and then 110. Slowly at first, and then with increasing vigor, the left wing began to recover and rose up to reverse the roll. A second later, the speed was 125 knots and increasing, the left wing returning back to level, and the aircraft was maneuverable again.

But the altitude was only 600 feet.

The nose was pointing down 35 degrees while the captain reversed his actions and began pulling hard to raise the nose and avoid the ground that was racing toward him at 7,500 feet per minute. The nose cooperated and began pulling out, trying to prevent the dark runway from taking them in. But the large and heavy airliner—more than 200 feet in length—could not muster the excruciating maneuver needed to escape the ground. Four seconds later, with the captain and copilot pulling as hard as they could on the controls, and every rivet, beam, and metal straining to prevent the worst—the white aircraft slammed into the ground. Within seconds it was engulfed in flames.

There were no survivors.

Why Did it Happen?

The entire flight lasted only 36 seconds and the aircraft crashed near the end of the runway from which they took off. Figure 15.2 is the flight profile. Like any accident, this tragic crash is a sequence of many abnormal events, each one not dominant enough to cause a catastrophe, but their combined effect did. We will start with the machine and its behavior, and then proceed to consider the pilot interaction with the automation.

The autopilot engaged six seconds after takeoff following repeated attempts by the captain. The altitude was 500 feet and the aircraft was catapulting into the sky at 6,000 feet per minute. As soon as it activated, the autopilot switched to ALTITUDE CAPTURE mode and began the curved maneuver to capture 2,000 feet.

Why 2,000 feet when the flight-test order called for a straight climb to 10,000 feet?

The post-accident investigation found that the autopilot's (reference) altitude was indeed set to 2,000 feet. One supposition for this discrepancy is that the captain and copilot forgot to reset the altitude to 10,000 feet; another is that they did intend to level off at 2,000, but coordinated this out-of-the-ordinary change to the test order using only hand signals (the cockpit voice recorder did not contain any mention of this 2,000 feet level-off modification). Since there was no record of any intercom contact with the flight-test engineer, who was sitting in the cabin away from the cockpit, regarding this change, nor any communication with the air traffic controller about such a major departure from the flight test, we will continue our discussion assuming that the crew *forgot* to set 10,000 feet.

If that's the case, where did the 2,000 feet come from? The only plausible explanation is that it was left over from the previous flight. And indeed, it turns out that the last altitude that the crew had entered into the flight control panel during the previous flight was 2,000 feet. For some reason, perhaps from being tired or for other reasons, they forgot to reset the altitude to 10,000 feet after landing and failed to notice this discrepancy during the preparation for the second takeoff.

The Autopilot

The captain tried several times to engage the autopilot. It did not engage and the captain had to push several times on the button to make it happen. The

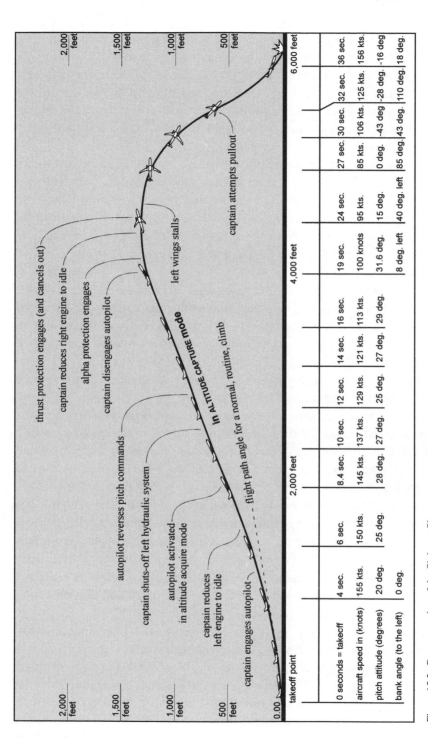

The following annotations appear on the flight profile diagram:

thrust protection engages (and cancels out)

captain reduces right engine to idle

alpha protection engages

captain disengages autopilot

in ALTITUDE CAPTURE mode

left wings stalls

captain attempts pullout

autopilot reverses pitch commands

captain shuts-off left hydraulic system

autopilot activated in altitude acquire mode

flight path angle for a normal, routine, climb

captain reduces left engine to idle

captain engages autopilot

takeoff point

Altitude reference lines: 2,000 feet, 1,500 feet, 1,000 feet, 500 feet

0 seconds = takeoff	4 sec.	6 sec.	8.4 sec.	10 sec.	12 sec.	14 sec.	16 sec.	19 sec.	24 sec.	27 sec.	30 sec.	32 sec.	36 sec.
aircraft speed in (knots)	155 kts.	150 kts.	145 kts.	137 kts.	129 kts.	121 kts.	113 kts.	100 knots	95 kts.	85 kts.	106 kts.	125 kts.	156 kts.
pitch attitude (degrees)	20 deg.	25 deg.	28 deg.	27 deg.	25 deg.	27 deg.	29 deg.	31.6 deg.	15 deg.	0 deg.	-43 deg.	-28 deg.	-16 deg.
bank angle (to the left)	0 deg.							8 deg. left	40 deg. left	85 deg. left	43 deg.	110 deg.	18 deg.

Figure 15.2. Reconstruction of the flight profile.

culprit was not the button, however. It was actually the autopilot's internal logic. When the captain tried to engage the autopilot, the copilot was exerting a slight nose-down pressure on his control in order to counter the high (25-degree) pitch attitude. There is a built-in logic in the autopilot that resists engagement of the autopilot (and also delays activation) when the pilot is making manual inputs. When the autopilot finally activated 2 seconds later, the pitch attitude was 29 degrees.

After the autopilot took control, it initially lowered the high nose-up attitude back to 25 degrees. And then 4 seconds later the autopilot reversed itself and progressively increased the pitch attitude—attaining almost 32 degrees by the time the captain intervened. If the captain had not disconnected the autopilot and assumed control, the autopilot, left on its own, would have probably continued increasing the pitch attitude beyond 32 degrees.

Why?

Post-accident analysis of the autopilot logic revealed a design flaw, which although it had existed for a long time, was neither detected nor encountered during software testing, simulator tests, or flight tests. This flaw had to do with control authority. When the autopilot transitions to ALTITUDE CAPTURE mode, it immediately calculates the rate of climb it needs to follow in order to close on and reach 2,000 feet. At the time the calculation was made, there were two strong engines producing full thrust.

And this is the underpinning of the entire accident, because shortly after the profile calculation was done by the autopilot, the captain reduced power on the left engine (to simulate an engine failure). At that point, the aircraft had only one engine producing thrust. The right engine could not produce enough thrust to keep the aircraft flying on the steep profile. But the autopilot did not care; it wanted to close the gap to 2,000 as fast as it could. The autopilot was not designed to progressively calculate the altitude capture profile; rather it was designed to follow the rate of climb that it computed initially. And follow this rate blindly, it did. The aircraft, nonetheless, could no longer fly that steep profile because its left engine was idle and not producing thrust. As a consequence, the aircraft began slowing down, falling off and deviating from the climb profile (see figure 15.3).

But as soon as the autopilot detected the deviation, it responded immediately by pitching up and increasing the aircraft's attitude in an attempt to climb back again. Yet the increased pitch attitude further reduced the aircraft's speed and caused an even greater deviation. Then the autopilot commanded an even higher pitch attitude. It was like a snowball effect—because every increase in pitch brought about an even greater deviation. Dumb and dutiful, the autopilot progressively commanded higher and higher pitch attitudes, well beyond the normal values, in a futile and ignorant attempt to make the aircraft fly a flight profile that had become impossible to attain. The autopilot was guiding the aircraft into a stall.

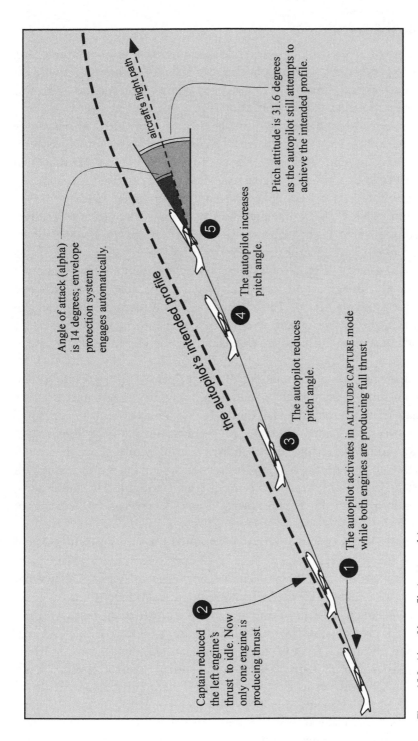

Figure 15.3. Aircraft's profile (not to scale).

Limited Protection

The flight control system, however, was designed with envelope protection modes to prevent the aircraft from entering a stall. To understand why the protection modes did not take effect, we need to go a bit deeper into what these systems can do and what they can't do.

Envelope protection systems provide a cocoon-like shell in which the autopilot (or the pilot) can do whatever he wishes to do—but once the system is about to go beyond the safety of the cocoon, the protection modes kick in and attempt to bring the aircraft back to safety. What this implies is that the envelope protection system splits the entire operational space (which is defined in terms of speed, pitch attitude, and other parameters)—into two separate regions: a normal operational region and an unsafe operational region. Figure 15.4(a) is an abstract description of the operating region and its division. The envelope protection system constantly monitors the location of the aircraft with respect to the unsafe region. The goal of the envelope protection system is to detect a transgression from the normal operational region into the unsafe region. And when that happens, the system takes control and tries to push the aircraft back into the normal and safe operational region.

Because envelope protection systems are in place to prevent the system from going unsafe, we want them to kick in *just before* the system crosses the boundary into the unsafe region. This is because it may take the system some time to reverse its actions and return to the normal region. This implies that there is an additional region, which I shall call the *buffer zone*. When the aircraft enters the buffer zone, the envelope protection system takes over and guides the system back before it reaches the unsafe region. The size of the buffer zone, which is always embedded in the normal region, depends on the dynamics of the aircraft.

Figure 15.4(b) is a graphic depiction of how the envelope protection system works. Point "a" corresponds to the aircraft operating in the normal region, and "b" is where the envelope protection system engages. Point "c" shows that the aircraft is already inside the buffer zone, but just skirting the unsafe region and is safely ushered out. In "d" the aircraft has returned to the normal region.

Envelope protection systems are critical for the sophisticated automation of today and the future because they provide an additional level of safety. Envelope protection systems bestow on us confidence that the system will not become unsafe and can always recover. Yet envelope protection systems are very difficult to design and should not be accepted naïvely by users as some kind of a fail-safe. There are several major issues that make the design of these Golems of our time rather complex: one, determining which cues truly and reliably indicate that the system is indeed crossing into an unsafe situation; two, establishing an appropriate buffer zone between the safe and unsafe

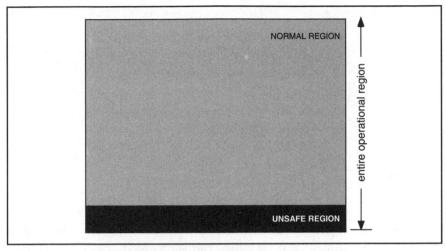

Figure 15.4 (a). Normal and unsafe operational regions.

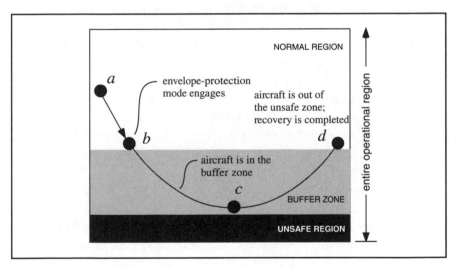

Figure 15.4 (b). Buffer zone and recovery.

regions of the operational space, which, as we mentioned earlier, has to do with the dynamics of the system. For example, for a really fast aircraft, the buffer zone must be very large—because it will take longer for the envelope protection system to bring the aircraft back to safety.

With this in mind, let us return to the particulars of the accident. The envelope protection system onboard the aircraft was designed to protect the aircraft from a stall. It does this by sensing the aircraft's *angle of attack*, which, generally speaking, is the angle between the nose-up (pitch attitude) and the aircraft's flight path (see figure 15.3). This angle of attack, termed *alpha* in aeronautical engineering lingo (and hence ALPHA PROTECTION mode), is quite indicative of stall. When this angle gets too high, stall occurs.

Nineteen seconds into the flight, when the pitch attitude reached 31.6 degrees and was increasing, the *angle of attack* exceeded 14 degrees. The guard on the mode transition to ALPHA PROTECTION mode became true and the transition took place (see figure 15.1). The ALPHA PROTECTION mode commanded the aircraft to lower the nose. Five seconds later, the second envelope protection mode engaged. The THRUST PROTECTION mode commanded full thrust on both engines, in an attempt to move the aircraft out of the unsafe (stall) region.

The captain, however, was faster than both protection systems: he had already commanded full nose-down attitude (in the same direction as the ALPHA PROTECTION mode) and then pulled back the right throttle, which canceled the full thrust that the THRUST PROTECTION mode was commanding. Even if the captain had not intervened manually, it is unlikely that the envelope protection system would have made a successful recovery. Why? Because there was simply not enough altitude for the airplane to recover. And here is a subtlety about the envelope protection system that is important to understand: the system sensed only what the aircraft was doing—it was not designed to monitor and take into account the aircraft's altitude.

At the time that the envelope protection system engaged, the altitude was only 1,300 feet. The aircraft simply couldn't switch from the situation in which it was about to stall to a full recovery, given such limited altitude. Had this problem occurred at 10,000 feet, the outcome might have been very different.

This issue of the time (and resulting loss of altitude) it takes the aircraft to recover and return to the normal operating region is an important consideration in the design of the buffer zone. Naturally, for a 200-foot-long, 325,000-pound aircraft, it takes quite some time to recover. The transition to the envelope protection mode occurred when the aircraft's angle of attack reached 14 degrees—but this was too late considering the aircraft's altitude (1,300 feet). For the aircraft to recover, given its low altitude, the transition to ALPHA PROTECTION mode should have occurred much earlier.

Note also that the quality and integrity of envelope protection systems depend on many factors, including (1) the internal events (e.g., speed, angle of attack, engine thrust) that are being sensed and used as conditions for transitioning into the protection mode; (2) the status of the environment (altitude, distance to obstacles such as buildings and mountains, and so on); and (3) the possible disturbances (in this case engine failure, loss of all hydraulic systems) and the resulting limits on aircraft performance. Therefore, the size of the operating envelope constantly changes during a flight. With it, there are constant changes to the size and shape of the buffer zone and the unsafe region.

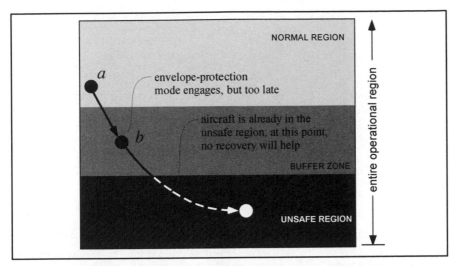

Figure 15.5. Shrinking of the normal and safe operational region (and expansion of the unsafe region) when altitude is considered.

Figure 15.5 shows the same recovery maneuver as seen in the previous figure. But here we take into account the aircraft altitude. Accounting for this factor shrinks the size of the operational region. The unsafe zone is bumped up, as is the buffer zone, demonstrating that for the recovery to succeed, the engagement of the envelope protection maneuver must begin earlier.

Erroneous Trim Setting

But even before the captain intervened and the envelope protection engaged in an attempt to save the situation, the aircraft, under the copilot's manual control, reached a 25-degree nose-up attitude. For comparison, the highest pitch attitude reached during the previous flight was only 14 degrees; during a routine takeoff on a commercial flight with passengers onboard it hardly ever exceeds 15–17 degrees. So the nose-up attitude indeed was unusual, and there was no lack of indications in the cockpit about this high-pitch attitude. Therefore, one of the immediate questions that arises is why did the copilot allow the pitch attitude to reach such a high angle in the first place?

There isn't a definitive answer to this question, but there are several explanatory factors in the way the aircraft was configured for takeoff: The longitudinal (pitch up or down) stability of the aircraft can be adjusted by setting the aircraft's trim (the pilots set the trim by moving the aircraft's horizontal stabilizer up and down). Trim adjustment is used to counter situations where the aircraft is loaded such that most of the weight is behind the wings, which will cause the nose to rise during flight. Figure 15.6 shows

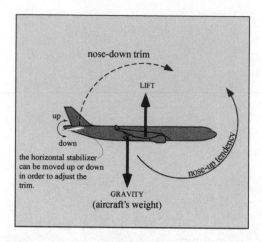

Figure 15.6. The aircraft's center-of-gravity is behind the center-of-lift. The resulting momentum causes a nose-up tendency. The pilots counter this effect by moving the aircraft's elevator (located in the tail of the aircraft). In the figure, the elevator is adjusted for a nose down effect.

that when the aircraft's center of gravity is behind the aircraft's center of lift, the pilots must adjust the aircraft's trim for nose-down (to counter the aircraft's tendency to lift up the nose).

During the test flight accident, the aircraft's center of gravity was behind the aircraft's center of lift, but the trim setting was set for a *nose-up* effect (2.2 degrees). It is not clear why the crew set the trim for a nose-up effect (the last setting used during the previous landing was 4.0 nose-up). But at any event, it created an even stronger nose-up tendency during takeoff and climb. (On their first takeoff, by comparison, the trim was set near zero degrees.)

The combination of a high nose-up tendency and maximum takeoff thrust caused the aircraft to pitch up aggressively during the climb. The copilot could have manually pushed the nose down to stop the aircraft from pitching up and prevented the autopilot from progressively commanding such excessive pitch attitudes. It may have been his familiarity with and confidence in the aircraft's autopilot and protection systems that made him less concerned about the steep climb profile.

Trust in Automation

Then comes the question of why the copilot did not disengage the autopilot when it commanded even higher pitch attitudes. Perhaps knowing that all this was part of an autopilot test, the copilot felt that they could go beyond the normal flight envelope. After all, that's what test flights are for. And maybe the copilot believed they should explore how the autopilot would correct such high pitch attitudes, even though the flight-test order did not call for such a demonstration. It is also possible that the copilot's failure to take action and disengage the autopilot may also have had something to do with a phenomenon called "trust in automation."

Most automated control systems are extremely reliable and safe. Over the course of human-automation interactions, just as in human relations, trust is built. Nevertheless, on rare occasions, automated systems fail. Over-trust in automation can bring users to the point of not recognizing the machine's inherent limitations. It can also lead to situations where the users fail to monitor and supervise the machine adequately. But worst of all, over-trust on automation can lull users into ignoring and dismissing cues that indicate the machine is failing to act as expected, and the situation is becoming progressively more dangerous. The sequence of events that led to the grounding of the *Royal Majesty* (chapter 8) is one sad example of this pervasive problem.

In aviation, over-trust on automation is a well-documented phenomenon. There were several aircraft accidents that involved inadequate monitoring of autopilots, and the National Transportation Safety Board (NTSB) began alerting the aviation industry to this problem as early as 1973, following the crash of an Eastern Airlines L-1011 aircraft in the Florida Everglades. (In that accident, the aircraft was on autopilot control, flying level at 2,000 feet and the entire flight crew was busy attending to a suspected landing gear malfunction. The autopilot, in ALTITUDE HOLD mode, disengaged unexpectedly. But since nobody was watching over the autopilot and monitoring the aircraft's flight path, the aircraft gradually descended into the ground.)

Several psychological experiments have shown that reliance on the decisions and actions of automated systems can make humans inattentive to information indicating that the automation is in trouble. Specifically, these experiments demonstrate that pilots have a strong tendency to trust automated systems, in spite of clear evidence from other cockpit displays that the automated system is misbehaving. In other words, pilots have a tendency, built on past experience, to *not* disengage the automation even when it goes wrong. Both the U.S. Federal Aviation Administration and several industry groups have publicly expressed concern about pilots' reluctance to disengage automated systems and take manual control of the aircraft. Further, although here we are discussing this issue in the context of flight crews and cockpit automation, anecdotal evidence indicates that the problem of over-trust is not unique to pilots. This issue is also a major concern in the design of futuristic automotive systems such as adaptive cruise controls and intelligent highway systems.

Crew Resource Management

When the aircraft was experiencing abnormal nose-high attitudes, the captain was busy with the flight-test tasks (turning the autopilot on, bringing the left engine to idle, and working the hydraulic system). However, neither the copilot nor the flight-test engineer said anything to the captain about the increasingly deteriorating situation.

The concept of Crew Resource Management emerged in the early 1980s in commercial aviation as many began to realize the importance of crew coordination and its impact on safety. The value of this concept was reinforced by several accidents, in which critical information (such as fuel status) was not shared and communicated among crewmembers. Lack of crew coordination, in which pilots acted alone and did not communicate clearly with fellow crewmembers, led to wrong decisions and catastrophic outcomes. Following several such accidents in which the post-accident investigation clearly indicated that a breakdown in crew coordination was a dominant factor, most major airlines began providing Crew Resource Management techniques to their pilots.

Dialogue between the crewmembers before the last test flight revealed that the two pilots and the flight-test engineer had a busy schedule that day, and they were tired. It also seems that although the captain and the copilot switched their roles just before the flight, they did not fully discuss or review the assignment of tasks. In particular, the copilot may have believed that once his takeoff was completed, and the autopilot engaged, he (the copilot) was relieved from his monitoring task. This may have affected the copilot's reluctance to disengage the autopilot when it attained such a high pitch attitude and explain his lack of communication with the captain. The corresponding problem was that the captain, who was occupied with bringing down the left engine to idle thrust and configuring the hydraulic system on the cockpit ceiling, was relying on the copilot to monitor the aircraft and the autopilot's behavior. The flight-test engineer also failed to alert the captain to the increasingly dangerous pitch attitude and the deterioration in airspeed. His first warning that something was very wrong came three seconds after the captain had already disengaged the autopilot and was well into his recovery attempts.

In Conclusion

This accident revolved around two issues: the problem in the autopilot's ALTITUDE CAPTURE mode and the recovery efforts. There was a serious deficiency in the autopilot's design. The autopilot had the authority to apply extremely high pitch attitudes. In this particular accident, the autopilot used its control authority to guide the aircraft along a capture profile that was no longer attainable. Following the accident, an analysis was conducted on autopilots used on other manufacturers' aircraft. Almost all autopilots had the same design problem, which rendered them vulnerable to the same kind of mishap. These findings prompted aircraft and avionics manufacturers to add software features to fix the problem—many of them opting for a software routine that limits the autopilot's pitch attitude in ALTITUDE CAPTURE mode.

Some engineers further argue that beyond the immediate solution that was implemented, a lack of integration between the autopilot and the autothrottle (that controls engine thrust) is the real culprit. Due to the piecemeal introduction of cockpit automation, the autopilot and the autothrottle are not fully integrated, leading to this and other problems that designers of autopilots, and ultimately pilots, must work around.

With respect to the recovery, there were two "agents" participating in these efforts: the captain, on the one hand, and the automatic envelope protection system on the other. After takeoff, the captain concentrated his efforts on the tasks prescribed in the flight-test order. His third task, shutdown of the left hydraulic systems, was completed 10 seconds into the flight when the speed was 135 knots and the pitch attitude was 27 degrees. Five seconds later he realized that something was abnormal and snapped, "What happened?" Three seconds later he disconnected the autopilot. Altogether, eight seconds elapsed between the captain's shutdown of the left hydraulic system and his disengagement of the autopilot.

One of the daunting questions that this accident raises is why it took so long for the captain to disengage the autopilot and begin the recovery. Given that the entire flight lasted about 36 seconds, 8 seconds constitutes almost a quarter of the time aloft. When we account for human reaction time (about 1.5-2 seconds in such situations), both for (1) recognizing and understanding the situation, and (2) disengaging the autopilot, it reduces this period to 4-5 seconds. It is unclear what went on during this period: Could it have been that the captain was preoccupied with yet another task, perhaps taking notes, showing something to the two pilot observers sitting behind him, or executing yet another task or procedure? Likewise, while he was occupied with executing the flight-test order, the aircraft's pitch attitude was progressively increasing well beyond normal pitch attitude. What kept the captain from sensing this acute nose-high attitude? Unfortunately, the accident report provides no answers to these perplexing questions.

What is obvious, however, is that the disappearance of the autopilot mode indicators prevented the captain from quickly recognizing and understanding the seriousness of the situation at hand. He could not tell what mode the aircraft was in, and that quandary may have contributed to the 4-5 second delay mentioned earlier. Yet when he disconnected the autopilot and began the recovery effort, the captain acted correctly and very quickly. Subsequent simulations have shown that if the captain had begun his recovery efforts just four seconds earlier, the same maneuvers that he so precisely executed would have prevented the accident and saved the day. This accident clearly demonstrates how the decision to disengage or keep using an automated system becomes one of the most critical decisions that a user must make while interacting with a safety-critical system, especially during an emergency. But hindsight, as we all know, is always 20/20.

Finally, it is important to note the unique interplay between the envelope protection system's engagement and the captain's recovery attempts. It is precisely in these critical moments, in which every second counts, that we see the potential for dangerous situations in which both the machine and the human are making simultaneous, yet uncoordinated, efforts for recovery. In the test flight accident described above, while the ALPHA-PROTECTION mode was engaged and active and while the THRUST PROTECTION tried to engage, the captain was interacting with the aircraft's controls alongside with the automation. We have already seen how "competition" between pilot and autopilot can lead to disaster (in an earlier discussion in chapter 3). The possibility of dual authority and, more poignantly, the possible partition between authority and responsibility, creates a serious potential for operational failure. Responsibility should always be accompanied with suitable authority. Therefore, if indeed the envelope protection responsibility is to rest with the automation, the pilot cannot manipulate the controls at the same time (yet the pilot should be given the authority to *completely* override the automation and take *full* manual control). These issues are far from simple and immediately raise critical questions. What should the pilot do when the machine takes action? What if the pilot's instinctive responses are in contradiction to the machine's actions? And then there is the ultimate question: who can, and who should—especially in these split-second situations—override whom?

An Important Lesson

The classic story of a computer that overrides the users, locks them out, and takes violent action is that of HAL. In 2001: *A Space Odyssey*, the super computer HAL-9000 appears to be sensitive, intelligent, and trustworthy. But then HAL begins to malfunction as the spaceship *Discovery* comes closer to its destination. Both ground control (on Earth) as well as the two (awakened) crew members realize that HAL is making errors. Ground control suggests temporarily disconnecting HAL, transferring control to another computer on Earth, and then repairing HAL's programs; the crew wants to override HAL and take manual control of the ship. But the computer is unable to accept the fact that its internal logic circuits are failing ("no 9000 computer has ever made a mistake or distorted information") nor can it accept a disconnect and an override; HAL wasn't pre-programmed for such a situation. The only thing HAL can do is to keep its pre-programmed task of continuing the mission at all costs. HAL begins to kill the astronauts on board in an attempt to continue the mission on its own. It is this acute problem of a machine that is given considerable authority, yet has limited ability to understand a situation, which produces the drama of *2001: A Space Odyssey*.

For all of us to avoid the HALs of the future, it is important to address such problems of authority, responsibility, and interaction head-on. It is the role of regulatory agencies (such as the Federal Aviation Administration, Nuclear Regulatory Commission, and Food and Drug Administration in the United States) to provide guidelines, requirements, and criteria for safe and reliable automated systems. For example, in aviation there are certification committees comprised of engineers, test pilots, and scientists that are tasked with developing design criteria for automated control systems. But for these committees to be effective, design guidelines and criteria must be based not only on in-depth understanding of how these machines work, but also on how humans can and should interact with them.

Modern-day machines, computers, and automated systems cannot think. As compelling as it may appear to a user who is working for days on end with a flawless computer or automated system, a machine cannot "read" an unexpected situation or scrutinize it. The autopilot described in this chapter continued increasing pitch attitude beyond reasonable angles with complete disregard of the situation and the consequences. It did what it was preprogrammed to do.

At its very essence, a computer is a system of states and transitions in-between. The computer senses an event and switches from one state to another. The description of this (pre-programmed) behavior is what you have in every model in this book—from the on/off light switch in chapter 1 (figure 1.2) to the model of the autopilot and envelope protection in figure 15.1. The reason a machine can fly an aircraft and guide a spacecraft to other planets, lies in the sophistication of what was programmed into the machine; namely, its map of states and transitions. Modern-day machines can only follow that map—nothing more and nothing less.

Chapter 16

Verification of User-Machine Interaction

(with Michael Heymann)

By now we have seen a multitude of problematic interfaces and systems and a variety of human-automation interactions that went sour. So instead of continuing on and presenting more and more automated systems that contribute to confusion, error, and tragedy, we stop here and ask a simple question:

What can be done?

Is there a way to identify interface design problems before they cause harm? Is there some design and evaluation methodology that can be brought to bear? Something that can be done before the user becomes confused and commits an error, before an accident takes place?

This chapter focuses on a methodology that allows us to verify that an interface is correct. We begin by evaluating several interfaces for a rather simple automotive system. Later in this chapter, we will use the same approach to verify an interface for a complex system.

Transmission

The machine in figure 16.1 is a simplified transmission system of a vehicle. Altogether, the transmission has eight states, which are arranged in three speed modes: LOW, MEDIUM, and HIGH. There are three internal states in the LOW mode (low-1, low-2, low-3), two in MEDIUM (medium-1, medium-2), and three in HIGH (high-1, high-2, high-3). When you drive this vehicle you switch among modes by manually "pushing-up" or "pulling-down" a gear lever. Within each mode, the transmission shifts up and down automatically based on throttle, engine, and speed values. In figure 16.1, you can see that the system initializes

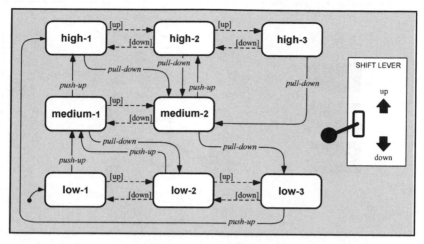

Figure 16.1. Transmission system of a vehicle and the driver's shift lever.

in low-1. When the driver increases speed, the system automatically transitions (when the conditions enabling *up* become True) to state low-2. If the driver decides at this point to shift up, the system will transition to medium-1.

The transmission is a semi-automatic system: The user shifts among the three primary modes, and the transmission switches automatically among internal states. The user's task in operating this transmission is to track the system along the three primary modes—LOW, MEDIUM, and HIGH. That is, the user must be able to determine the current mode of the machine and also to predict the next mode. This task requirement is akin to the type of questions users usually ask about automated systems: "What's it doing now?" "What's it going to do next?" and "Why is it doing that?" Note, however, that the user is required to track *only* the three primary modes and not every internal state of the machine (e.g., there is no need for the driver to distinguish between states medium-1 and medium-2 inside the MEDIUM mode).

Proposed Display

So by now we have the system (modes and transitions) and task requirements that specify what the user needs to do (unambiguously track the three modes); now the question is what kind of a display will work here? For starters, we want the display to be correct. That is, with this display and user model, it should always be possible to perform the task. Second, we strive for a simple display. Intuitively, the simplest display possible is the one proposed in figure 16.2. This display has three indicators—LOW, MEDIUM, and HIGH—which correspond to the three primary modes.

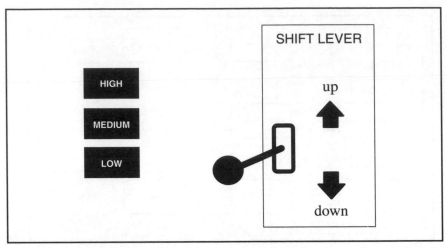

Figure 16.2. Proposed display showing three mode indications (HIGH, MEDIUM, and LOW), and the shift-lever.

Is this display correct for the task? To answer this question, we need to overlay the proposed display on top of the machine model (figure 16.3[a]). Starting from the top, we note that the HIGH indicator covers three internal states (high-1, high-2, and high-3). Note that all the three *down* transitions out of high-1, high-2, and high-3 go to MEDIUM. Therefore it is sufficient to tell the driver (in the user manual) that any *down* transition from HIGH will take the system to MEDIUM.

We continue along these lines and mask (or abstract away) all internal states and internal transitions, but leave in the transitions between the primary modes. The resulting model, which is presented in figure 16.3(b) describes the actual behavior of the transmission as viewed through this proposed display. Ideally, this model should constitute the user model for the transmission, and contain all the information that the user must know to operate the transmission correctly. The user model is supposed to be a correct abstraction of the underlying machine. However, when we look carefully at figure 16.3(b) we see a problem. There are two transitions out of the LOW mode—one going to MEDIUM and the other to HIGH. The same event (*up*) could take us into two different modes. The consequence is that the driver will not be able to predict whether the system will transition to MEDIUM or HIGH. The display is non-deterministic. And since there is no way to resolve this ambiguity, we must conclude that this display is *not* correct.

Alternative Display

The problem of non-determinism occurs because the proposed display abstracts too much information. It is over simplified. An alternative display that

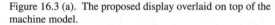

Figure 16.3 (a). The proposed display overlaid on top of the machine model.

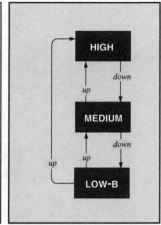

Figure 16.3 (b). Observed machine behavior through the (proposed) display.

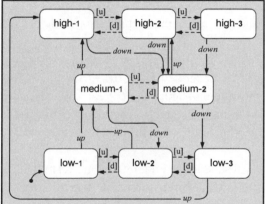

Figure 16.4 (a). Machine model of the transmission system.

Figure 16.4 (b). Alternative display and user model.

may eliminate the non-determinism is depicted in figure 16.4(b). It has two LOW modes (LOW-A and LOW-B) and two automatic transitions in-between.

Is this a good display?

Well, by intuitive inspection it seems quite reasonable. It is still a rather simple display (although it now has *four* mode indications), and the non-determinism appears to be rectified. But it can be seen that upon manual *push-up* from LOW-A, the system transitions to MEDIUM, while on *push-up* from LOW-B, the system goes to HIGH. Therefore, for this display to be correct, it is necessary that when the display shows LOW-A, the transmission must be in either low-1 or

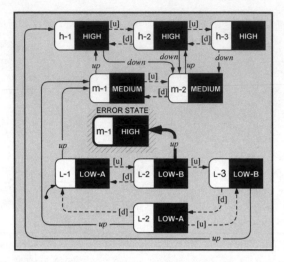

Figure 16.4 (c). Composite of the machine and the alternative display.

low-2, and when the display shows LOW-B, the transmission will always be in low-3.

To verify that this alternative display is indeed correct for the task, we take the machine (figure 16.4[a]) and the user model (figure 16.4[b]) and create a composite model. In the composite model of figure 16.4(c), each internal state (low-1, low-2, and so on all the way to high-3) is paired up with its corresponding display indicator. Now let's make a run through this machine, user model, and composite model.

The machine initializes in internal state low-1 and the user model initializes in LOW-A. The resulting composite configuration shown in figure 16.4(c) is low-1/LOW-A. When condition [u] becomes True, the machine automatically transitions to low-2 and the user model moves to LOW-B; the new composite configuration is low-2/LOW-B. So far the user model runs in synch with the machine model.

At this point, let's manually shift up to MEDIUM. Now look what happens: The machine, according to figure 16.4(a), will transition to internal state medium-1. The user model (figure 16.4[b]), however, transitions to HIGH mode. The new composite configuration is medium-1/HIGH, which, of course, is a contradiction. The display and user model says that we are in HIGH, but in fact the underlying machine is in MEDIUM (medium-1). The display and user model are deceiving!

So we realize that with this display and user model, the driver cannot predict the next mode of the machine. We conclude that this alternative display is also *not* correct for the task.

Error State

The composite state medium-1/HIGH is an error state. It is a case in which the display is no longer a correct abstraction of the underlying machine. This is a problem because it will lead to erroneous user-machine interaction. Given this display, there is nothing we can do to alleviate the problem; no additional training, no better user manuals, no warning signs, no procedure, nor any other countermeasures will allow the user to recognize the true mode of the system.

The problem with the transmission display is akin to the problem in the machine-machine interfaces of the *Royal Majesty*. There, the interface said that the GPS unit was sending accurate satellite-based data, when in fact the unit was sending dead reckoning data. The interface between the GPS unit and the radar-map/autopilot lied and the consequences were dire. Here, the interface to the transmission system tells the driver that the current mode is HIGH, when, in fact, the actual machine mode is MEDIUM. The display deceives, and this, of course, is a very serious design problem. Unfortunately, such pathogenic interfaces exist in modern automated systems. These erroneous interfaces are outright dangerous. It is therefore critical to identify and eradicate them before they cause harm.

Verification

The step-by-step process for evaluating a system and providing proof that the system satisfies (or doesn't satisfy) some criteria is known as *verification*. With respect to automated systems, we are interested in evaluating the interface and providing proof that it does not have error states. In the case of the transmission system we have done this step-by-step process manually, but as you can imagine this process can be implemented as a software tool and applied to larger systems.

Verification has been used for many years in the design of hardware systems and to some extent in software. The topic of verification came to the public's attention when Intel's infamous "floating-point-division" bug was discovered in November 1994. At first, the chip manufacturing company was trying to minimize the problem by saying that it affected a selective kind of computation that only a few people used. Eventually the company conceded that the problem had widespread ramifications, forcing it to recall all Pentium chips and pay out approximately $500 million. The cause of the bug was a logical error in which a look-up table was erroneously indexed. Following this fiasco, Intel and other manufacturers invested heavily in verification methods. Today, verification is a hot topic in software design with the objective of identifying

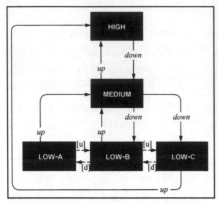

Figure 16.5 (a). The correct display.

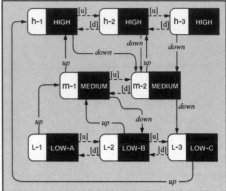

Figure 16.5 (b). No error states.

flaws in the (computer) code. In this chapter we see how the idea of verification can be extended and applied to interface design.

Good Transmission

Now we turn back to the transmission system. After concluding that the alternative display of figure 16.4(b) is also incorrect, we proceed to search for a better display. After several iterations, we come up with the display of figure 16.5(a), which exhibits no error states (figure 16.5[b]). Note that although there are two *down* events emanating from MEDIUM (figure 16.5[a]), the resulting non-determinism is not a problem for us because the task requirements specify that we should be able to discern between three primary modes (LOW, MEDIUM, HIGH) and not between sub-modes such as LOW-B and LOW-C. We therefore conclude that this new display is correct for the task. (For additional discussion on the process of verifying interfaces and coming up with a correct and also succinct interface, see the endnotes for this chapter.)

Verification of a New Autopilot

We now shift from automotive systems to aircraft systems. We will apply the interface verification methodology to evaluate a new autopilot for a new aircraft; this autopilot's behavior and some of its modes are similar to the autopilots discussed earlier. As was demonstrated in chapters 14 and 15, the altitude capture maneuver, which is common to all autopilots today, is a source of many potential interaction problems. In this chapter we will concentrate on evaluating pilot interaction with this autopilot during the altitude capture maneuver.

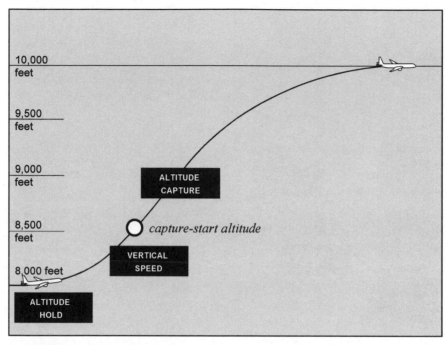

Figure 16.6 (a). The sequence of mode changes during an altitude change (from 8,000 feet to 10,000). The black rectangle indicates the current mode of the autopilot as displayed to the pilots.

Changing an Altitude

An altitude change involves a sequence of mode-switching on the part of the autopilot. The sequence begins when the airplane is level; it then climbs, levels-off, and then flies horizontally again. You know the sensation, when you sit in row 27B and hear the engines rev up and feel the aircraft nose up into a climb, and then the engine whines down as the aircraft levels off.

Figure 16.6(a) shows a typical climb to altitude profile. We start at 8,000 feet with an air traffic control instruction to "climb to and maintain 10,000 feet." The pilot sets the new reference altitude (10,000 feet) into the flight control panel, and then engages VERTICAL SPEED (which is one of several modes used for changing altitudes). The aircraft responds by pitching up and climbing toward the new altitude. At 8,500 feet, the autopilot switches *automatically* to ALTITUDE CAPTURE mode and begins the gradual maneuver to capture the 10,000-foot (reference) altitude.

Now we are going to examine what will happen when the pilot makes changes to the altitude setting. Changes in the altitude setting while the aircraft is performing this capture maneuver can come about either because the pilot has decided to modify the altitude, or, more likely, because air traffic control has again instructed the pilot to change the altitude. Figure 16.6(b) shows what happens when air traffic control instructs the pilot to "stop climb at

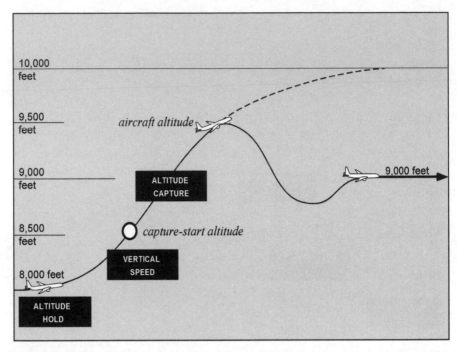

Figure 16.6 (b). While in ALTITUDE CAPTURE mode, the pilot changes the reference altitude to 9,000 feet. The autopilot responds by capturing the new altitude.

9,000 feet" while the aircraft is in ALTITUDE CAPTURE: the pilot enters this new reference value into the flight control panel, and the autopilot re-engages in VERTICAL SPEED mode. Then the autopilot guides the aircraft down, resulting in an eventual capture of 9,000 feet. So far so good.

In figure 16.6(c), air traffic control instructs the pilot to "descend back to 8,000 feet." After the pilot enters this new 8,000-foot reference value into the flight control panel, the autopilot also re-engages in VERTICAL SPEED mode, but then instead of descending, it keeps on climbing. The autopilot not only ignores the new setting, but also continues climbing indefinitely. Unless the pilot does something, the aircraft will just continue to climb on and on. This is indeed a strange and unexpected behavior, especially when compared with the aircraft's response when the altitude setting was changed to 9,000 feet.

Machine Model

We turn now to the machine model to understand the underlying logic of the autopilot that leads to this indefinite-climb behavior. Figure 16.7(a) is a model of the autopilot's behavior in VERTICAL SPEED and ALTITUDE CAPTURE modes. Beginning in ALTITUDE CAPTURE mode, we see that upon pilot interaction with the altitude reference value, the autopilot ends up in two different VERTICAL

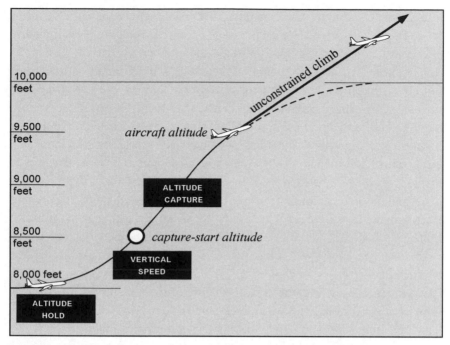

Figure 16.6 (c). While in ALTITUDE CAPTURE mode, the pilot changes the reference altitude to 8,000 feet. The autopilot responds by NOT capturing this new altitude. Instead, the autopilot continues on an inconstrained climb.

SPEED behaviors. The difference between the two behaviors ("climb to set altitude" and "unconstrained climb") has to do with a condition. And this condition is the altitude at which the autopilot switched to ALTITUDE CAPTURE mode—called the *capture start altitude*—which in our example occurs at 8,500 feet. Here's how this condition determines the outcome:

- If the new altitude is *above* the capture-start altitude, the autopilot *will* capture the new altitude. (In figure 16.6[b], the new altitude is 9,000 feet, which is indeed higher than 8,500, and the autopilot captures the new altitude properly.)
- However, if the new altitude is *below* the capture-start altitude, the autopilot *fails* to capture the new altitude. (In figure 16.6[c], the new altitude is 8,000 feet, which is lower than 8,500, and the autopilot climbs unconstrained.)

These two behaviors of the autopilot are a result of a contradiction between commands that are given to the autopilot. When the pilot engages the VERTICAL SPEED mode, he or she also has to set the rate of altitude change (e.g., 1,000 feet per minute) and also the direction (e.g., + for climb or - for descent). The contradiction occurs when the pilot sets the autopilot in vertical speed mode, say for a climb at a positive rate of 1,000 feet per minute, and then sets the altitude to a value that is below the current aircraft's level. What should the autopilot do? Should it honor the rate of climb and ignore the altitude, or do the reverse? In a way, this is a default issue because the autopilot, without any clear instruction from the pilot, must decide one or the other. It cannot do both. It turns out that the autopilot's default logic is to ignore the altitude, and continue climbing unconstrained, and hence comes the VERTICAL SPEED (unconstrained climb) behavior. This behavior, as disturbing as it may sound, is not unique to this particular autopilot. This problem has been around for many years in any autopilot that has a vertical speed mode (most of them do), and is called "kill the capture" in pilots' jargon. Pilots receive instruction in how to avoid the problem and it is also mentioned in the aircraft's flight manual.

Beyond the technicalities of the autopilot's logic, note what is going on here in terms of the underlying structure of the system. First, a change in (altitude) reference value causes a mode change (from ALTITUDE CAPTURE to VERTICAL SPEED); and we already well know that this structure of mode and reference value confuses users (see chapter 12). Second, the same pilot action (setting a new altitude reference value) results in two different autopilot behaviors (capture of an altitude or unconstrained climb). But as we shall see next, there is a much more serious problem lurking beneath this autopilot design.

Interface and User Model

Figure 16.7(b) is a user model of the autopilot. It describes the information provided on the interface and what the pilot knows about the system's behavior (from training and the aircraft's flight manual). The autopilot interface tells the pilot what the current mode of the autopilot is—there is a display indication for ALTITUDE CAPTURE and an indication when the autopilot is in VERTICAL SPEED mode. However, recall that this mode, depending on pilot interaction, can produce two different behaviors—namely, VERTICAL SPEED (climb to set altitude) and VERTICAL SPEED (unconstrained climb). And since this behavior is taught to pilots and mentioned in the aircraft's flight manual, we depict these two behaviors in our user model by splitting the VERTICAL SPEED indication into two states: VERTICAL SPEED (climb to set altitude) and VERTICAL SPEED (unconstrained climb).

Until now, the aircraft flight manual gave a fully accurate description of how the autopilot works. But from here on, it simplifies (or abstracts away) some of the information. Specifically, the manual mentions that the autopilot will transition from ALTITUDE CAPTURE to VERTICAL SPEED (unconstrained climb) if the pilot sets the altitude below the *aircraft's current altitude*. And this is what you see in figure 16.7(b). While this behavior is indeed true when the autopilot is in VERTICAL SPEED (climb to set altitude), it is only partly true when in ALTITUDE CAPTURE (compare figures 16.7[a] and 16.7[b]). In ALTITUDE CAPTURE mode, the

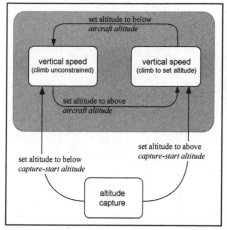

Figure 16.7 (a). Machine model.

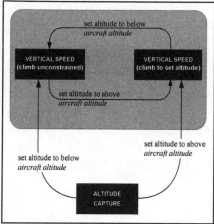

Figure 16.7 (b). User model.

autopilot will go into VERTICAL SPEED (unconstrained climb), only if the pilot sets the altitude to below the *capture-start altitude.*

So the difference between the description in the aircraft's flight manual and the actual autopilot behavior centers around the critical value (*aircraft's altitude* or *capture-start altitude*). The flight manual says that while in ALTITUDE CAPTURE, setting the altitude to a value *below* the *aircraft flight altitude* will trigger a transition to VERTICAL SPEED (unconstrained climb), but in fact the autopilot will trigger this transition only if the altitude is set below the *capture-start altitude.* Why is this subtlety omitted from the aircraft flight manual? We can assume that the people who wrote the manual either did not know about this subtle difference, or, and this is the more likely case, they wanted to simplify things for the pilot. Because with this abstracted description, instead of having to remember two critical values (*aircraft altitude* and *capture-start altitude*), there is only one critical value to remember (*aircraft current altitude*).

Composite Model

Now that we have a description of the machine's logic and the abstraction in the user model, we proceed to verify whether the user model is correct for the task. And the pilot's task here is to reliably predict whether the aircraft will capture the new altitude (e.g., 8,000 feet) or not. Figure 16.8(a) is the machine model and figure 16.8(b) is the user model. The two models look the same, but there is a subtle difference that we noted earlier: in the machine model of figure 16.8(a), the critical altitude that determines whether we switch to VERTICAL SPEED (climb to set altitude) or VERTICAL SPEED (unconstrained climb) is the

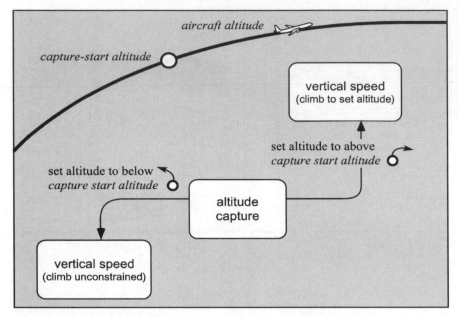

Figure 16.8 (a). Machine model.

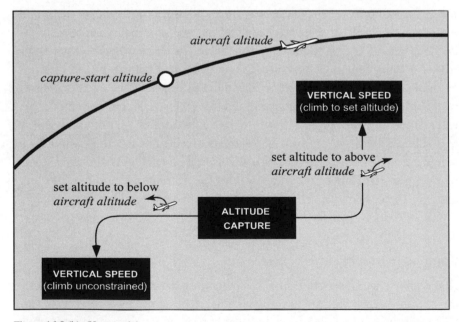

Figure 16.8 (b). User model.

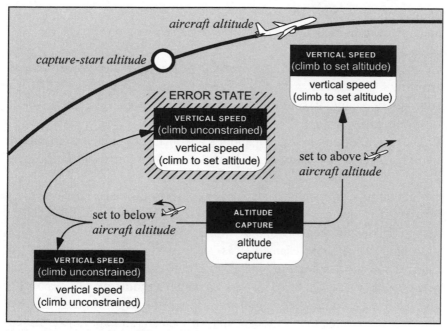

Figure 16.8 (c). Composite model (from user's point of view).

capture-start altitude; in the user model of figure 16.8(b), the critical altitude is the current *aircraft altitude*.

Figure 16.8(c) is the resulting composite. As usual, we look at this composite model from the user's (in this case pilot's) point of view. When we are in ALTITUDE CAPTURE mode and the pilot sets an altitude that is *above* the aircraft altitude, the aircraft will capture that altitude. This is what the pilot expects and all is well. However, when the pilot sets an altitude that is *below* the aircraft altitude, the machine's behavior becomes unpredictable: If the newly set altitude is *below* the aircraft as well as *below* the capture-start altitude, all is well and the aircraft will go into an unconstrained climb. But if the newly set altitude is *below* the aircraft and *above* the capture start, the pilot expects that the aircraft will go into an unconstrained climb, when in fact, the aircraft will capture the newly set altitude. This is an error state, and we therefore must conclude that this interface and user model is incorrect.

Here is an analogy that may help illustrate the problem: Say you are taking a state board examination. To prepare for this career-critical test, you sign up with a tutor who tells you, perhaps to encourage you to work harder, that the passing grade is 80 percent, when in fact the actual passing grade is 70 percent. Months later you take the exam: If you score above 80, you come out happy, and sure enough in a week's time comes the official congratulations letter from the state board. If you scored 60, the letter from the board confirms that you flunked and have to retake the test. But if you come out scoring 75 you go home with sadness on your face—only to discover a week later that you actually passed!

A score of 75 (or any score between 70-79 for that matter) constitutes an error state. You assume that you flunked, when in fact you passed. There is a contradiction, and it surely confused you, although here there was a happy ending.

The essential point here is not about state board examinations, shameless tutors, or aircraft autopilots, although the latter is an important application. The essence is about how we go about verifying that an abstraction (or simplification) of a system is indeed correct for the task. Like it or not, every interface is an abstraction. The more complex the machine, the more abstract the interface—and the more important the verification.

In light of the above example, it is possible to distinguish among three categories of abstractions: the good, the bad, and the ugly. An abstraction that tells the pilot that when he or she sets the target altitude to a value *above* the current aircraft altitude, the aircraft will capture the target altitude, is a *good* one. It is akin to telling the student that 80 percent and above is a passing grade. Although it is not the complete story, it always holds true. An abstraction that tells the user that he or she is in one mode when in fact the machine is in another mode, is *bad* abstraction. Because of this contradiction, the user is surprised and can make an error. But there is also the *ugly* case. And the ugly case has to do with the fact that an error state is a direct result of non-deterministic behavior. You can see the non-deterministic structure in figure 16.8 (c), where the transition "set to below aircraft altitude" forks into two different outcomes. Sometime resetting the altitude to *below* the current aircraft altitude leads to capture, sometimes resetting the altitude to *below* the current aircraft altitude leads to unconstrained climb. Now the pilot is not only surprised, but also confused. Sometimes the system works just as expected and sometimes not. But the pilot will never be able to figure out what is really going on.

The Impact of Non-deterministic Interfaces

An abstraction that inflicts non-deterministic behavior on the user is elusive. Not only the interface is confusing, but it also prevents users from identifying the source of confusion because the problem may disappear the next time a user interacts with the machine. We have all experienced a situation in which after calling and calling, a technician finally comes to your home to fix a problem, but then the machine works just fine! It may happen also when you drive to your mechanic's shop with a weird noise in the car. In aviation, there are many examples where non-deterministic problems in interfaces plague a system. Pilots attribute the resulting incidents to lack of awareness, their own errors, complacency, and a variety of other guilt-trip explanations—when in fact the problem was embedded and hidden in the machine. Sometimes, such problems are written up as a malfunction, such as a failure of the autopilot system. However, when technicians try to fix the autopilot, it doesn't show any faulty modes and the pilot cannot replicate the failure. For lack of evidence, the system is declared safe and operational. The only problem is that such latent flaws may lead to accidents.

It is also important to note the long-term effects of such design flaws on users. When encountering non-deterministic and incorrect interfaces, users usually concoct some explanations in an attempt to clarify the confusing behavior. These explanations are called "folk models," and in most cases they are only partially correct and therefore quite brittle, which in turn only breeds more confusion.

Finally, there is an implicit assumption on the part of designers, managers, and end-users that any kind of interface problem can be fixed, either by mentioning the potentially dangerous condition in the user manual, or perhaps by designing a procedure to work around the problem. This assumption is deeply ingrained in complex and safety-critical domains such as aviation, the military, and nuclear power. This is partly because after a system is fielded, it becomes extremely difficult and very expensive to recall it in order to change the interface or the underlying logic. But in this particular autopilot design, as in many others, such fixes will *not* solve the problem. For the pilot to determine whether the aircraft will or will not capture the altitude, he or she needs not only a better flight manual, but also to know the value of the capture-start altitude. However, this critical altitude information, which varies with every altitude change, is not provided on any display in the cockpit. To solve this problem, either the interface or the underlying autopilot behavior must be changed. There is no Band-Aid "fix" to the problem.

Verification of User Interfaces

It is precisely because of the high cost of recall that verification of user interfaces and early identification of design problems is so important for current and future automated systems. So when it comes to the design of safe and efficient user interfaces in medical equipment, avionics systems, and other safety-critical systems, verification is not only a good design practice and important for the bottom line, but also a way to identify design problems before they manifest themselves in serious mishaps and cause harm.

In user interface design, there are several formal and strict criteria that make an interface incorrect for the task. In the transmission and autopilot examples I described one of these—an error state—represents a contradiction between the machine model and the user model (with respect to the user's task requirement). That is, the interface tells the user that the machine is in one mode (VERTICAL SPEED to unconstrained climb), when in fact the machine is in another (VERTICAL SPEED to altitude setting). And since part of the user task requirements is to differentiate between these two outcomes, an error state is born. Naturally, a correct interface is one that does not exhibit error states.

Augmenting

Another verification criterion is an *augmenting* state. An augmenting state is a situation in which the user is told that a certain mode or state change is possible, when in fact the machine has no such mode or it cannot transition into a mode (it may be disabled, for example). To find an augmenting state, you need look no further than the elevator in your office building or the thermostat on the wall. Next time you ride an elevator, look carefully at the control panel. There are buttons for the various floors, and then there are the "open door," "close door," "alarm," and "emergency" push buttons. Ever try to press the "close door" button? In most elevators it doesn't work, unless you are a fireman or an elevator technician with special keys. The rest of the time, even though the button is there and may even light up, it is disabled.

The most common case of a placebo button, one that directly affects many of us, is the office thermostat. Many office thermostats are disconnected. Worse, some thermostats are in place to give you the impression that you can alter the office's temperature, when in fact it is "fixed" and cannot be changed unless you have special access to the system. It turns out that many landlords are fed up with constant tampering with the temperature setting—which eventually breaks the heating/air conditioning system and results in high maintenance costs. Quietly they ask their heating ventilation and air conditioning technicians to "fix" the thermostat.

Restricting

The third criterion that we shall mention here is a restricting state. A restricting state represents a situation in which the user is unaware that certain event(s) can trigger mode or state changes in the machine. The user model abstracts away these events and the resulting mode changes. But if the user accidentally or unintentionally presses a button, all of a sudden the system switches modes—to the dismay of the user. One embarrassing example is the landing gear lever of a certain aircraft. A pilot was told that while the aircraft is on the ground (with the landing gear extended) the gear handle is disabled (i.e., it is impossible to retract the gear). But it turns out that in a certain aircraft configuration, the gear lever is "hot" (enabled). On one occasion, a pilot accidentally bumped the gear handle with his hand—the nose gear retracted and the aircraft's nose hit the ground.

In military fire-control systems there are many internal locks (interlocks) to prevent the system from launching missiles inadvertently and/or without authorization. There is a rather convoluted sequence of events and conditions before a missile can be launched. In one such system, it was found that in addition to the designed sequence with all its safeguards and key entries, there was another sequence (which turned out to be a short-cut with no safeguards) that also led to missile launch. A correct interface is the one that does not include restricting states.

In Conclusion

Interfaces and related user manuals are always a reduced, or abstracted, description of the machine's behavior. No interface provides a complete description of the underlying behavior of the machine. Therefore, a major concern of designers is to make sure that these abstracted interfaces and user manuals are indeed correct for the task. Many of the intrinsic problems that exist in today's interfaces lend themselves to the occurence of *error*, *augmenting*, and *restricting* states. The verification approach discussed here is a viable method for reducing the presence of interface flaws in consumer products, automated devices, and automated control systems—in particular flaws that are embedded in safety-critical systems.

While in this chapter we verified the transmission and an autopilot system in a graphical way, the underlying approach and methodology can be incorporated into a software tool and applied to larger and more complex systems. Modern verification tools include fast computers and efficient algorithms to search for error, augmenting, and restricting states.

Chapter 17

Beyond the Interface

(with Meeko Oishi and Claire Tomlin)

T he previous chapter was about a systematic methodology for verification of interfaces, making sure that the interfaces are correct for the task. We now take it one step further—analyzing and verifying the underlying system. I will use the same presentation style as before: illustrating a situation that we are all familiar with, and then moving on to analyze a more complex automatic system.

Red Light Violation

Consider the following scenario: you're driving toward an intersection, and you can see the green traffic lights looming in the near distance. But as you approach the intersection, the light turns yellow. What do you do? Should you stop, or perhaps drive through? As we all know, the answer depends on many factors as well as the context of the situation—the car's distance and speed, whether there's a car behind, if you're in a hurry, if there are cars waiting at the intersection; or perhaps this is the middle of the night and the intersection is empty and nobody is looking. Let's begin by focusing on two main issues: the distance from the intersection and the car's speed.

And to sharpen the issues even more, consider here an intersection that has a video camera and a large sign that says "Red light violation, $281 minimum fine."

First we analyze the two available options for the driver: braking before the intersection or proceeding through the intersection. Figure 17.1(a) shows a region from which braking is safe. Given any combination of speed and distance within the dark gray region, the driver can always safely brake. But how do we know that? We know the car must stop at the white line before the intersection; at that point, the speed is 0. From there we go backward and

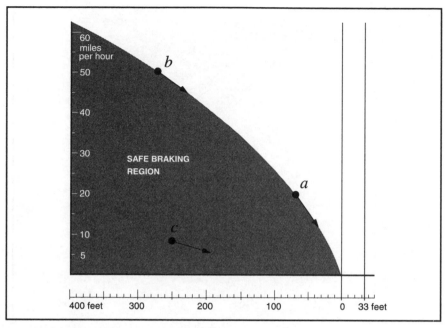

Figure 17.1 (a). The safe breaking region. (The horizontal axis shows distance, in feet, from the intersection and the vertical axis shows the car's speed in miles per hour.)

calculate, based on the car's maximum braking performance, the stopping distance for every speed (from 1 to 60 miles per hour).

The yellow light is 4 seconds long, and we take into account that it takes the (average) driver 1.5 seconds to react and press on the brakes. So, for example, at point *a*, the car's speed is 20 miles per hour and if you apply maximum braking, you *will* stop at the white line; the same is true when you apply maximum braking at point *b* (speed 50 miles per hour). Naturally, for every point inside the safe-braking region, for example point *c*, maximum braking is not required—the driver can brake lightly and still be able to stop before the intersection.

Now we turn our attention to the other option—driving through the intersection. Let us assume for now that the driver, once he or she observes the yellow light, simply maintains current speed. How do we figure out the safe region for driving through? Just as before, we start from the intersection and work our way back.

Contrary to common belief, it is *legal* to be caught inside the intersection when the light turns red and cross to the other side with the red light above you. What is *illegal*, however, is to enter the intersection when the light is red (see the discussion in the endnotes about California's driving regulations). Therefore, we calculate from the white line at the entry of the intersection,

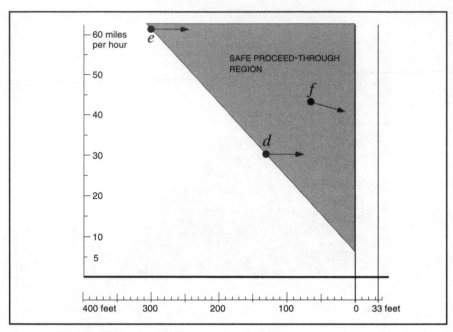

Figure 17.1 (b). Safe proceed-through region.

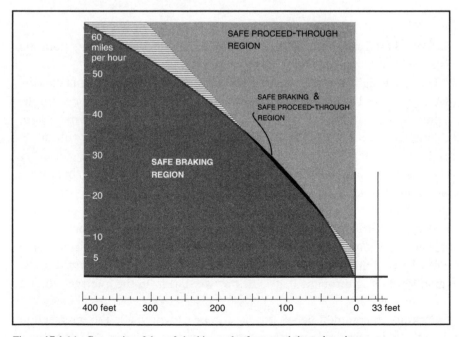

Figure 17.1 (c). Composite of the safe braking and safe proceed-through regions.

assuming that we want to pass the white line at a minimum speed of 7 miles per hour (because we do need to vacate the intersection before it turns green for the crossing traffic). In figure 17.1(b), we see that when the speed is 30 miles per hour and the distance from the intersection is 130 feet (point d), we can safely drive through. From point e (speed 62 miles per hour, distance 300 feet) we can also drive through and pass the intersection in time. For every point inside the safe proceed-through region (e.g., point f), maintaining the current speed is not required—the driver can even reduce speed and still cross the intersection before the light turns red.

Composite Graph

Now that we know the consequences of either stopping before the intersection or driving through it, we can turn to the decision itself. How do we know, when the yellow light comes on, whether we should brake or proceed through? To analyze this dilemma, specifically in situations where it is not so obvious which option is safe and legal, we need to consider the "safe braking" region and the "safe proceed-through" region together.

The composite shown in figure 17.1(c) divides the entire operational region into four sub-regions. The dark gray region is the "safe braking," the light gray is the "safe proceed-through," and the narrow black region is where they overlap. If you are in the black region, you can either brake or proceed through, and you'll still be safe and legal.

But what are the hatched areas?

These are sub-regions of the operational space from which you can't safely brake or proceed through the intersection. Here's the problem: if you stop, even at maximum braking, you will find yourself entering the intersection in red; if you proceed-through, the light will turn red before you reach the intersection. The point is that given the two options, brake or drive at constant speed, there is just nothing you can do. If you are in the hatched region when the light turns yellow, you will commit a violation and get a ticket.

So now you're thinking to yourself, wait a minute, what about "gunning" through the intersection. Won't we avoid the red light? Well, we all know that accelerating to cross an intersection that is about to turn red can be rather dangerous, but regardless, let's consider this acceleration option. In figure 17.2, we expand the "proceed through" region to include the acceleration option. If the driver is in the white region, for example at point g, and he or she accelerates, it is possible to pass through and vacate the intersection in time.

Note that by accounting for accelerations, the hatched regions in figure 17.2 have shrunk, but they were not eliminated. Therefore, regardless what we do, there are still the regions where we are just too fast to stop in time and too far away to accelerate through the intersection. The point is that there are

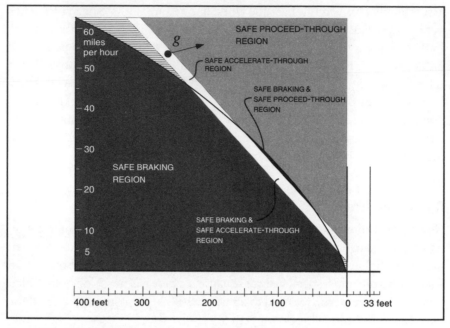

Figure 17.2. With acceleration.

combinations of speed and distance from which it is *impossible* to be safe. This is one of those "damned if you do and damned if you don't" situations. The hatched regions represent a lock, because if the driver is there when the light turns yellow—there is no way to escape.

Possible Remedies

The above problem is called the "yellow interval dilemma" and is well known to traffic engineers. The unsafe (hatched) region at the top of the graph is called "the dilemma zone." Now let's consider some of the viable design options for eliminating the unsafe regions. For a start, we can limit the speed to 35 miles per hour (see figure 17.1[c]). That way nobody will reach the unsafe (hatched) region at the upper end of the curve; and if they did, it is because they were speeding (now if you are caught you may get two tickets—one for speeding and one for a red light violation). But there are problems with this speed limit option because sometimes, for example outside of town, 35 miles per hour is just too slow.

Another option is to increase the duration (interval) of the yellow light, which will increase the size of the "proceed through" region and eliminate the unsafe region. Interestingly enough, it turns out that this solution is only partially successful, and the reason for this is due to the human factor: once drivers notice that there is ample time, they tend to ignore the yellow light and proceed through the intersection even when they are far away from it. Short of increasing the yellow light interval, one way that traffic engineers deal with the problem, albeit not directly, is to delay the green light to the crossing traffic. That is, for about two seconds, all sides of the intersection are red. The solution increases overall safety by adding a buffer zone, but does not eliminate the yellow light problem (and the unwarranted red-light violations).

Automatic Landing System

In the example above we analyzed the intersection and the car's performance to understand possible actions on the part of the driver; but in an implicit way, it was also a verification of the yellow light interval. By creating a composite of two graphs, we identified regions that allow the driver safe braking and safe passage and regions that don't. The same approach can be used for analyzing and verifying more complex systems—in particular, automated control systems, where we want to verify the design of the automation itself as well as human interaction with it.

In most automated control systems, and in particular those that are used in safety-critical systems, the user has the authority to break away or disengage the automation when unsafe situations arise. One case in point is the automatic landing system in modern airliners. As the name implies, this system flies the approach, makes the landing, and then steers the aircraft to the end of the runway. Automatic landing systems are commonly used in bad weather, specifically, in a condition called "zero-zero" in aviation lingo, which means that the out-of-the-cockpit visibility is zero and the clouds or fog reach all the way to the ground.

In these severe visibility conditions, only the automatic landing system is permitted to make a landing. Therefore, automatic landing systems are designed to be extremely accurate and have built-in redundancies to reduce the likelihood of a system malfunction. Furthermore, automatic landing systems go through rigorous testing and retesting on a monthly basis. It is the only component of the automatic flight control system for which we cannot rely on the pilot as a backup in case the system is in error. Why? Because in zero visibility the pilot cannot see the runway at all, and therefore he or she cannot judge if the autoland system is doing its job properly. Nevertheless, the pilots are required to monitor the automatic landing system, making sure that it switches among its internal modes properly; that it maintains appropriate speed and pitch attitude; and is tracking the glide-slope (which the pilots can see on their displays). In making an automatic landing, the autopilot is in full control of the situation. But there is one option that is always available to the pilot—to discontinue the approach and make a go-around.

A go-around is a maneuver that every seasoned airline passenger has experienced at one time or another. During the approach for landing, all of a sudden the aircraft pitches up and the engines roar, and within a few seconds the aircraft begins to climb rapidly. Usually the reassuring voice of the captain comes on the public-address system and informs us that we are "going around" for yet another approach and landing. From the passengers' point of view, the go-around is perhaps not a very pleasant experience. On the pilots' side, go-arounds are well-practiced maneuvers with the intention of taking the aircraft away from an unsafe landing. Sometimes, a go-around is requested by air traffic control—perhaps the aircraft has come too close to another aircraft on the approach, or there is debris, a vehicle, or another aircraft on the runway.

Going Around

In modern automated aircraft, the pilots execute a go-around by pushing a small lever located on the throttles. From then on, the automatic flight control system does the rest—pitching up the aircraft, advancing the engines to full thrust, and flying the aircraft to a higher altitude (usually 2,000 feet). From our point of view, a go-around is a disturbance event because it may come *any time* during the approach. Performing a go-around places considerable workload on the pilots: first, it is usually a surprise and they need to break quickly away from their focus on landing. From there on, flaps must be set, landing gear raised, a reference airspeed maintained, and checklists performed. Sometimes the pilots also need to formulate new plans and make quick decisions, then consult charts and enter information into the computer (not to mention briefing the weary passengers).

Although the pilots can always disengage the autopilot and execute the go-around manually, it is recommended, and in some airlines mandatory, that this go-around maneuver be performed automatically. The reason is that the maneuver is quite difficult to perform optimally, that is, with a minimum loss of altitude, especially when the aircraft is close to the ground.

Figure 17.3 shows the sequence of mode switching in making an automatic landing and the consequence of engaging the go-around mode: Initially, the AUTOLAND mode is engaged and the airplane is flown automatically toward the runway. At 50 feet above the runway, the system transitions to FLARE mode and begins the curved maneuver to touchdown. After the airplane lands, the system goes into ROLLOUT mode and steers the airplane on the runway's centerline until the pilots bring the airplane to a stop. A go-around may be

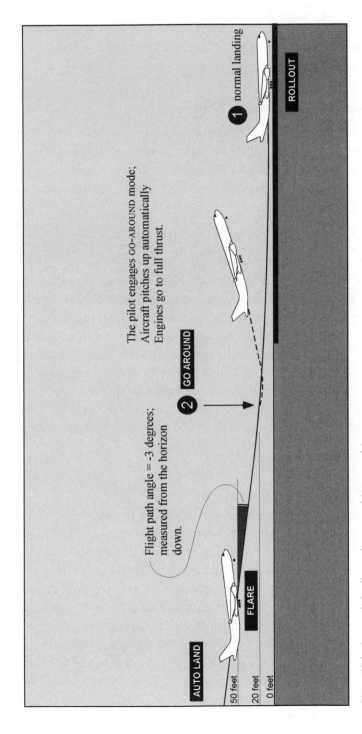

Figure 17.3. Automatic landing and go-around (not to scale).

initiated any time during the approach to landing. Figure 17.3 shows a go-around that is initiated when the aircraft is at an altitude of 20 feet.

Verification of the Go-Around Maneuver

We want to verify the aircraft responses during a go-around maneuver. In particular, we wish to explore the consequence of engaging a go-around when the aircraft is close to the ground. Why? Because at that vulnerable point in time, the speed is at its lowest as the airplane is making the delicate transition from flying to landing. But how do we go about performing this verification? Very much as we did in the analysis of the intersection, we begin by considering the available options. There are two in this case: either the airplane lands or the airplane makes a go-around.

Unlike the intersection example, where car performance data, such as maximum braking and acceleration are readily available, information about AUTOLAND performance is proprietary information. This information is kept behind locked doors as a trade secret and is therefore unavailable publicly. For that reason, the information used in this analysis is based on aeronautical engineering textbooks and general knowledge of how modern autoland systems work. As usual, we are interested here in the approach and methodology for verification of an automatic system, and not so much in the details of a specific autoland system.

Safe Landing

The funnel-like shape in figure 17.4(a) shows us the region from which an autopilot can make a safe landing. There are three variables that combine to create this three-dimensional shape: the aircraft's altitude above the runway, the aircraft's speed, and the aircraft's flight-path angle (which is the angle at which the airplane descends toward the ground—see the inset at the top of figure 17.4[a]). In principle, the shape is computed in the same way the regions in the yellow light example were computed. We start from touchdown, where the flight path angle should be between 0 and -2 degrees and work our way back. (If the angle is greater than zero, the airplane will not be able to land; if the angle is less than -2 degrees, the aircraft's tail will hit the ground.) For each altitude, from 0 to 60 feet, we compute the speed and flight-path angle that the autopilot needs to maintain such that eventually the aircraft will make a safe landing.

This is exactly what we did in calculating the safe-braking region in the yellow interval dilemma. For example, at an altitude of 60 feet, if the flight-path angle is

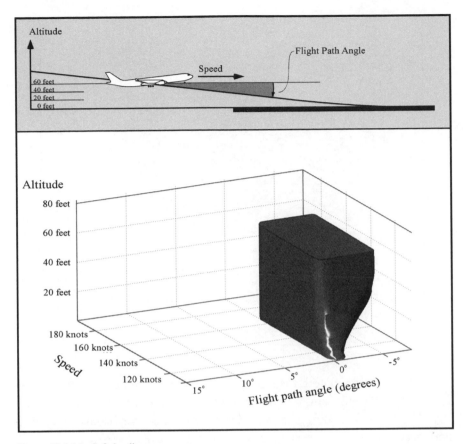

Figure 17.4 (a). Safe landing.

between 0 and -7 degrees and the aircraft's speed is between 120 and 170 knots, the autopilot will eventually make a safe landing. However, if the speed is below 100 knots the autopilot will not be able to make a safe landing.

Safe Go-Around

Figure 17.4(b) shows the safe region for executing a go-around. This region is rather large because unlike the safe landing region that funnels down to the runway with a tightly constrained angle and speed, the go-around can be executed safely at a variety of flight-path angles and speeds. But as you can see in figure 17.4(b), the shape has a wedge-like cutout at low speeds and negative flight-path angles. Why? Because it takes the go-around maneuver several seconds in order to pull the aircraft out of the descent and avoid hitting the ground. What figure 17.4(b) shows us is the combinations of altitudes, speeds, and flight-path angles from which a safe go-around is always guaranteed.

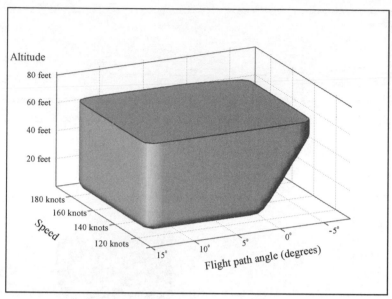

Figure 17.4 (b). Safe go-around.

We have identified both the safe region for landing and the safe region for go-around; now we want to verify that while the autopilot is controlling the airplane for a safe landing, the pilot can always break away from the landing and make a go-around to evade danger. To do this, we need to consider the safe "landing" region and the safe "go-around" region together. And for the sake of illustration, let us consider what will happen when the pilot engages the go-around mode and the aircraft is 20 feet above the ground. Figure 17.5(a) is a slice of the safe-landing region at 20 feet above the ground, figure 17.5(b) is the same for the safe go-around region, and figure 17.5(c) is the composite graph.

Composite Graph

The composite graph of figure 17.5(c) shows three emerging sub-regions: the light gray is the safe go-around region, the black is where safe go-around overlaps with safe landing, and the hatched is also part of the safe landing region. Because a go-around is usually triggered by a disturbance event that can occur at any time, the black region is where we always want to be: from here the autopilot can make a safe landing, and if a go-around is needed, the aircraft will be able to break away from the landing and escape safely.

So what about the hatched region? The hatched area is problematic, and here's why: under normal conditions the autopilot will try to make the landing when the flight-path angle is close to 0 degrees, but under less than normal

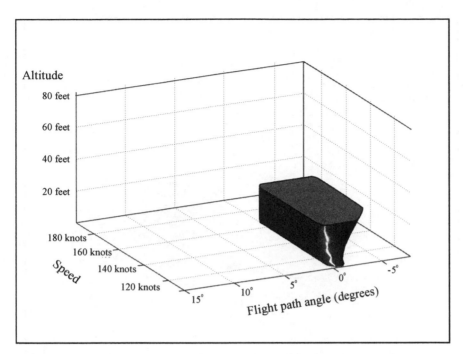

Figure 17.5 (a). Safe landing region at 20 feet.

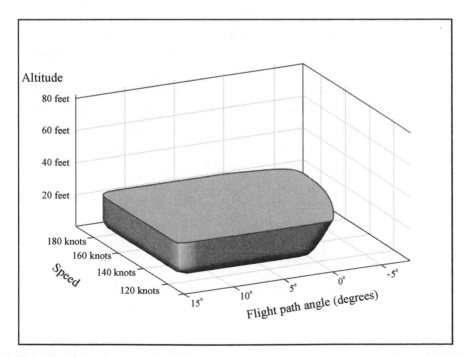

Figure 17.5 (b). Safe go-around at 20 feet.

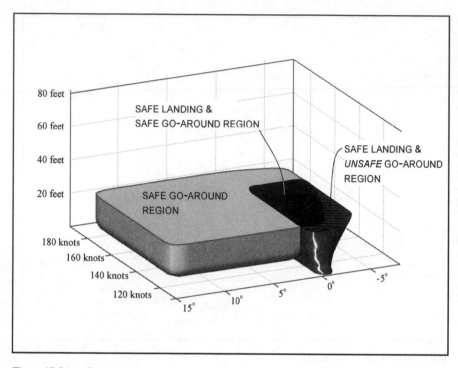

Figure 17.5 (c). Composite.

conditions, such as gusts or a strong tailwind, the autopilot may be operating in the hatched region. Nevertheless, the autopilot can handle the situation and still make a safe landing.

But we are not just concerned about safe landing; we also want to ensure a safe go-around. And here lies the problem: if the autopilot is operating in the hatched region, the aircraft will not be able to execute a safe go-around; neither from this altitude of 20 feet, nor from any lower altitude. In this example, if the autopilot is operating in the hatched region and a go-around is initiated, the aircraft may stall.

Automation Locks

There is a certain similarity between the yellow interval dilemma and the automatic landing example. In both cases, there are regions where the system is no longer safe; no matter what the driver or the autopilot does, they will not be able to escape the situation. In these regions the human–machine system finds itself in a lock.

But there also is an important difference between the two examples, which has to do with the locus of control: in the case of the yellow light, the driver is *manually* controlling the car. When we operate machines in everyday life, there are many situations in which we find ourselves in such a lock. Sometimes while driving on a highway the traffic is jammed and safe braking distances are hardly maintained. We all know that if someone ahead slams on the brakes there will be a chain accident, and because the lanes to the right and left are tightly packed, there is no escape. We are cognizant of this, yet we find ourselves again and again in these situations. There are many similar situations on the road, sometimes resulting from our actions or from inherent design deficiencies (e.g., poorly designed on-ramps). And we tend to accept the risks.

However, when it comes to an automated system, it is the machine itself that enters the potentially unsafe region from which recovery is difficult or impossible. Initially, the automation is in control, the user is monitoring the machine, and everything looks fine. And then, for example, air traffic control instructs the pilots to make a go-around. But when the user takes manual control or commands the automation to escape, the system is in a state from which recovery is difficult. This is an automation lock. Over the years, such "automation locks" have manifested themselves in several incidents and accidents.

China Airlines Boeing 747

On February 19, 1985, a China Airlines Boeing 747-SP, flying from Taipei, Taiwan, to Los Angeles, California, experienced an engine malfunction over the Pacific. (There are four engines on a Boeing 747 and the aircraft is designed to operate safely with three.) The aircraft was flying at 41,000 feet when the outmost right engine (number 4) got stuck in a low thrust setting. The aircraft began to slow down and the crew was trying to diagnose the problem and initiate remedial action.

Before and during the engine problem, the aircraft was on autopilot control. As soon as the outmost right engine malfunctioned, the aircraft started to roll to the right (due to asymmetric thrust—two good engines on the left side of the aircraft vs. only one and a half good engines on the right). But then the autopilot immediately countered by rolling the aircraft to the left. Meanwhile, the captain was focusing on trying to regain airspeed and was reluctant to disengage the autopilot and take manual control of the aircraft (as was recommended in his training and procedures for such a situation). A minute and a half later, the malfunctioning engine failed completely and was no longer producing any thrust. Now the autopilot had to work harder to roll the aircraft to the left in order to keep the wings level. The amount of left roll required to keep the aircraft flying level was coming close to the autopilot's control limits.

But inside the cockpit, except for the decreasing airspeed, the only feedback that would have depicted the worsening asymmetric situation and the autopilot's excessive attempts to counter the turn to the right, was the pilot's control wheel increasing left-wing-down deflection (about 6-10 degrees). However, this was an area that was not included in the captain's regular instrument scan, and since he was not "hands on" the wheel, he was not aware of this deflection.

The autopilot actions to counter the aircraft's tendency to roll to the right introduced a turning motion (side slip) and increased the aircraft's drag, which, in turn, further reduced the aircraft's speed. The captain reacted by switching the autopilot from the flight management computer control to a more simple pitch-hold mode and rotated a knob to begin a descent. Meanwhile, the autopilot was losing its struggle with the asymmetric thrust and the right wing was dropping. The aircraft began to gradually pitch down and descend while the autopilot was still fighting to bring up the right wing. Nevertheless, the autopilot did not provide any warning or indications to let the crew know of the alarming situation. Seconds later, the autopilot could no longer sustain the load of the wing and gave up—the right wing dropped rapidly, passed 45 degrees and the aircraft entered a downward spiral.

The large aircraft, with 251 passengers and 23 crewmembers onboard, plunged down out of control as the captain was trying to regain manual control of the aircraft and the flight engineer was attempting to restart the failed engine. The aircraft was in the clouds and this hindered the captain's ability to regain control of the aircraft. When the aircraft emerged from the clouds at 11,000 feet, the captain could see the horizon and was finally able to stop the dive and level the aircraft at 9,500 feet. The crew declared an emergency and made a safe landing at San Francisco Airport. One passenger and one crewmember suffered serious injuries during the 31,500-foot dive. The airplane suffered structural damage in the tail section.

It is important to note that at the time of the incident, the aircraft had already been 10 hours in the air crossing several time zones. The crew's ability to obtain, assimilate, and analyze the data from the flight displays and engine indicators as well as recall information from their training and procedures was probably impaired by the effect of monotony, boredom, and fatigue. In light of these factors, the captain's reluctance to disengage the autopilot and hand-fly the aircraft as soon as the engine malfunctioned was probably not because of over-trust in the automation per se, but perhaps because of his need to rely on the automation while being fatigued.

Simmons Airlines ATR-72

On October 31, 1994, a large commuter aircraft with 64 passengers, two flight attendants, and two pilots, was flying in bad weather to Chicago. The aircraft, flying at 16,000 feet, was in the clouds and had encountered icing conditions. As the aircraft approached Chicago's O'Hare airport, it was instructed to enter a holding pattern and maintain 10,000 feet. The crew knew they were flying in icing condition, but because the wings of the aircraft were far behind the cockpit, they could not visually see the accumulation of the ice on the wings. When ice builds up on an aircraft's wing, it changes the wing profile (or cross section) such that it is no longer smooth and efficient. The immediate implication is that the wing cannot produce as much lift as normal and the plane can therefore stall earlier.

The autopilot was engaged throughout the flight and when air traffic control instructed them to "descend and maintain 8,000 feet," the flight crew selected VERTICAL SPEED mode and began a descent to 8,000 feet. As the ice attached itself to the wings, it formed different shapes; specifically, the ice on the right wing of the aircraft was accumulating faster, causing the right wing to drop down, and rolling the aircraft into a right turn. Meanwhile, the autopilot was constantly countering (the right turn) by commanding a left turn. As this was going on, the crew had no idea of the deteriorating situation; they could not see the wing and the autopilot system did not indicate or alert them that it was working extremely hard to keep the aircraft's wings level. As the aircraft was descending through 9,500 feet, the autopilot could no longer compensate for the accumulation of ice on the right wing and counter the right roll. A second later the autopilot disengaged. This was by design, because there are conditions in the autopilot logic that trigger an automatic disengagement when the autopilot is working beyond a pre-determined operating range.

When the autopilot disengaged and quit, the right wing, no longer held up by the autopilot, immediately dipped down, corkscrewing the aircraft into a brutal right turn. The crew, who just seconds before had been sitting passively and monitoring an aircraft flying on autopilot, all of a sudden had a very vicious and strangely behaving aircraft on their hands; the accumulation of ice on the wings dramatically changed the way the aircraft behaved (and altered its handling characteristics). The nose dropped down to 15 degrees below the horizon; and the aircraft rolled rapidly to the right—the right wing reaching 77 degrees down. From there on the crew constantly tried to gain control over the aircraft. But the aircraft was rolling to the right and pitching the nose down to 60 degrees below the horizon. The aircraft was losing altitude fast. Twenty-seven seconds after the autopilot disengaged, the aircraft crashed in a soy-bean field.

The National Transportation Safety Board and most aviation experts agree that the flight crew's actions, before and after the autopilot quit, were consistent with their training and knowledge and in line with what is expected from a competent crew. It was extremely difficult—given the abruptness and severity of the unexpected roll and the changing characteristics of the aircraft (due to ice accumulation)—for the pilots to recover from this icing encounter and subsequent loss of control.

In Conclusion

In both of the above cases, the autopilot did not clearly indicate to the pilots the increasingly troublesome situation, nor did it give a warning that it was about to give up. It just did. As a consequence, the flight crews found themselves in situations from which recovery was difficult and at times impossible. Moreover, the immediate and unexpected transition from being a passive observer (monitoring the autopilot, sometimes for hours) to actively controlling a disabled aircraft that behaves unpredictably, was extremely difficult for the pilots. Dealing with these adverse situations and regaining control required split-second responses—a tall order, in particular when there was no time to ascertain the current status of the aircraft and assess the situation.

With increased automation and the shift of authority toward the machine, such "automation locks" pose serious threats to safe and reliable operation. They also highlight an important aspect of human-automation interaction, which has to do with the way we perceive automated machines. We, as humans, have a hard time accepting the possibility that an automated machine that is given full control will blindly enter into an unsafe and potentially disastrous situation. These automation locks, which exist in modern-day systems, bring us again face-to-face with HAL. Just as HAL, the ever-reliable machine, locked the astronauts out of control in *2001: A Space Odyssey*—in these two examples the autopilot, usually very reliable, locked the pilots into a situation from which recovery was extremely difficult.

While I do not wish to imply that it is necessarily the case that human operators can always recover from failures that automation can not, sometimes this is indeed the case. Had the pilots in the above mishaps been informed in time of the impending problems, they could have manually stabilized the aircraft and most likely recovered. Furthermore, the automation is usually designed to control the system under pre-specified operating conditions that do not include all irregular circumstances that might occur as a result of environmental impediments or internal failures. These impediments and failures are frequently too many and too diverse to account for and to guard against. The human operator, if suitably informed, could decide whether to continue and rely on the automation, or assume control manually. Thus, suitable indications and annunciations regarding departures from normal operating conditions are extremely important when designing automatic controls of safety-critical systems.

Before closing, it is important to mention that automation locks can be addressed in several ways: one, by identifying those critical situations in which the user will not be able to bring the automated system back to safety; two, by improving the design and thereby reducing the size of unsafe regions; three, by developing more sophisticated control systems (for example, the autopilot) that will resist entry into unsafe regions; and most importantly, by providing the user with timely and appropriate feedback about the dynamic behavior of the control system and departures from normal operating conditions. We need interfaces that know when to abstract-away information and when to display additional information—and how to do this in a way that adapts to the user's needs, interaction style, and the context of the situation. Developing such "intelligent" interfaces requires a level of sophistication that does not exist today. In many ways, interface technology lags behind advances in information technology and the ever-powerful machine. Needless to say, there are many challenges ahead when it comes to interface design and human–automation interaction.

Chapter 18

Conclusions and Observations

"You become responsible, forever, for what you have tamed."

—Antoine de Saint-Exupéry, *The Little Prince,* chapter xxi

I n the early nineteenth century, with the rise of the industrial revolution in England, workers led by one Ned Ludd broke into workshops and ransacked machines and equipment. The "Luddite" movement spread rapidly throughout the industrial heartland of England in 1811, resulting in the destruction of many wool and cotton mills before the British government suppressed it harshly. In 1813 the British Parliament passed a law making "machine breaking" a capital crime, and executed 17 men soon thereafter. By 1817, the Luddite movement had ceased to be active in England. Although the image of the heroic "machine-wrecker" fighting against the goliath machine captures the imagination, most historians agree that the revolt was primarily against the changing economic structure. Machine breaking was a way of putting pressure on employers and an effective method of "collective bargaining by riot and destruction." In any event, over the years the term "Luddite" has become synonymous with anyone who opposes industrial progress—or currently, technological advances.

This book is not a Luddite attack on automation, nor does it argue that we need to limit the fielding and application of automated control systems. On the contrary, we need to embrace the machine. No matter how you look at it, machines and automated systems are part of our daily lives. Further, their widespread application, authority, and power is only expected to grow in the coming years. Automated systems will be with us for the long haul.

For us to be able to embrace (and at the same time tame) the machine, we first need to form ties with it. That is, we need a better understanding of how machines work and how to interact with them. From a baby's night monitor, to a teenager's MP3 player, all the way to the robotic arm on board the International Space Station, automated systems have internal states, modes, parameters, reference values, initial settings, defaults, and history settings. We, as users, need to understand these characteristics of machines and have an appreciation for what they can and cannot do. This understanding is important not only for our immediate benefit, but also in order to have an educated opinion about the directions we, as a society, should take in designing and fielding the automated systems of today and the future. That is one important message of this book.

The second message of this book is about HAL's legacy. One of the prominent challenges that faces us in the design and use of automation is how to make the interaction between humans and automated systems smooth, cooperative, efficient, and, most of all, safe. We certainly want to avoid automation that behaves contrary to the user's intent and expectations. We should not accept interfaces that mask and obscure critical information, and especially those that are non-deterministic and incorrect.

Any interface that is non-deterministic and incorrect is just as dangerous as a HAL-like computer that goes berserk—because, from the user's point of view, it means that it is impossible to fully control the machine. And a machine that misbehaves, especially in an unforgiving and high-risk environment, is an accident waiting to happen. We no longer have to read fiction or go to the movies to witness machines turning against users. This reality is here today— we already operate and rely on automated control systems that can get us into a situation from which it is extremely difficult to recover.

It is easy, at this point, to put the blame on engineers, designers, and all those involved in making, fielding, and operating these machines. However, this is not a fair and constructive approach, because designers of automated systems are not a malicious lot, and in most cases a considerable amount of effort and resources are devoted to making automated systems usable, efficient, and safe. Nevertheless, it is true, as we have seen throughout this book and through our own (sometimes unhappy) experiences with certain devices, computers, and automated systems, that serious problems do creep in and go undetected.

Taming HAL

Throughout this book we have encountered many approaches and concepts for looking at human-automation interaction. We also employed an array of evaluation and verification methods to analyze interfaces and user interaction. These concepts and methods come from a variety of disciplines: discrete mathematics, psychology, computer science, human factors, control theory, cognitive science, and industrial engineering. They allow us to better understand the machine, to appreciate user-machine interaction, and actually to see with our own eyes the kinds of problems that plague human-automation interaction. In general, we have seen four main categories of human-automation design problems.

The first category, which really goes beyond the interface, yet affects it directly, concerns the integrity and safety of the underlying system. If there are flaw or gaps in the designer's understanding of how the automated system behaves, then there will be a corresponding flaw in the design of the interface. Specifically, before any serious evaluation and analysis of interfaces can begin, the underlying system must be analyzed to make sure that *all* unsafe regions are accounted for, and that there are no hidden unsafe regions. The issue here is about having an accurate and sound description of the machine.

The second category is about incorrect interfaces. In chapter 16 we encountered an interface (for an aircraft autopilot) that was incorrect. Given the interface indication, it was impossible to determine whether the aircraft was capable of capturing a new altitude. Such incorrect interfaces represent the most severe case of interface design problem. In the case of the *Royal Majesty* (chapter 8), we found a similar problem in the interface between two machines. No matter what kind of interface we are dealing with (machine-machine or human-machine), an incorrect interface is a design flaw that can lash out and lead to confusion and error.

The third category of interface design problems pertains to interfaces that are technically correct, but practically incorrect. One example was the design of the alarm mode in the wristwatch from chapter 3. On the face of it, the interface was correct—all the information was there—the user could examine the display very carefully after pressing the light button and observe that the alarm was disarmed. However, when one considers that the light goes off in 3 seconds, the alarm symbol is tiny, and the user in most cases is half asleep, the interface "appears" non-deterministic to the user. Therefore, for practical purposes, the interface is incorrect.

The fourth category of interface design problems belongs to systems such as the cordless phone (chapter 6), TV/VCR (chapter 7), and the entertainment system (chapter 9). Although these systems are technically and practically correct, they do create interaction problems because their design violates principles of user interaction (such as population stereotypes, consistency, and navigation and orientation problems).

I believe that the application of the insights, concepts, principles, and methods discussed in this book will allow those of us who are involved in the design of devices, as well as those who use, operate, and manage automated systems, to detect and identify the kind of interaction and interface design problems that exist in automated systems. (See also the last section of this chapter for a summary of 35 key issues of human-machine and human-automation interaction.)

Several Caveats

Before I conclude, there are several important caveats about this book that you ought to be aware of: first, this book addresses human interaction with automated systems by focusing on the information content that is required for correct, efficient, and safe user interaction. This book does not address the issue of what should (and what should not) be automated, but rather how human-automation interaction should be implemented. Second, this book does not directly address the issue of how the information is presented to the user on the graphical user interface, how the information is perceived and processed by the user, and what actions this processing may prompt.

Throughout this book, the term "task requirements" is used as the source for the kind of information (states, modes, operational regions) the user needs to know in order to operate the machine. There are varieties of analytical methods for deriving these task requirements. There are also design guidelines—some of them general to any human-machine system, some more specific for automated systems—that help designers define the user's task requirements and the extent of the user's control of the system's components. Last but not least, designing any interface requires an extensive amount of usability testing and sometimes experimental work to make sure that the proposed design is indeed usable and suitable for human use.

In Conclusion

Machines, devices, and automated control systems are all around us. We like them. This is obvious from the excitement in the tone of almost every news commentary and TV program about a new machine, robot, spacecraft, or any other automated device. We are fascinated by these machines because they are powerful, dutiful, precise, unemotional, and extend our abilities beyond our human reach. Yet at the same time, we have a deep fear of being betrayed and hunted by the very machines that we create and nurture. I have a feeling that this coexistence of fascination and fear is what fuels the popularity of stories about rampaging automatons like Frankenstein, the Golem, the dinosaurs from *Jurassic Park*, the *Terminator*, and yes, HAL.

Automated systems are efficient and reliable, and can improve our lives tremendously. It is also true that the same automated systems can run away and cause us grave harm. And we are all aware of the fact that there will only be more and more of them in the near and distant future. Richard Feynman, a Nobel Prize laureate and one of the most influential scientists of the twentieth century, was among a large team of scientists involved in the development of the atom bomb. After World War II, he turned his attention to academic research, education, and peaceful applications of atomic power. In his last book, *What Do You Care What Other People Think?*, Feynman wrote that throughout his life and in applying his work he was greatly inspired by a saying that he once received from a Buddhist monk:

To every man is given the key to the gates of heaven;
the same key opens the gates of hell.

I believe that the same applies when it comes to designing and applying automation.

The choice is ours.

Observations About Human-Automation Interaction

The following is a summary of 35 key issues we have uncovered and discussed in this book:

1. Time delays in response to user interaction (such as in ceiling fans and computers) can confuse users, especially when there is no positive and immediate feedback as to what the system is doing. (chapter 1)
2. In almost every control system, from the simple three-way light bulb to an aircraft autopilot, there exists an underlying structure where the same event leads to several different end states (or modes). Therefore, unless you know the current state, you will never be able to predict where you will end up. (chapter 1)
3. An interface (and all the related information provided in the user manual) is always a highly abstracted description of the machine's behavior. Making sure that this abstraction is correct, efficient, safe, and suitable for the user is a fundamental and critical aspect of interface design. (chapter 2)
4. There is a very fine balance between simplification of the interface (by abstracting-out any superfluous information) and over-simplification of the interface until it becomes non-deterministic. The interface design challenge, from the point of view of the information provided to the user, is to make sure that the abstraction is correct and as simple as possible. (chapters 3 and 16)

5. It is possible to describe the underlying machine (machine model) and the information provided to the user on the interface and in the manual (user model) using the same language. This allows for comparisons between the models and identification of design problems. (chapter 3)

6. A non-deterministic interface is one where the same user action can lead to more than one outcome. Such ambiguous interfaces, where users cannot determine what will be the outcome of their interaction with the system, are the gremlins that plague many of today's systems. In high-risk environments where there is no easy recovery, or when actions are irreversible, such interfaces are not only confusing and frustrating, but also dangerous. (chapters 3 and 7)

7. Automatic transitions take place when one or more conditions that guard the transition become True. The user has no direct control over these transitions because the events that trigger them are computed inside the machine, are time-dependent, or are affected by the environment (e.g., temperature). (chapters 3 and 4)

8. When the consequence of an automatic transition is important for the user, and especially when events that trigger the automatic transitions are not easily identifiable, the user should be provided with positive feedback at the time of the transitions and, when necessary, an advance warning. (chapters 4 and 16)

9. In automated control systems, the onset of a potential problem occurs when there is a significant divergence between what the users expect and what the machine is actually doing. We saw this dynamic and ongoing divergence, and the disastrous results, in the case of Korean Air Lines Flight 007, in the *Royal Majesty*, and in the case of the accident with the blood-pressure machine. (chapters 4, 8, and 12)

10. When users interact with automated control systems, small errors due to forgetfulness and inattention can be magnified because of a powerful and persistent machine—and can lead to disastrous consequences. (chapters 4, 8, and 12).

11. Population stereotypes are conventions that we learn over time. These conventions and experiences combine to form an "internal model" which shapes the way we interact with machines and how we expect them to respond. The design of user interfaces should carefully consider these stereotypes and try to support them (rather than violate them). (chapters 6, 7, 10, and 12)

12. We intuitively understand and are well accustomed to flipping switches of all sorts. But when these switches are affected by some external event, time delay, or additional automatic transitions, interacting with them can easily become confusing. (chapters 6 and 14)

13. Designers of devices should make every effort to ensure that user interaction is consistent throughout menus, sub-menus, and all forms of interaction. Preferably, this consistency should be in line with a related population stereotype. (chapters 6 and 7)

14. Automatic control systems have modes and reference values. In most cases the mode is a discrete state (e.g., ALARM-ON) and the reference values are continuous parameters (e.g., volume). (chapter 6)

15. There is a subtle difference between mode engagement and mode activation. Just having a mode engaged (e.g. ALARM-ON) does not necessarily mean that the mode is active and producing the desired behavior. (chapters 7 and 9)

16. A system is usually comprised of several components or machines working concurrently and exchanging information. Many of the principles of human-machine interaction, specifically those dealing with the abstraction of information, are also applicable to the design of machine-machine interfaces. (chapter 8)

17. In large and complex automated control systems where the user is physically removed from controlling the underlying process and is relegated only to monitoring, supervising, and perhaps entering data, the consequences of users' mistakes (as well as internal failures in the system) are not easy to pinpoint and identify. In these automated systems it becomes even more critical to provide the user with timely, correct, and sufficient feedback. (chapters 4, 8, 15, and 17).

18. Every automated system has modes, reference values, initial settings, timed events, defaults, and history settings. These features have a direct impact on user interaction and user expectations. It is important that these settings, which sometimes change automatically, are described to the user and are positively indicated. This is especially critical for defaults (e.g., the GPS defaulting to DEAD-RECKONING mode in the *Royal Majesty*). (chapters 4 and 8)

19. To help users cope and interact efficiently with "walk-in interfaces" such as ATMs, digital kiosks, and Internet sites, the underlying structure of the human-machine interaction must be made apparent to the user. (chapter 9)

20. Magic acts rely on misdirection created by the differences between what is conveyed to the audience and what is hidden from them. The magician purposefully provides the audience with extraneous information and tries to conceal the important information to make sure the end result is a surprise. Interface design is in the business of un-magic.

21. We strive for interfaces that are both correct for the task and also minimal in terms of number of modes, display indications, and events that the user must monitor and track. (chapter 11)

22. It must be recognized that users *will* make mistakes while interacting with automated control systems. It is possible, however, by employing a variety of evaluation and analytical techniques, to reduce the likelihood of error. Nevertheless, no system is infallible. Therefore, considerable effort should be made to help users quickly identify the consequence of a mistake and provide an efficient and safe means of recovery. (chapters 12 and 15)

23. A safety analysis of a system begins by identifying unsafe states and regions, then employing a variety of design features (such as interlocks, interface indications, and procedures) to block viable paths to these unsafe regions. (chapters 13 and 17)

24. User interaction with devices and machines, especially in safety-critical systems, is frequently governed by procedures, which provide the user with a unique sequence of actions that must be followed in order to complete the task safely and efficiently. Many procedures, especially those that are used while interacting with highly dynamic systems (such as aircraft or nuclear power plants) and those that are performed during emergencies, involve timing constraints and are essentially a race against time. (chapter 13)

25. Although procedures provide guidance to the user, promote safety, and, in case of a multi-person crew, enhance crew coordination, their use as a Band-Aid to fix a faulty and less than robust design is not a viable solution. Humans are not machines, and as a result, either inadvertently or on purpose, a procedure sequence, at one time or another, may go unused or bypassed. (chapter 13)

26. The relationship between a mode and its reference value is delicate and fragile. When implemented, this relationship should be properly displayed, explained to the user, and every effort should be made to avoid violating population stereotypes and other customs of interaction. (chapters 12 and 14)

27. In some automated systems, modes and reference values are combined, such that a change in a reference value can trigger a mode change and vice versa. Such designs must be carefully evaluated because they tend to confuse users and become a source of repeated mistakes. (chapters 12 and 14)

28. Modern automated systems are highly reliable. As a consequence, users tend to depend on these systems—often without understanding and recognizing their inherent limitations. Lulled into a false sense of security, over time, users may neglect proper monitoring and supervision of the machine. Worst of all, over-trust in automation can lead users to ignore and dismiss cues that indicate that the automated machine is not performing well and is becoming unsafe. (chapter 15)

29. The old platitude of "just hit the button, sit back, and the rest will happen automatically," which was prevalent in the 1960s and 70s as automation began to enter our daily lives, doesn't work anymore when it comes to safety-critical systems such as medical equipment and aircraft. We cannot afford to sit back and be ignorant of what is going on. We need to know how it works and what is going on. This will become even more critical as future automated devices (such as adaptive-cruise controls) will become an everyday reality. (chapters 9, 11 and 16)

30. Envelope protection systems are critical for the sophisticated automated systems of today and the future. They provide an additional degree of safety. Envelope protection systems give the impression of an overall guarantee of safety, a kind of global safety net. Yet envelope protection systems are very difficult to design and should not be taken for granted by users. (chapter 15)

31. Modern-day machines, computers, and automated systems cannot think. As compelling as it may appear to a user who is working for days on end with a flawless computer or automated system—a machine cannot "read" an unexpected situation or scrutinize it. The machine can only do what it was pre-programmed to do. At its very essence, a computer is a system of states and transitions. Modern-day machines can only do what is specified by this system of states and transitions—nothing more, nothing less.

32. Some of the accidents detailed in this book—the *Royal Majesty*, the blood-pressure machine incident, the flight-test crash—are poignant reminders that the decision to keep on using, or to disengage, the automation can be one of the most important decisions a human operator can make, particularly in critical and emergency situations. (chapter 15)

33. Verification is a rigorous methodology for proving that a system satisfies (or doesn't satisfy) a given criterion. Such verification methods, which can be incorporated into a software tool and applied to larger and more complex

systems, can be used to verify the correctness of interfaces. Several interface and system characteristics—such as the lack of error states, of augmenting states, and of restricting states—provide the criteria for this kind of verification. (chapter 16)

34. When designing user interfaces, we assume that the underlying system works as specified. However, in many complex systems this is not always the case. Therefore, in order to develop interfaces that are safe and correct—the design, evaluation, analysis, and verification can only take place after it is certain that the underlying system behaves as specified. (chapter 17)

35. "Automation lock" is a condition in which the automation drives the system into an unsafe region from which it is difficult and at times impossible to recover. Automation locks are particularly dangerous when users rely too much on the automation and tend to over-trust it to the point of not monitoring it and dismissing cues that the automation is not working properly. (chapter 17)

Notes

Introduction

Arthur C. Clarke was still writing *2001: A Space Odyssey* while Stanley Kubrick was building the set and shooting the movie. The novel, script, and the movie proceeded simultaneously, with feedback in each direction. The novel was published in July 1968, shortly after the release of the movie. As a result, there is a much closer parallel between book and movie than is usually the case, but there is one major difference. In the novel, the destination of the spaceship *Discovery* was Japetus, one of Saturn's many moons. In the movie, however, the destination is Jupiter. (The change from Saturn to Jupiter made for a simpler and shorter story line—Jupiter is closer to Earth—and, according to a later account by Arthur C. Clarke, the special effects department could not produce an image of Saturn that Stanley Kubrick found convincing.) For a scientific treatment of the relationship between the science fiction shown in the movie and today's technological facts and the state-of-the-art in computer science and engineering, see *HAL's Legacy* by David Stork (MIT Press, 2000).

Golem (GO-lem) is the Hebrew word for a "shapeless mass." In the Bible, a form of the word (galemi), is used to describe a fetus before taking human shape, according to one interpretation, and mankind before creation, according to another—"Thine eyes did see my substance (galemi), yet being unperfect" (Psalms 139:16). The word is also used to denote anything imperfect or incomplete; unconscious Adam, initially a body without a soul, is referred to as a Golem. According to Jewish mysticism (Kabala), the creation of a Golem is achieved by using combinations of letters from the Hebrew alphabet (see *Golem*, by David Wisniewski, Clarion Books, 1996).

The original tale of a rabbi creating a Golem to defend his congregation was told of Rabbi Elijah of Chelm, Poland. Over the years, the story's locale had shifted to Prague and focused on Rabbi Judah Loew ben Bezalel (1513-1609). A renowned scholar and mystic who wrote extensively about religious matters, much of his fame rests upon his supposed creation of the Golem to defend the Jews of Prague. For a literary treatment of the Golem story see Gustav Meyrink, *The Golem* (Dover, 1986). A more folk-tale version was written by Isaac Bashevis Singer (Farrar, Straus & Giroux, 1992); see also *The Amazing Adventures of Kavalier & Clay* by Michael Chabon (Picador, 2001) about the Superman-Golem. For centuries, the story of the Golem served as a cautionary tale about the limits of human abilities and the potential harm of runaway creations. Probably the first link between the Golem and modern day computers was made by Norbert Wiener's thought-provoking book on life, computing, and automation—titled *God and Golem, Inc.* (MIT Press, 1988).

The terms *automa*, *automaton*, and *automate* have been used since the seventeenth century to describe mechanical systems that work on their own ("taken with admiration of watches, clocks, dials, automates" [*Oxford English Dictionary*, second edition, 1989]). A much older use of the word has a similar albeit somewhat contrasting meaning: it was used to denote an action of a biological body (e.g., amoeba), of which the causes of action lie in the body itself.

The definition of the term *automatic*, when applied specifically to machinery, is "self acting under conditions fixed for it" (*Oxford English Dictionary*, second edition, 1989). Since in this book we are interested in user interaction with an automatic machine, we use the term "automatic" to denote mechanized event(s) triggered without direct user involvement.

Automation is a relatively modern term that first appeared in 1948. It was coined by Delmar S. Harder, an engineer and vice president of the Ford Motor Company, and was initially used to describe a system for the automatic handling of parts. The idea behind such automation was to increase efficiency and reduce manual demands on workers. One of the early usages of such automation was the development of a machine to transfer hot, heavy coil springs from a coiling machine to a water tank—a job that before the use of automation, was called "man-killing" because the worker had to reach down, lift the hot and heavy part to chest height, turn around and put it in the water tank, all within seconds. (The first plant that was designed around automation concepts was Ford's stamping plant in Buffalo, New York. It began operating in September 1950.)

Another definition of automation, one that focuses on the ongoing influence of automation on individuals and society, is "the execution by a machine agent (usually a computer) of a function that was previously carried out by humans." The implication is that what we define as automation evolves and changes with time, and what was considered previously as automation (automatic starters, automatic washing machines) is now simply considered a device or a machine (Raja Parasuraman and Victor Riley, "Humans and automation: use, misuse, disuse, abuse" in the 1997 issue of the *Human Factors Journal*, volume 39).

A broader definition of the term automation is "the application of automatic control to any branch of industry or science" (*Oxford English Dictionary*, second edition, 1989), which is a more applicable definition of the term in the context of this book. Throughout this book I will use the term "automated devices" to denote the general class of systems that have automatic features, switching from one mode to another on their own, and "automated control systems" to denote automatic mechanisms that govern and control a dynamical process (e.g., a cruise control system that governs the car's speed and an autopilot that controls the aircraft flight).

Automatic teller machines (ATMs) may perhaps one day be called "tellers" or "teller machines" because there won't be bank tellers of the human kind. There are about a million ATMs in the world today. Their overall reliability is high, but they do sometimes fail, causing erroneous transactions. On February 15 and 16, 1994, customers using ATMs for Chemical Bank were billed twice the amount they actually withdrew. The error affected some 150,000 withdrawals and transfers between accounts made at ATM machines and stores. Customers got their cash and receipt as normal, but the computer that processed the transaction erroneously sent it to another computer that processed paper checks, which, in turn, deducted the (same) amount from the customer's account. The error only affected debits. People who made deposits did not receive extra money (Jeff Hecht, "Bank error not in your favor," *New Scientist*, March 5, 1994; and *Fatal Defect* by Ivars Pereson [Vintage Books, 1994]).

The Therac-25 is a radiation machine for cancer treatment. Built in the early 1980s, it was one of the first radiation machines to be controlled by a computer. The machine had two modes: a LOW energy mode (200 rads), and a HIGH x-ray mode (25 million electron volt). In the LOW energy mode, which is used for treating superficial tumors, the beam was aimed directly at the patient's body. In HIGH energy mode, which is used to treat tumors in deep tissues, the electron beam was directed toward the patient through a large and thick metallic (tungsten) shield.

The machine was controlled by a computer and the technician would enter the desired mode and different settings through the computer screen. Due to a design flaw, both in software and also in the interface, there were several sequences of user (technician) actions that would result in the patient receiving the HIGH energy beam, and without the tungsten shield, instead of LOW energy. This amounted to 8,000 to 25,000 rads (depending on the initial dosage setting), which was somewhere between 60 to 125 times more than intended. Between 1985 and 1987, 6 patients were mistakenly overdosed. All of them died. For a detailed account of this series of accidents and a comprehensive discussion of approaches and methods for safety analysis of software, see *Safeware: System Safety and Computers* by Nancy Leveson (Addison Wesley, 1995).

On October 31, 1994, a large commuter aircraft carrying 64 passengers, 2 flight attendants, and 2 pilots was flying in bad weather to Chicago. The aircraft, flying at 16,000 feet, was in clouds and encountered icing conditions. As the ice attached itself to the wings, it formed different shapes; specifically, the ice on the right wing of the aircraft was accumulating faster, causing the right wing to drop down, and pushing the aircraft into a right turn. As the aircraft was descending through 9,500 feet, the autopilot could no longer counter the accumulation of ice on the right wing. When the autopilot quit, the right wing, no longer held up by the autopilot, immediately dipped down, corkscrewing the aircraft into a brutal right turn. The crew, who just seconds before had been sitting passively and monitoring an aircraft flying on autopilot, all of a sudden had a very vicious and strangely behaving aircraft on their hands; the accumulation of ice on the wings dramatically changed the way the aircraft behaved (and altered its handling characteristics). From then on the crew constantly tried to gain control over the aircraft, but they could not recover. Twenty-seven seconds after the autopilot disengaged, the aircraft crashed. (For more details on this accident, see chapter 17.)

The quote from the airline pilot about his apprehension and concerns to fully interact with the automatic landing system was recorded during a series of interviews conducted between 1992 and 1994 as part of a study on the use of checklists and procedures in automated cockpits. (See Asaf Degani and Earl Wiener, *On the design of flight-deck procedures*, NASA Contractor Report number 177642, 1994—see especially Appendices 4 and 5. This and similar comments made by experienced airline pilots about their uneasiness while interacting with automated systems led to the initiation of a detailed study of pilot interaction with an automatic landing system, which is discussed in chapter 17.

Peter Hancock's statement regarding humans and technology comes in an article published in the *Human Factors and Ergonomics Society Bulletin* (volume 42, issue number 11, November 1999). The Human Factors and Ergonomics Society lists about 5,000 members from a variety of diverse fields such as psychology and engineering. The Society's mission is to "promote the discovery and exchange of knowledge concerning the characteristics of human beings that are applicable to the design of systems and devices of all kinds."

Chapter 1

Figure 1.1 is a black-and-white reproduction of a painting (synthetic polymer on canvas) by Anatjari Tjampitjinpa, an Australian painter. It is said to represent a series of journeys made by his aboriginal ancestors. Reminiscent of ancient desert treks, the painting plots the location of soaks, rock holes, and wells, in the language of Tajukurrpa—the religious "dreamtime" activities of the Aborigines. The painting appears in *Spirit Country: Australian Aboriginal Art from the Ganter Myer Collection* by Jennifer Isaacs (1999, Hardie Grant Book, page 40).

The notion of travels from one water hole to another, from one station to another, thereby describing paths in a landscape, is a natural way to record human activities. The same notion is also behind the theory of finite-state machines, with its graphical notation of states, transitions, and events, which is used in figures 1.2, 1.3, and 1.4. The theory, which is more about abstractions than actual machines (finite-state abstractions would have been a much more appropriate name), was developed in the early twentieth century to answer theoretical questions in mathematics and received a boost in the 1930s with the work of the English logician and mathematician Alan Turing. Turing's pioneering work and his idea of a "state" as a basic concept in representation of systems, set the foundation for the development of computing machines. Later, the concept of states was employed by Claude Shannon in developing information theory. The work of Shannon and others is the theoretical foundation behind the communication protocols that link computers together and make the Internet a reality.

The theory of finite-state machines is concerned with (mathematical) descriptions and models that serve as abstract representations of physical systems and phenomena. It describes a system as a network of states and transitions, which is not unlike, for example, the depiction of stations (states) and (transition) lines on a subway map. Finite-state machines have been used to describe and analyze systems in practically every field of science—from physiology to electrical engineering, and from biology to linguistics. This is achieved by viewing a system (e.g., a machine, a living organism) through its way of behavior, and not through its physical composition. One of the most important contributions of this theory is the ability to describe, using the same language, very different systems, and then analyze them together. The significance of this is that by using the description language as a kind of a "common denominator," it allows us to compare and see the interaction between two (or more) systems. We will certainly take advantage of this throughout this book to analyze human interaction with machines.

The classical textbook on finite-state machines, still used today in the classroom for introducing computer-science students to thinking in abstract ways, is *Introduction to Automata Theory, Languages, and Computation*, by John Hopcroft, Rajeev Motwani, and Jeffrey Ullman (Addison-Wesley, 2000). A lighter treatment of this topic can be found in a variety of math textbooks. A good start is *Discrete Mathematics with Applications*, by Susanna Epp (2nd edition, PWS Publishing Company, 1995); look for section 7.2 (pages 357–369).

Since Turing's seminal work, several extensions and modifications to this way of describing and modeling systems have been implemented. The research work of Edward Moore at Princeton University and George Mealy at Bell Laboratories, published in the mid-fifties, provided a way to make the description language more applicable to how information systems and modern computers behave. In chapter 5, we will use one recent development, called *Statecharts,* that embodies many of these extensions and modifications.

Chapter 2

Abstractions is an ongoing research topic in computer science. The objective is to describe complex systems in an abstracted way so that one can perform various types of tests on the computer code. The use of the London Underground map as a vivid example for the concept of abstraction can be found in a book about mathematical methods for system analysis, *Using*

Z: Specification, Refinement, and Proof by Jim Woodcock and Jim Davies (Prentice Hall Europe, 1996, pages 5-8).

Mr. Henry Beck was the designer of the (abstracted) London diagram. He worked as an electrical draftsman for the Underground company (London Transport), which at the time was suffering from lack of customers and incurring heavy losses. As a commuter himself, he realized that the problem with the existing maps was that they were too accurate. He used his experience in drawing circuit diagrams to turn the geographical map into an abstract diagram. He created a clean representation by using only vertical, horizontal, and diagonal lines. Station symbols also come from mechanical and electrical-engineering notations.

When Mr. Beck submitted his highly abstract sketch idea to the publicity department at London Transport, officials rejected it as "too revolutionary." They were concerned that the map was not geographical, because he enlarged the downtown area and compressed the outer areas. Henry Beck sent it back to them a year later. This time the publicity manager changed his mind and ordered 750,000 copies made. Some accounts tell us that Beck was never actually paid for the original job, some mark the price at five guineas. Whatever it was, it is a meager price given that the diagram is reproduced over 60 million times each year by companies other than London Transport. The successors to the original Underground diagram are still used by Londoners today, and the abstraction principles that Beck developed are used in the New York, St. Petersburg, Sydney, and many other underground maps. For the history of the diagram and the intricate human story of pride and obsession on the one hand, and corporate intrigue on the other, see *Mr. Beck's Underground Map* by Ken Garland (Capitol Transport, 1994).

Growing up as a child in London, I spent hours staring at the diagram of the Underground, wondering how these lines connect and imagining trains moving back and forth in the diagram. One time, while returning home (we lived in the suburb of Wembley) from one of my first unaccompanied ventures on the Underground, I made a mistake, and instead of taking the Bakerloo line, I took the Metropolitan line. I got off at Preston Road station with a fright, assuming, based on the diagram, that I would have a very long walk to our home near South Kenton Station. To my delightful surprise, I was home in a jiffy!

Indeed, the abstracted map is not always perfect. For example, out-of-town visitors using the diagram to get from Bank station to Mansion House (figure 2.6[a]), have to switch several trains and spend quite some time underground. When they emerge out of Mansion House station, they find themselves 200 yards down the street from the location they'd started at (Bank Station). According to the BBC information web page, there are also reports of tourists getting on at Covent Garden and traveling one stop to Leicester Square. The diagram hides the fact that this is about 300 yards by foot, and is the shortest distance between any two stations in the Underground network.

Chapter 3

In this chapter I discussed the topic of non-determinism in a fairly broad and loose sense. Strictly speaking, a system responds non-deterministically whenever its output cannot be predicted. This was, for example, the case in the soft drink machine where the outcome *root beer* could not be predicted by the customer who has no insight into the machine's innards. On the other hand, the alarm clock problem was not strictly non-deterministic: the user could have turned on the light to see the clock's current mode. Alternatively, he or she could have examined the display very carefully after pressing the "light" button and observed that there was no shockwave symbol and the alarm was disarmed. So, in a strict sense, the interface for the digital watch was okay. But in a practical sense it was not—the traveler acted quite rationally but still was surprised by the clock. When it comes to user-machine interaction, the usability test is not just an issue of strict formalities but of practice. The proof of the pudding is in the eating.

I wore the digital watch described in this chapter for several years before I was finally able to figure out why it would *sometimes* fail me on wakeups. A somewhat similar problem in another digital watch is told by Donald Norman in *The Design of Everyday Things* (Basic Books, 2002):

> I had just completed a long run from my university to my home in what I was convinced was a record time. It was dark when I got home, so I could not read the time on my stopwatch. As I walked up and down the street in front of my home, cooling off, I got more and more anxious to see how fast I had run. I then remembered that my watch had a built-in light, operated by the upper right-hand button. Elated, I depressed the button to illuminate the reading, only to read a time of zero seconds. I had forgotten that in stopwatch mode, the same button (that in normal, time-reading mode would have turned on a light) cleared the time and reset the stopwatch. (page 110)

The aircraft accident mentioned in the beginning of the chapter took place on April 26, 1994, in Nagoya, Japan. After the copilot inadvertently engaged the GO-AROUND mode, which automatically increases engine thrust to maximum power and initiates an aggressive climb, the copilot pushed the control wheel down to force the aircraft to descend. The captain also instructed the copilot to push the control wheel down and make the landing.

As the copilot was pushing the wheel to force the aircraft to land, the autopilot—engaged in GO-AROUND mode—was countering by adjusting the control surfaces to make the aircraft climb. The crew continued with the landing, unaware of this abnormal situation (there was no warning to alert the crew directly to the onset of this abnormal condition). By the time the captain recognized the abnormal situation, and took control of the aircraft, he was unable to recover in time. The aircraft pitched high, stalled, and crashed tail-first on the runway. Of the 271 passengers and crew, only 7 passengers survived.

The captain and copilot assumed that if they would exert heavy nose-down forces to the control wheel, they would be able to override the autopilot (in GO-AROUND mode) and continue the descent to the runway. This was true for most autopilots, but on this particular model, it was different. The feature that allows the pilot to override the autopilot was disabled when the aircraft altitude was less than 1,500 feet. (The intent was to prevent the pilots from inadvertently pushing or pulling on the wheel and thereby overriding the autopilot). It also appears that the aircraft manual had unclear descriptions of this autopilot feature, which contributed to the pilots' assumption that the manual override always works. Subsequently, all autopilot models were modified such that this manual override feature is never disabled.

Chapter 4

The discussion in this chapter is primarily based on the report of the International Civil Aviation Organization (ICAO) investigative team and transcripts of recorded conversations within the Soviet Air Defense System. Additional information was obtained from the article "Closing the File on Flight 007" by Murray Sayle (*The New Yorker*, December 13, 1993, pages

90–101), a chapter by David Beaty titled "Boredom and Absence of Mind" in his book *The Naked Pilot: The Human Factors In Aircraft Accidents* (1995 Airlife Publishing: Shrewsbury, England), as well as videotapes of a press conference in Moscow several days after the disaster featuring the Red Army's chief of staff and a large and detailed diagram of the flight route, the first interception over Kamchatka, and the final attack on Flight 007.

Although the inertial navigation system with its triple redundant computers was hailed as the technological solution that would eliminate navigation errors, similar navigation snafus were no news to the commercial aviation community. There were more than a dozen similar incidents that were investigated (and Lord only knows how many never got reported, to save face), in which flight crews selected INERTIAL NAVIGATION mode but did not detect that the navigation system was not steering the autopilot. Other incidents occurred in situations in which flight crews failed to re-engage INERTIAL NAVIGATION after temporarily using HEADING mode to fly around thunderstorms. Most of the incidents involved track deviations of less than 60 miles. One incident, however, involved a 250-mile off-track deviation, and another (110 miles) almost resulted in a mid-air collision between an off-course Israeli El-Al Boeing 747 and a British Airways 747 over the Atlantic Ocean.

The presence of military reconnaissance aircraft, flying close to commercial air routes and collecting electronic and communications information, is not a new practice; the first intelligence-gathering reconnaissance missions of the Cold War era were flown during the airlift to Berlin in 1946. The practice is still going on today. On April 1, 2001, a U.S. Navy aircraft was flying a communication-gathering mission over the South China Sea. A Chinese fighter aircraft was sent to intercept and track the Navy EP-3 reconnaissance aircraft. The fighter aircraft came too close to the reconnaissance aircraft and collided with it. The Chinese fighter crashed in the sea and the pilot died; the U.S. Navy EP-3 aircraft made an emergency landing on the Island of Hainan. The crew was released after long and tense diplomatic negotiations between China and the United States. The aircraft was dismantled and then brought back to the United States.

Commercial aircraft straying into military zones, causing confusion, and at times strife, has occurred in the past. In 1978, MiG fighters strafed a Korean Air Lines plane that mistakenly wondered into Soviet airspace. The aircraft made an emergency landing on an icy lake in the Kola Peninsula (in northeast Russia, close to Finland), and fortunately only a few passengers were injured. In 1985, two years after the shooting down of Korean Air Lines 007, a Japan Airlines Boeing 747 also had its autopilot in HEADING mode and strayed 60 miles into the same (Siberian) airspace. By this time, the Soviets had made changes in their rules of engagement and were more cautious. The aircraft was escorted out.

The shooting down of Korean Air Lines Flight 007 was used by both the Soviets and the United States for political propaganda and internal politics. President Ronald Reagan called it a "heinous act" and used it to press Congress to continue with the arms race and the mystifying "Star Wars" weapons system. Yuri Andropov, in turn, denounced the event as "a sophisticated provocation masterminded by the US special services with the use of a South Korean plane" and drove the USSR into a counter arms race that eventually broke the Soviet Union's economic back, giving rise to the reforms of his successor, Mikhail Gorbachev. Cold War politics aside, the disaster shook every nation on both sides of the Iron Curtain and became an international tragedy. It eventually led Reagan to offer the world's civil aviation operators free use of the then military-only, Global Positioning System (GPS), a constellation of more than three dozen orbiting satellites that allows precision navigation for ships and aircraft. Many have argued that if KAL 007 had the highly accurate global positioning system, such a blunder could not have happened. But as we will see in chapter 8, which details the story behind the grounding of the cruise ship *Royal Majesty*, which was equipped with a global positioning unit, accuracy is not the only factor that must be considered in the design of highly automated navigation systems.

Chapter 5

The *Statecharts* description language, which embodies these three characteristics of automated systems (concurrency, hierarchy, and synchronization), was conceived in the early 80s to deal with the growing complexity of automated systems. It has since been used to describe a variety of systems such as avionics, automotive, medical technology, as well as for specifying and understanding computer security systems. *Statecharts* is similar to the finite state machine description that we have been using in the previous chapters, but extends the description to include concurrency, hierarchy, and synchronization. In this book we will stick with the original *Statecharts* language as described by David Harel and Michal Politi in their book *Modeling Reactive Systems with Statecharts* (McGraw Hill, 1998). *Statecharts* (and its many derivatives) are commonly used in the process of designing and specifying computer code, and is one of the components of the Unified Modeling Language (UML), a structured approach for software development (see *The Unified Modeling Language Reference Manual* by James Rumbaugh, Ivar Jacobson, and Grady Booch published by Addison Wesley, 1999).

The leap from *Statecharts* as a specification language for code generation into using it for user interfaces was taken by Ian Horrocks (*Constructing the User Interface with Statecharts*, Addison-Wesley, 1999). Horrocks, a professional designer working for British Telecom, shows a systematic methodology for specifying interfaces using the *Statecharts* language. His is a book about implementation—how to specify the behavior of the user interfaces and how to generate from this specification a well-structured code.

The notion of hierarchy casts some light on the topic of abstraction, which, as I have mentioned earlier, is the essence of any interface design. By going up in a hierarchy of states and super-states one can abstract information that resides within a super-state, and by going down one can refine. For a down-to-earth treatise on these issues, see a small book, titled *Science of the Artificial* (MIT Press, 1996), by the late Nobel prize winner, Herbert Simon.

Chapter 6

The technical term for the dual switch we discussed here is a *three-way* switch. The reason it is called a three-way, even though there are only two switches, is that for this arrangement to work, each light switch must have three terminals (see the figure on page 291); one "traveler" wire goes from switch A to switch B, another from switch B to A, and then there is a "common" wire that runs from the electrical source to the switches and from there to the light fixture. The light fixture is ON when there is a continuous pathway from the electrical source (via the common wire) to the light. When either switch creates a gap in that pathway, the light fixture is OFF.

On a more general level, the dual, triple, or multiple switch poses an interesting quandary: how do we know what should be the state of the light fixture, for any number of switches? What if we have four switches—how can we tell?

Let's work it out together; it is rather illuminating. We already know how a system with two light switches works, right? Here is a table for a system with two switches:

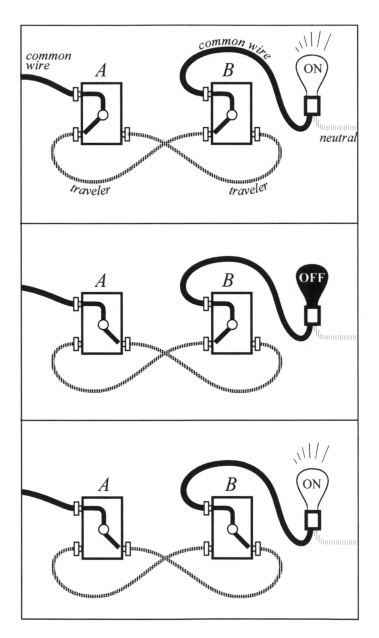

Three configurations of switches and light.

*down, down = **off***
*down, up = **on***
*up, up = **off***
*up, down = **on***

Now let's consider a system with three switches. When all three switches are down, the light fixture is OFF.

*down, down, down = **off***

Flip up the first switch, and the new configuration is

*up, down, down = **on***

From here, we can go on and complete the "switching" table.

*up, up, down = **off***
*up, up, up = **on***
*down, up, up = **off***
*down, up, down = **on***
*up, down, up = **off***
*down ,down, up = **on***

To make it easier on the eye, we replace "down" with 0 and "up" with 1. Here is the same table in a more digital form:

*0, 0, 0 = **off (0)***
*1, 0, 0 = **on (1)***
*1, 1, 0 = **off (2)***
*1, 1, 1 = **on (3)***
*0, 1, 1 = **off (2)***
*0, 1, 0 = **on (1)***
*1, 0, 1 = **off (2)***
*0, 0, 1 = **on (1)***

Now let's go and sum up the values for each row. In the first row, the sum is 0 (0+0+0 = 0) and the light is off. In the second row, the sum is 1 (1+0+0 = 1) and the light is on; in the third it is 2 (1+1+0 = 2) and off, and so on and on. Do you see a pattern here?

If you look carefully you will note that every time the sum is an *odd* number the light is ON; every time we have an *even* number (and 0), the light is OFF. It turns out that this rule works no matter if we have two, three, four, five, or hundred switches. With this little insight, we can always tell what should be the state of the light, no matter how many switches we have.

The notion of "population stereotype" surfaced in examination of training incidents and accidents that occurred to U.S. aviators during World War II. At the time, there were no standards, or any appreciation of convention, for the design of switches and levers in aircraft. There were occasions where pilots, during training and also in combat situations, would pull up the flap lever instead of the gear lever, turn off the generator switch instead of the landing light switch, and takeoff with flight controls locked. Paul Fitts and his colleagues at the U.S. Army Aero Medical Laboratory started working on these and other related problems and resolved most of them by the end of the war.

Fitts and his colleagues strongly believed that "it should be possible to eliminate a large proportion of so-called 'pilot-error' accidents by designing equipment in accordance with human requirement"; and with that conviction as well as extensive theoretical and experimental work, they laid the foundation of what we know today as human factors and ergonomics. (The classical article on the topic is by Paul Fitts and R. E. Jones, "Analysis of factors contributing to 460 'pilot-error' experiences in operating aircraft controls," which appeared in a book by Wallace Sinaiko, *Selected Papers on Human Factors in the Design and Use of Control Systems* [Dover Publications, 1961]). The initial investigations into the notion of capture error were conducted by James Reason in England (*Human Error,* Cambridge University Press, 1990) and Donald Norman in the United States (*The Design of Everyday Things,* Basic Books, 2002).

"Red sky in the morning, sailor's warning; red sky at night, sailors delight," is a fairly good predictor of weather patterns in the Northern Hemisphere. In mid latitude, such as 20^0-40^0 North, the storm systems generally follow the jet stream as it blows from west to east. When a sailor goes to sea and sees red sky in the morning, it means that the sun, rising in the east, is casting its rays on clouds that are coming from the west. These clouds may be an indication of a coming storm. When we see red sky in the evening, it means that the sun, now setting in the west, is casting colorful rays on clouds that are in the east. These clouds, perhaps storm clouds, are pushed by the wind further and further away—and therefore one can expect a safe sailing day (*National Geographic,* June 2001, p. 1). The phenomena and the resulting rule-of-thumb was probably known and used by early Mediterranean sailors. According to St. Matthew, Jesus said: "when it is evening, ye say, it will be fair weather, for the sky is red. And in the morning, it will be foul weather today, for the sky is red and lowering" (16:2-3). Jesus makes the statement following his stay at Tyre and Sidon, two prominent maritime cities.

Chapter 7

On the issue of setting the time on your VCR, some new VCRs and TVs come with a feature that adjusts the clock automatically. This feature works by scanning TV channels for a time signal that is broadcast by many public stations as well as some commercial network stations. It sounds like the perfect solution to the nagging problem of setting the time in VCRs and TVs, one that will eliminate forever the inconvenience of going through menus, modes, and long user manuals. But as it turns out, the new feature brought with it its own set of problems. In one case, technicians in a San Jose, California, TV station did not check on the accuracy of the signal for about a year. Consequently, the time signal slipped by 24 minutes—affecting thousands of VCRs in the area.

In another case, a Los Angeles station fed this time signal for the nationwide Fox network; however several affiliates across the country failed to adjust the signal for their local time zones. As a consequence, clocks on the East Coast were automatically set to Pacific Time. Another source of confusion was when shows were taped for a later broadcast, yet technicians forgot to remove the time signal. Automatic clocks all over the country started to dance. The point is that human error and the potential for confusion was not eliminated, it was only shifted from the user to station technicians. In addition, there is the fact that we all tend to rely on these automated devices to the point of not checking or monitoring what is going on. And all of us, lay people or professionals are in the same boat when it comes to over trusting automated systems (as we shall see in chapter 8). The auto-clock story is told by Kim

Vicente in his book *The Human Factor: Revolutionizing the Way People Live with Technology* (Knopf Canada, 2003; see the auto-clock story in chapter 7).

Chapter 8

Sailors are a superstitious lot, and so are executives of cruise lines. The reputation of a grounded ship does not sail very far. And hence the good ship *Royal Majesty*, once the jewel of the fleet, was eventually sold to Norwegian Cruise Lines, where she began sailing under the name *Norwegian Majesty*. In 1999 she voyaged to Bremerhaven, Germany, where she was hauled up and sliced in the middle. A 112-foot midsection was inserted, giving her an overall length of 680 feet and erasing all physical signs of her embarrassing grounding.

In August of 2001, the Asia Rip (AR) buoy (which was confused by the crew of the *Royal Majesty* with the entrance buoy to the Boston traffic lanes), was "decommissioned" and removed from location. The long and heavy chain that connected her to the wreck below was cut, and the buoy was towed back to shore. Notices to Mariners, issued by the U.S. Coast Guard and dated August 28, 2001, instructs mariners and navigators to cross out Asia Rip from all charts and piloting books.

The factual account for this chapter came primarily from a 1997 National Transportation Safety Board (NTSB) report about the accident and a variety of maritime publications. (The full report, titled *The Grounding of the Panamanian Passenger Ship Royal Majesty on Rose and Crown Shoal near Nantucket, Massachusetts on June 10, 1995,* can be obtained from the NTSB or downloaded from their web site.) The sequence of events and interactions between the bridge officers and the captain is based on the crews' testimonies following the accident. The description of the routine sea duties and some of the non-technical aspects of this story come from my own experiences as a bridge officer on naval and commercial vessels.

The Global Positioning System is a constellation of 24 satellites orbiting the Earth. The GPS unit obtains its distance from at least three satellites, and then triangulates its exact position on the Earth. Nevertheless, a three-satellite fix can only work if the GPS unit is synchronized with the atomic clocks onboard the satellites. Since it is impractical to install an atomic clock in every GPS unit (it would drive the cost of every unit to above $50,000), the designers of the GPS used an ingenious idea to correct for any timing errors between the (regular) clock in the GPS unit and the atomic clocks of the satellites: the distance to a fourth satellite is sought and a little trick of geometry (and algebra) is used to calculate the fix as well as to synchronize the GPS's internal clock with the atomic clocks in the satellites. The implication is that every GPS unit (yes, even your $100 hand-held unit) is also an atomic-accuracy clock. As mentioned earlier in the notes of chapter 4, one of the reasons for making the GPS system, which was developed and maintained by the U.S. military, accessible to commercial operators, was the navigation error that led to the downing of KAL 007. There are many similarities between the two accidents, which suggest that these navigational snafus are not only about the accuracy of the data, but also about user interaction with automated devices.

Chapter 9

Walk-in interfaces are pervasive and widespread. They are encountered on a daily basis by all of us wherever we turn, and not just by avid Internet surfers. We all use the telephone to call our bank, our doctor's office, the nearby box-office, or our insurance agent. We no longer hear a human voice on the other side of the line, but rather an automated answering system that guides (or misguides) us to our destination. This also is a "walk-in interface"—vocal rather than graphic. These automated voice answering systems often drive us up the wall. They tend to provide us with multiple and sometimes unclear options, offer menus that are annoyingly deep and multilayered, and when we finally understand where we missed the turn, give us no option for recovery. And when we just want to speak to a human on the other side of the line—not a chance. The system is set up to discourage us from reaching them. How often do we get frustrated and ultimately hang up just to retry our luck again?

There are four guiding principles that we can identify in a good walk-in interface: *Orientation, guidance, clarity of options* and, finally, *recoverability*. Orientation deals with where we are. A good interface must enable us to know, at all times, exactly where we are in the system (or interface). This is precisely the principle that was violated in the airliner entertainment system that we described earlier. The initial state remained obscure and the passenger had no idea what to do.

Guidance is about "where do we go from here?" or "how do we get there?" A good walk-in interface must enable us to know the answers to these questions from the interface or from our population stereotypes. It must not assume that we will learn by trial and error. If the designer wants to spare the trained and experienced users the added overhead, an override switch can be provided as an option for eliminating the guidance tips.

Clarity of options refers to description of choices. What can we, or must we, do next? The options must be clear and well defined. In walk-in interfaces, the user must not be assumed to have prior knowledge. He must be guided unambiguously by the interface. The internet visitor information system violated this second principle in that it did not specify clearly the implications of the various termination options.

Recoverability deals with what recourse is provided after making a mistake or taking the wrong turn. A good interface should minimize the damage. It must not leave us hanging there helplessly. It must not delete our previous work when we return from an unwanted detour and must guide us smoothly back to course. To conclude, while walk-in interfaces slowly mature and are getting progressively better, they are still frequently problematic for us. Designers are learning from gained experience. But not every mistake must be first made just to be corrected later.

Chapter 10

The strong, albeit reversal, relationship between the design of displays and magic is described in eloquent detail by Edward R. Tufte in his book *Visual Explanations: Images and Quantities, Evidence and Narrative* (Graphics Press, 1997). See chapter 3, "Explaining Magic: Pictorial Instructions and Disinformation Design" for a tour de force on this topic.

As another aspect of this theme, note that the first part of the chapter deals with how web designers can exploit population stereotypes to the advantage of the web-page owner rather than the user. The web page is designed to serve its owner, who is benefiting from "exploiting" the user. This role reversal is significant, since we have always assumed that an interface is designed to serve the user, and suddenly . . . woops—it's the owner who is being served! But indeed, this is what, for example, advertising is all about. We do not turn our TV on just to watch the "new and improved suds campaign," we came to watch our favorite police drama. But the advertiser foots the bill of the programs we like. Ultimately, the interface designer must serve the owner without losing the user—not a simple challenge.

With respect to the problem with the bill of sale, the reader may be interested to know that there is a way out of the perpetual lock, albeit not an elegant one. Click another "department" in the e-boutique—doesn't matter which—video, posters, books, whatever. Go there; add an item even though you have no intention of buying it—say, a poster—to the shopping cart. When you get to the shopping cart page, remove the poster, and then proceed to checkout and you will find yourself at the white form-filling page. All your previously entered information is there, including the problematic email. So update the e-mail carefully, and continue on to the bill of sale and finally, after a long detour, pay.

Chapter 11

Gregory Bateson was the son of the British biologist William Bateson (who coined the term "genetics" and was one of the pioneers of the field), and the husband of the renowned (yet very controversial) anthropologist Margaret Mead. His scientific work crossed the disciplinary boundaries of biology, anthropology, psychology, education, and ecology. One of his notable contributions is the notion of "double bind" in psychiatry, resulting from his study of alcoholism and schizophrenia at the Veterans Administration Hospital in Palo Alto, California. Appointed by Governor Jerry Brown in 1976, Gregory Bateson served on the Board of Regents of the University of California until his death in 1980. In his last two years he was Scholar-in-Residence at Esalen Institute in Big Sur, California, where he had a lasting influence on many innovators in diverse fields such as dance, mythology, and psychotherapy.

Bateson played a major role in the early formulation of cybernetics—the science of communication and control in animals and machines. An interdisciplinary field of study, cybernetics (derived from the Greek word for a steersman on a boat), was first introduced by the mathematician Norbert Wiener and grew out of the work of Claude Shannon on information theory and the related engineering discipline of control theory (see endnotes for chapter 1). Bateson was instrumental in introducing cybernetics concepts into the work of social and natural scientists. His seemingly simple definition of information, which on consideration is quite insightful, is based on his work in the area of communication and cybernetics. (For a down-to-earth and very clear treatment of many of his scientific principles see *Mind and Nature: A Necessary Unity*, 1979, Dutton.)

The cruise-control incident described in this chapter took place in Atlanta, Georgia, several years ago. I was the driver. Driving north on Interstate 85, in heavy traffic and rain, I engaged the cruise control and it activated at a speed of 40 miles per hour. Once the rain ceased and the traffic opened up, I pressed on the gas pedal and drove on as usual for about half an hour (completely forgetting that the cruise control was engaged in the background). As I came off the highway at North Druid Hill exit, I let go of the gas pedal and the car began to slow down. Approaching the large intersection at the top of a small hill, the car jerked forward unexpectedly—I ran the intersection at 40 miles per hour and in *red* lights. It was my lucky day.

Although uncommon in Europe and in most of the world, almost every mid-sized car and above sold in the United States and Canada comes equipped with a cruise-control system. Along the multilane and mostly straight highways that connect distant cities in the United States (the Florida Turnpike and I-95 come to mind), cruise control is an important aid for the modern traveler. There is a variety of interfaces for cruise-control systems on the market.

Most small- and mid-sized cars do not have any display indications about the modes of the cruise control. In most Japanese-made cars, and in some American-made cars, there are cruise ON and also cruise ENGAGED indications. But as the evaluation details, there is still a missing indication: the interface does not display to the driver when the cruise control is in CANCELED mode. As a consequence, the driver cannot tell what will be the response of the system after he or she presses "resume": sometimes, "resume" will cause the system to be ENGAGED, sometimes not. The only way to know is by trial-and-error.

Ralph Teetor, an engineer and a successful businessman, is the inventor of the cruise control. Oddly enough, Ralph Teetor was blind (from a work-related accident when he was a boy) and of course could not (legally) drive. Mr. Teetor mentioned on numerous occasions that the inspiration for his invention came while driving with his patent attorney, who was the jerkiest driver with whom he had ever driven. The story goes that the attorney had a tendency to slow down while talking and speed up while listening. The resulting accelerations and decelerations so annoyed the blind engineer that he set out to devise the "Speedostat."

By 1941 he had an operational prototype, but not much enthusiasm from any car manufacturer. Then the United States entered into the Second World War, and the government imposed a national speed limit of 35 miles per hour in order to conserve fuel and tires. People found it boring and difficult to maintain a constant speed during long travel. The invention became practical for many—especially for truck drivers. The initial design of the Speedostat was quite different from today's: using a dial on the dashboard, the driver would set the desired speed. The driver would then accelerate manually (foot on gas pedal), until reaching the set speed. At that speed, the driver would feel a resisting pressure from the gas pedal, against which he would hold his foot to maintain the set speed. To occasionally increase the speed beyond the limit, such as while passing another car, the driver had to press through the resistance on the gas pedal.

Customer surveys and early road tests made it clear that drivers of both cars and trucks wanted a "lock in" feature that would allow them to take their foot off the gas pedal. The inventor resisted because of safety concerns, namely that the drivers may become distracted, panicky, or fall asleep while using the "Stat." After the major car manufacturers demanded it, Teetor designed a small electric motor that would hold the speed and relieve the driver.

The first systems appeared in the late 1950s on the Chrysler Imperial, New Yorker, and Windsor models with the marketing name "Auto-pilot." In 1959, Cadillac offered the "cruise control" as optional equipment on all their cars, and the new name stuck. Later, both Chrysler and Cadillac asked for an automatic acceleration that would function after the driver braked and then wanted to return to the pre-set speed. Ralph Teetor flatly refused to add the "resume" feature out of his concern that it would only bring accidents. Chrysler engineers added the resume feature anyway.

In the end, it was another gasoline crisis that finally sealed the success of the invention. During the oil embargo of 1972-1973, the U.S. government imposed a speed limit of 55 mph. Car manufacturers made the cruise control an optional accessory in almost every car sold in the United States. Ralph Teetor's life story is told by his daughter, Marjorie Meyer-Teetor, in a book titled *One Man's Vision* (Guild, 1995).

Today, cruise-control systems are evolving at a fast pace. On the near horizon are adaptive cruise-control systems, which are devices that automatically adjust the speed to avoid collision with the car ahead or any other object it can detect. The system has a radar mounted on the car's grill, and the radar can detect objects up to 500 feet away. With data inputs from the radar and the car, the adaptive cruise control automatically decelerates to maintain a safe distance from the lead car. If the situation is dire, the system has full authority to apply breaking. Once safety is assured, the system re-engages automatically and resumes to the set speed.

Chapter 12

The incident described in this chapter took place several years ago at a large teaching hospital. The incident is detailed in a chapter, titled "Automation in anesthesiology," by David Gaba, M.D. which appeared in a book titled *Human Performance in Automated Systems: Current Research and Trends* by Mustapha Mouloua and Raja Parasuraman (Erlbaum, 1994, pages 57-63). The incident was not an isolated case, similar incidents have occurred in the past and the problem is known to many anesthesiologists. After the incident was investigated, the hospital management decided to remove this particular model of blood-pressure machine from all surgeries and wards.

In writing this chapter, I have relied on several academic publications on the topic of human factors in medicine. The book *Human Error in Medicine* (edited by Marilyn Bogner and published by Erlbaum, 1994) provided background information, as well as *Under the Mask: A Guide for Feeling Secure and Comfortable During Anesthesia and Surgery* by Dr. James Cottrell and Stephanie Golden (Rutgers University Press, 2001).

Chapter 13

The aircraft accident described in this chapter occurred several years ago. The factual information is based on the cockpit voice recorder and flight data recorder. The non-factual description and the painting of the scenes are based on my own flying experience. The actual evaluation of this procedure and the synchronization problem is somewhat more complicated than presented here, yet the results are the same. In writing this chapter I drew on previously published work conducted with my former advisor, Professor Earl Wiener, on the use and design of procedures. (See Asaf Degani and Earl Wiener, *The Human Factors of Flight-Deck Checklists: The Normal Checklist,* NASA Contractor Report number 177549, published in 1990; and *On the Design of Flight-Deck Procedures,* NASA Contractor Report number 177642, 1994.)

Problems in arming spoilers for landing have occurred in the past and contributed to many incidents and to a few accidents: In 1999, an American Airlines MD-80 aircraft crashed while landing on a wet and slippery runway in Little Rock, Arkansas. The American Airlines pilots also forgot to arm the spoilers before landing. Once they landed, as much as they tried, they were unable to stop the aircraft before the end of the runway. The aircraft overran the runway and broke in half, killing the captain and ten of his passengers. The full report on the American Airlines Flight 1420 accident can be obtained from the National Transportation Safety Board (NTSB) or downloaded from their web site (*American Airlines Flight 1420, Runway Overrun During Landing, Little Rock, Arkansas, June 1, 1999*; NTSB report number AAR-01/02).

Premature deployment of spoilers has also occurred in the past. On July 5, 1970, a McDonnell Douglas DC-8-63, operated by Air Canada, was making an approach to Toronto-Pearson International Airport. Sixty feet above the runway, the aircraft all of a sudden began to sink rapidly. The right outboard engine was torn off the aircraft in the subsequent heavy landing. The crew initiated a go-around and climbed to 3,000 feet. Then, a large piece of the right wing separated from the aircraft. The DC-8 stalled and crashed, killing all 109 people on board. The investigation report declared that the probable cause of the rapid descent while the aircraft was close to the runway was premature deployment of spoilers. As a result of this accident, the U.S. Federal Aviation Administration issued an Airworthiness Directive cautioning pilots against in-flight operation of ground spoilers by requiring the installation of a warning placard in the cockpit and the insertion of an operating limitation in the aircraft's flight manual cautioning pilots not to use ground-spoilers in flight.

The general topic of procedure correctness is rather involved and beyond the scope of this book. But generally speaking, a procedure is a (conditional) sequence of actions aimed at driving a system safely from an initial state (e.g., lowering the gear) to a desired end state (e.g., gear is locked in place and spoilers are armed). Since the user's actions frequently interact with the system's and the environment's responses, the user's actions are conditional on these responses. To this end, one requirement of a correct procedure is that there will be no known situations in which an automatic event or a disturbance will unexpectedly drive the system to an unsafe state (e.g., as in the case of premature spoiler deployment in mid-air). For an additional example and discussion of correct procedures, see Asaf Degani, Michael Heymann, and Michael Shafto, "Formal Aspects of Procedures: The Problem of Sequential Correctness," a short paper that appeared in the 1999 *Proceedings of the 43rd Annual Meeting of the Human Factors and Ergonomics Society.*

When timing constraints are involved in the procedure's execution, the correctness of a procedure is more subtle, because a specified sequence of actions might become unattainable—thereby disrupting the sequence (and possibly forcing the user to skip over an essential operation). It is also important to note that when executing a procedure that involves a dynamical system, the user and the system are in a race against time, an issue that becomes paramount when considering emergency procedures such as in-flight fires.

Finally, just like in interface design, making sure that the procedure is correct and safe is only the first step. For the procedure to be correct and also useful and suitable for the user, the procedure must be further refined. Consideration must be taken so that the procedure steps are clear to the user. Namely, that the user actions prescribed in the procedure leave no room for ambiguity. Likewise, that information such as values and state of the system are easily obtainable from the interface and are not subject to interpretation. Additional considerations are about the relationship between the sequence of the procedure and the layout of a control panel, how the procedure fosters coordination in a multi-person crew, and how the status of the procedure (awaiting a response, completed, needs to be executed again) is shared among the crew members.

Chapter 14

This year, 2003, marks the centennial of powered flight. On December 17, 1903, after many unsuccessful trials, Wilbur and Orville Wright made a series of engine-powered flights and realized one of humanity's greatest dreams—to fly. The famous picture showing Orville flying several feet above the ground, and Wilbur running in chase, have us all suspended between our primal connection to the earth and our desire to reach for the stars.

The aircraft speeding incident described in the chapter took place several years ago. I took the liberty of simplifying some of the details regarding the autopilot and flight management computer, so as to make it more understandable and readable. The full details exist in my Ph.D. dissertation, *Modeling Human-Machine Systems: On Modes, Error, and Patterns of Interaction* (Georgia Institute of Technology, Atlanta, 1996) and in an article titled "Modes in Human-Machine Systems: Review, Classification, and Application," by Asaf Degani, Michael Shafto, and Alex Kirlik, which was published in 1999 in the *International Journal of Aviation Psychology* (volume 9, issue 2, pages 125-138).

The term "automation surprises" was coined by Nadine Sarter and David Woods in the early 1990s following their work on the human factors of automated cockpits. This work has received a lot of attention and helped designers and pilots to better understand the problems involved in operating complex automated systems. (See the chapter titled "Automation Surprises," by David Woods, Nadine Sarter, and Charles Billings, in the *Handbook of Human Factors and Ergonomics,* published in 1997 by John Wiley.) Another related term that is sometimes used to describe such human-automation interaction problems is called "mode error." The term was coined by Donald Norman in the early 80s to categorize a special type of human error and is described in *The Design of Everyday Things* (2002, Basic Books). The importance of feedback to the pilot about the state of the automation is stressed in a classic article by Donald Norman, published in 1990, titled "The 'Problem' with Automation: Inappropriate Feedback and Interaction, Not 'Over-Automation'" (*Philosophical Transactions of the Royal Society of London,* B-327, 585-593).

There are several other (speed-related) problems with the interface to the autopilot described in this chapter. Look again at figure 14.6 and note that while the phone's design has the automatic transition only in one direction (from OFF to ON), the speed source in the autopilot has an automatic transition in both directions. What's the significance of this? Well, it makes the user interaction more vulnerable to error, because the user may get confused going in either direction—from manual to automatic or from automatic to manual operation.

And indeed, there have been several reported incidents in which pilots engaged the VERTICAL NAVIGATION mode and entered a series of new speed limits into the flight management computer, anticipating that the computer would change the speed automatically at specific waypoints during a descent. Then air traffic control instructs the crew, for example, to hold at a certain altitude. The pilot switches to ALTITUDE HOLD, but in his or her mind, still assumes that all the speed limits so laboriously entered into the computer will be honored. However, since the autopilot is not in VERTICAL NAVIGATION mode, the speed input no longer comes from the flight management computer. Instead, as we already know, engagement of the ALTITUDE HOLD MODE causes a transition to the flight control panel as the source for the speed reference value. But since there is no speed value there (the pilot did not enter anything), the autopilot defaults to the aircraft's speed at the moment of mode switching, which, in most cases, is very different from the speed limits entered by the pilots into the computer. If the crew does not catch this discrepancy in time, it can easily result in a speed violation.

Chapter 15

This chapter is based on a preliminary report that was published on July 28, 1994, four weeks after the aircraft accident, as well as on several trade publications and articles on the topic. The accident report was reprinted in its entirety in the *Aviation Week and Space Technology* magazine (1995, volume 72, issues 14, 15, 16, 20, 21, 22). As mentioned in the chapter, the preliminary report leaves gaps in our understanding of the accident (the final report was never made publicly available). I have simplified some of the technical details of the autopilot, aerodynamics, and envelope protection system's logic so as to make the chapter more readable. In describing the accident, I took the liberty of adding non-factual information from my own familiarity with automated aircraft.

Following the test-flight accident, many procedures and practices for flight tests were re-examined and changed. In addition, new certification rules for conducting test flights specify that in order to test the ALTITUDE CAPTURE maneuver, the maneuver should only be attempted at an altitude of at least 8,000 feet, which provides ample altitude for recovery.

The problem of over-trust on automation became a research agenda in the late 1980s and early 1990s. Considerable work has been done to understand this problem in the context of aviation and medical systems. The problem, however, is not unique to medicine and aviation—it exists in every domain where automated control systems are supervised by humans. The interested reader can find additional information in the work of Kathleen Mosier and Linda Sitkaka as well as in a recent article by John Lee and Katrina See, "Trust in Automation: Designing for Appropriate Reliance" (which will be published in the *Human Factors Journal* in 2004). The importance of the decision to disengage or keep using an automated system is discussed in an article by Raja Parasurman and Victor Riley, "Humans and Automation: Use, Misuse, Disuse, Abuse," which was mentioned in the endnotes for chapter 1.

Crew Resource Management is now a mature concept. When first introduced, it signaled a major shift in emphasis away from the maverick, individualistic pilot who has the "right stuff," to a sensitive and well-attuned manager of a crew. The shift in emphasis was traumatic to many pilots who were brought up in the old school and could not adjust to the new. However, after 20 years, it is clear that a cultural change has taken place in the aviation industry, and indeed, it was for the best. The concept was initially dubbed as *cockpit* resource management, and was later modified to *crew*, when others, from surgical teams to maritime bridge officers and power-plant technicians, began adopting and using the concept in their training and operations. The initial work on Crew Resource Management was conducted by Hugh Ruffell Smith and John Lauber (see *Cockpit Resource Management* by Earl Wiener, Barbara Kanki, and Robert Helmreich, published by Academic Press, 1997).

A very emotional debate about the role of the computer and the pilot in controlling modern aircraft has been going on for years in the aviation community. The debate only intensifies around the topics of envelope protections and who (machine or pilot) has final authority. This debate has many implications beyond aviation, because automated machines are now implemented in many safety-critical systems, such as medical technology and nuclear power. There are several excellent publications and books on the human factors of cockpit automation: Earl Wiener and Renwick Curry, "Flight-deck Automation: Promises and Problems" (published in the journal *Ergonomics*, volume 23, in 1980), and Charles Billings' *Aviation Automation: The Search for a Human-Centered Approach* (published in 1996 by Erlbaum). The most current book on the topic is Thomas Sheridan's *Humans and Automation* (published in 2002 by John Wiley and Sons).

Chapter 16

Semiautomatic transmission systems, such as the one presented here, can be found in older buses and construction equipment. Modern transmission systems are somewhat more sophisticated: they have an automatic feature that down-shifts between modes. Think of this as a kind of protection system: if the user fails to down shift when the bus is already at the highest state (e.g., high-1) and the speed is still dropping, an automatic transition into MEDIUM will take place so as to prevent the vehicle from stalling. A similar transition is also present between MEDIUM and LOW. Note that it is practically impossible for the driver to predict the onset of such automatic transitions, because they are based on a variety of conditions that are hidden from the driver. In a situation like this, it is possible to provide the driver with some indication just before the automatic transition takes place. The driver can then decide to increase speed or shift down (either manually, or by letting the transmission do it automatically).

The autopilot example presented in this chapter is based on an article by Asaf Degani and Michael Heymann entitled "Formal Verification of Human-Automation Interaction," which appeared in the *Human Factors Journal* (Volume 44, issue 1, pages 28-43). The *Wall Street Journal*, in an article titled "Cold, Hot, Cold, Hot: Employees Only Think they Have Control" (January, 15, 2003, page B-1), reports that Heating Ventilation and Air Conditioning (HVAC) technicians admit to what office workers suspected for years: that many office thermostats are fake. The estimates for the percentage of "fixed" thermostats ranges from 2 percent according to one technician to 90 percent according to another. Such thermo-fraud has been going for some 40 years—only increasing when utility prices go up.

Beyond just verification of a given (e.g., prototype), it is possible to extend the concepts of error, augmenting, and restricting states to a systematic methodology for generating correct interfaces. That is, a process that takes into account the machine's behavior and the user's task requirements and generates an interface solution. Further, it is possible to use this process not only to generate a correct interface, but also one that is the simplest possible. Such a methodology exists, and we have already used it in this book. In particular, the methodology and algorithm for generating correct and succinct interfaces follows the same steps that we took in redesigning the cruise-control system in chapter 11. (For a detailed discussion on this topic see Michael Heymann and Asaf Degani, *On Abstractions and Simplifications in the Design of Human-Automation Interfaces*, NASA Technical Memorandum number 211397, which was published in 2002). For a shorter and lighter treatment of this topic, see Asaf Degani and Michael Heymann, "Analysis and Verification of Human-Automation Interfaces," which appeared in the Proceedings of the 10th International Conference on Human-Computer Interaction, June 22-27, 2003.

Chapter 17

The examples, analysis, and methods presented in this chapter are based on a recent report on analysis and verification of automation systems and interfaces (see Meeko Oishi, Claire Tomlin, and Asaf Degani, *Safety Verification of Hybrid Systems: Applications for User Interface Design*; NASA Technical Memorandum, 2003). As for the yellow light interval, see an article titled "A Review of the Yellow Interval dilemma," by C. Liu, R. Herman, and D. Gazis, which appeared in a journal called *Transportation Research A*, volume 30, issue number 5, 1996. With respect to entering an intersection, the California Vehicle Code (of 2002) states the following rules (which I edited here for comprehension):

- A driver facing a yellow light is, by that signal, warned that the green light period is ending or that a red light will be shown immediately thereafter (regulation 21452(a)).
- A driver facing a red light alone shall stop at a marked limit line (regulation 21453(a))

Since the regulation says nothing about entering the intersection on yellow and encountering a red while the car is inside, the situation is open for interpretation. The common interpretation used by police officers and judges is that you can enter the intersection when the light is yellow (regulation 21452(a) does not prohibit entering—it only says that you are warned of an impending red) and therefore it is okay if the light changes to red while you are inside the intersection.

Many of the examples discussed in this book have distinct, or discrete, modes as well as continuous parameters (reference values). Such systems, which have both discrete and continuous dynamics, are called *hybrid systems*. In the last decade or so, a theory and methodologies for analysis and verification of such systems have been developed. The theory extends the classical notion of a finite-state machine to add continuous dynamics such as time, speed, pressure, and more. These methods are used in the analysis of systems that have complicated dynamics, such as collision avoidance systems for aircrafts and also for cars. As shown in this chapter, these methods can also be extended to the analysis and verification of human-automation interaction and interface methods.

There is an implicit assumption, especially when it comes to designing interfaces, that the underlying system works per specifications. Assuming that the underlying system behaves as specified is imperative for the analysis and verification of interfaces, because otherwise it is impossible to analyze an interface and come to any meaningful conclusions. Therefore, one of the objectives of this chapter was to highlight the fact that interface analysis and verification can *only* take place after there is an assurance that the underlying system behaves as specified and all the unsafe regions are understood and accounted for. In other words, the model of the system must be sound.

To further illustrate this point, say that you are asked to develop a new system for aiding drivers. Specifically, you want to provide guidance to the driver about whether he or she should stop at a yellow light or proceed through an intersection. You can think about a display on the dashboard that will indicate GO if the car is in the safe "proceed through" region and STOP if in the "safe braking" region. But if at certain speeds and distances from the intersection, the driver will commit a violation regardless of whether he or she stops or goes, then, of course, the interface is useless.

But even if you perform the above analysis and identify the unsafe region, just indicating it to the driver is not enough. Why? Because it's too late! In many respects it is like telling a diver who has already jumped off the platform that there is no water in the pool. I hope you'll recognize that in these situations the problem is beyond the interface. The only way to deal with the problem is to provide an advisory system that monitors the car's speed and distance with respect to the unsafe region. When the driver is *about* to enter the unsafe region, the system can warn the driver to change speed and avoid entry.

For a more detailed description of the Boeing 747 incident, see *China Airlines Boeing 747-SP, 300 nautical miles northwest of San Francisco, California; February 19, 1985.* This accident report was published in 1986 by the National Transportation Safety Board (NTSB report number AAR-86/03). I obtained the information regarding the commuter aircraft accident from the NTSB accident report (*In-flight icing encounter and loss of control, Simmons Airlines, doing business as American Eagle Flight 4184. Roselawn Indiana, October 31, 1994.* NTSB Report number AAR-96/01) as well as several trade publications on this accident.

Chapter 18

The Luddites claimed to be led by Ned Ludd, also known as "King Ludd" and "General Ned Ludd." It is believed that he was responsible for the destruction of two large knitting machines that produced inexpensive stockings, and his signature does appear on a "workers manifesto" of the time. Whether or not Ludd actually existed is unclear; it is also well documented that acts of machine wrecking took place prior to Ludd's appearance on the scene.

Some historians argue that the Luddites, most of them skilled knitters and weavers, were enraged by the use of unapprenticed workmen, wage reductions, and the inexpensive wool products that were coming out of factories—undercutting those produced by traditional (small) workshops that were becoming obsolete. Often characterized as violent, thuggish, and disorganized, historians believe that the Luddites were mostly normal people in the community who were protesting against forced introduction of damaging changes in their lives. Although the aim of the protest were factories, warehouses, and machines, the impetus for the riots was the abolition of pricing defined by custom and practice (and the introduction of what we would today call "free market"). (See H. E. Hobsbawm, *Labouring Men: Studies in the History of Labour*, published in 1964 by Basic Books.)

A variety of task analysis techniques exist today and many of them are applicable to human interaction with automated systems (see Kim Vicente's *Cognitive Work Analysis: Toward Safe, Productive, and Healthy Computer-Based Work*, published by Erlbaum in 1999, and *A Guide to Task Analysis* by Barry Kirwan and L. K. Ainsworth, published by Taylor & Francis in 1992). As for guidelines for the designing of human interaction automated control system, see Charles Billings' *Aviation Automation: The Search for a Human-Centered Approach* (published in 1996 by Erlbaum).

In addition to his long career and many contributions in physics, Richard Feynman also served on the committee investigating the explosion of the space shuttle *Challenger*. He caught the public's attention with his simple and elegant demonstration of a small rubber "o" ring in a glass of ice water, which helped all of us understand how the freezing temperature at the day of launch (January 28, 1986) made the large "o" ring of the rocket hard and brittle and allowed a fuel leak that enabled the explosion.

Richard Feynman recognized, taught, and wrote that scientific knowledge is the enabling power to do either good or bad. He was also very concerned about scientists' social responsibility and strongly advocated that scientific concepts and ideas should be made available to the lay audience and not kept behind an academic shroud. As for the Buddhist quote, see "The Value of Science," a public address given in the 1955 meeting of the National Academy of Sciences, which is reprinted as the concluding chapter in his last book (*What Do You Care What Other People Think?: Further Adventures of a Curious Character*, W. W. Norton, 2001).

Index

Aboriginal art, 13, 285-6
abstraction, 6, 31, 117, 255, 278, 280; in autopilot user model, 249, 252; hierarchy and, 290; non-deterministic interface and, 161-2, 240, 278; in underground map, 24-6, 286-7. *See also* non-determinism
active *vs.* engage mode, 97, 280; in blood-pressure machine, 154, 156, 157-60
advertising, Internet, 135-40, 142, 141, 295
Air Canada crash (1970), 298-9
air-conditioner unit, 73-8; hierarchy in, 73-6; synchronization in, 76-8. *See also* climate-control system, in car
aircraft: automated control systems for, 2, 201-3; computers on, 4, 202, 208, 209, 210, 270, 300; lighting systems, 19-20; personal entertainment systems on, 121, 122-7; Soviet fighters, 52, 53; U.S. reconnaissance, 50, 52, 53, 65, 289; in World War II, 293. *See also* autopilots; procedures, in aircraft, *and specific aircraft*
aircraft, protection systems for: *See* envelope protection systems
aircraft flight manual, 249, 288, 299
aircraft landing systems. *See* instrument landing system (ILS); landing systems, aircraft
aircraft test flight. *See* test-flight accident
air traffic controller, 219, 262
ALARM mode: in clock radio, 95-7, 287-8; in digital watch, 40, 43, 45, 47, 276
ALPHA-PROTECTION mode, 217, 222, 230, 236
ALTITUDE CAPTURE mode, 206, 208; autopilot verification, 244-53; interface and user model, 247-9; machine model, 246-8; non-deterministic interfaces and, 252-3; test-flight accident, 221, 226, 234-5, 301
ambiguity, in human-machine interactions, 38, 240, 278. *See also* abstraction; non-determinism
American Airlines Flight 1410, 298
"Analysis and Verification of Human Automation Interfaces" (Degani and Heymann), 302

anchor, initial setting as, 124, 127, 132
Anchorage Airport, 50
Andreessen, Marc, 127
Andropov, Yuri, 52, 289
anesthesia (anesthesiologist), 164, 166, 168-9, 171-2, 298. *See also* blood-pressure machine
annunciator panel, 182, 183
antenna, GPS, 102, 103
antilock braking system (ABS), 218
artificial horizon display, 179-80, 183
Asia Rip (AR) buoy, 112, 113, 294
ATMs (Automated Teller Machines), 36, 37, 121, 134, 284; default conditions in, 126; PIN for, 57
augmenting state, 254
autoland. *See* landing systems, aircraft
automated control systems, 1, 2, 7, 120, 237; understanding, 274-5. *See also* autopilots
automated devices, frustration with, 1-2. *See also specific devices*
automated systems, complexity of, 4
automated systems, humans and. *See* human-automation interaction
automated tellers. *See* ATMs
automatic flight control systems, 201-3. *See also* flight management computer; instrument landing system (ILS)
automatic pilot. *See* autopilot
automatic transitions, 59, 62, 279, 301-2; in cruise control, 151, 152, 161; in digital watches, 40, 43
automation: defined, 284; over-trust in, 120, 232-3, 281, 282, 294, 301
"Automation in Anesthesiology" (Gaba), 298
automation locks, 254-5, 267-73, 282, 297; in aircraft accidents, 269-72; in automobiles, 268-9
automation surprise, 206, 300
automobiles (cars): audio system settings, 125; braking in, 151, 152, 218, 256-7, 269; display icons in, 92, 160; non-deterministic interface in, 48; red-light violation and, 256-60, 268-9, 302-3; standardization in, 92-3; transmission system for, 238-43, 244, 301-2. *See also* climate control system; cruise control systems

Justin -

It has been a pleasure
having you in class for
the past two years. I am
confident that you will
thrive at Cornell. Best of
luck and stay in touch.

John Khalgir

PAVLOV'S DOG

METRO BOOKS
New York

An Imprint of Sterling Publishing
1166 Avenue of the Americas
New York, NY 10036

METRO BOOKS and the distinctive Metro Books logo are trademarks of
Sterling Publishing Co., Inc.

Interior design and illustrations: Jason Anscomb, Rawshock design
Photo credits: Shutterstock.com 13, 14, 85, 127, 164.

ISBN 978-1-4351-6131-3

For information about custom editions, special sales, and premium and corporate purchases,
please contact Sterling Special Sales at 800-805-5489 or specialsales@sterlingpublishing.com.

Manufactured in China

2 4 6 8 10 9 7 5 3 1

PAVLOV'S DOG

GROUNDBREAKING
EXPERIMENTS IN PSYCHOLOGY

ADAM HART-DAVIS

METRO BOOKS

New York

Contents

Introduction

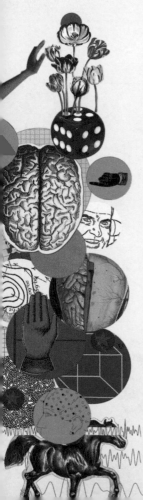

Can the human mind ever hope to understand the human mind? This may not be an impossible goal, but it is beset with difficulties, which is perhaps why psychology was such a latecomer among the sciences. There have been chemists and physicists (known as "natural philosophers") for hundreds of years, but the first psychologist announced himself less than 150 years ago.

People certainly thought about minds and behavior long before then; the ancient Greeks Plato and Aristotle wrote about the "psyche," which is where the word "psychology" comes from. Originally the word meant life, or breath, and later spirit or soul (Psyche being the goddess of the soul in Greek and Roman mythology). Now we use *psychology* to refer to all aspects of the human mind.

And what about that mind? How does it work and can we ever hope to understand it? In the sixteenth century the French philosopher René Descartes argued that the body and brain are machines, but we need a mind as well, to think, feel, and make choices or decisions. This view is known as Cartesian dualism and it pervades almost all aspects of psychology. Even as very young children we feel as though they are thinking, conscious "selves" inside our bodies. But the more we learn, the less plausible this seems to be, and right from the start psychology has struggled with this problem.

The first person to call himself a psychologist was Wilhelm Wundt, who founded a psychological research laboratory in Leipzig, Germany, in 1879. He is known as the father of experimental psychology (a branch of the science that focuses on empirical evidence gathered through experimentation rather than theory). The first textbook on the subject, William James's great *The Principles of Psychology*, appeared in

1890. The noted naturalist Charles Darwin didn't consider himself a psychologist, but he was captivated by what factors proved intelligence and made a 40-year study of the humble earthworm, and at the turn of the century Edward Thorndike explored the degree to which animals could learn and reason.

The early twentieth century, saw the beginnings of "behaviorism," which sought to explore only observable events using rigorous experimental methods and resist speculating about the subjective or invisible. In hindsight there were many ethical breaches in this period, and a number of studies remain highly controversial, but much was learned and the field psychology was greatly expanded by the likes of Ivan Pavlov and his discovery of classical conditioning.

Postwar, Jean Piaget's carried out groundbreaking studies of the cognitive development in children and Leon Festinger coined the notion of cognitive dissonance. The 1960s saw Stanley Milgram's obedience studies gain worldwide attention and in the 1970s Donald G. Dutton and Arthur P. Aron asked if there was a link between sexual attraction and fear. The science of psychology grew and grew, influencing numerous areas of our everyday lives.

In the following pages these great experiments, and many more besides, will carry us through the history of psychology as well as propelling us on a journey deep into the understanding of ourselves.

CHAPTER 1: Beginnings: 1848—1919

The idea of psychology was barely conceived during the nineteenth century, but Charles Darwin's groundbreaking explorations roused a great curiosity in animal behavior and what it might teach us about human beings. And after the publication William James's *The Principles of Psychology* (1890) this curiosity was to flourish and expand into a whole new science.

Over the following decades, Edward Thorndike would investigate whether animals could learn, and Ivan Pavlov would be awarded the Nobel prize for demonstrating that reflex responses could be trained and conditioned. These studies and others paved the way for the rise in interest in perception, behavior, and thought.

1881

THE STUDY

RESEARCHER:

Charles Darwin

SUBJECT AREA:

Animal behavior

CONCLUSION:

Earthworms display a basic
degree of intelligence

ARE WORMS INTELLIGENT?

DARWIN'S INVESTIGATIONS INTO THE INTELLIGENCE OF EARTHWORMS

Worms have no ears or eyes, so how do they function in their environment; is it through learning or instinct?

Charles Darwin was a superb naturalist, and his studies encompassed all sorts of animals, from tiny barnacles to giant tortoises. He was first encouraged to look at earthworms by his uncle Josiah Wedgwood in 1837, and continued to observe their behavior over the next 40 years in his garden at Down House in Kent. He described his findings in detail in the last book he wrote before his death in 1882: *The Formation of Vegetable Mold through the Action of Worms, with Observations of their Habits*. He called it "a small book of little moment," but thousands of copies were sold within a few weeks of publication.

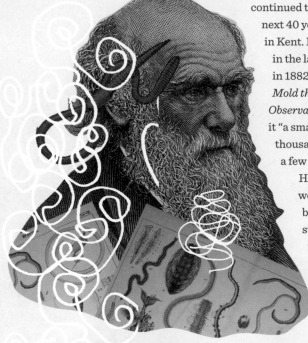

He was fascinated by the fact that worms bring soil to the surface, thus burying other things, which is why stones sink into the ground. He installed in his garden a worm-stone, which is still there. He went by railroad to Stonehenge and drew diagrams showing that some of the great fallen stones had sunk between 4 and 10 inches into the ground.

A family affair

Charles Darwin was a family man and loved being in the garden with his children. He used them as his research assistants, lining them up along the flower beds, and instructing them to note which bees were on which flowers when he blew his whistle. This unusual approach also helped Darwin to gather a lot of data in a short time.

He also enlisted his children to assist with his earthworm studies. He kept a number of worms in flowerpots, and asked his children to try and stimulate them. They tried shining lights on the worms, but having no eyes the worms paid no attention until the lights were extremely bright, and even then only when the light shone on their back ends.

The children blew whistles, shouted at them, and played the bassoon and the piano, but the worms showed no interest. However, if the worms were actually placed on the piano, they reacted immediately when a key was pressed; presumably they could feel the vibration coming through the instrument, even if they could not hear the note.

Instinct or intelligence?

The feature that struck Darwin most forcibly, however, was the apparent intelligence shown by the worms. They had the habit, outside, of pulling leaves into the mouths of their burrows.

> *Worms seize leaves and other objects, not only to serve as food, but for plugging up the mouths of their burrows; and this is one of their strongest instincts.... I have seen as many as 17 petioles of a clematis projecting from the mouth of one burrow, and 10 from the mouth of another. Hundreds of such plugged burrows may be seen in many places, especially during the autumnal and early winter months.*

What surprised him most was that they almost always pulled the leaves in by their tips—and since the worms had no eyes, he wondered how they could they find the tips of the leaves. He reasoned that if the worms acted entirely through instinct or chance, they might pull them in randomly. But otherwise they must use intelligence. He pulled 227 withered leaves from worm burrows; 181 of them (80 percent) had been pulled in by their tips, 20 had been pulled in by the base, and 26 by the middle.

He and his son Francis tried cutting the tips off some leaves, and later observed that the worms had instead pulled the majority of them into their burrows by their stalks. They also did various experiments with other leaves and pine needles, and concluded that the worms always seem to choose the easiest option.

To test his theory further in a controlled experiment, Darwin cut pieces of stiff writing paper into elongated triangles, so that they were similar in shape to leaves. He then used tweezers to pull these triangles into a narrow tube. When they were pulled by the apex—the narrowest corner— they went straight in with the sides curling up to form a neat cone. When they were pulled by a point away from the apex, the task was harder and more of the triangle folded back within the tube.

Next they rubbed dozens of these paper triangles with fat to prevent them from disintegrating in the dew, and scattered them around his lawn. Over several nights Darwin observed that 62 percent of the paper triangles that the worms had pulled into their burrows, were pulled in by the apex, and the proportion was even higher for narrower triangles.

Darwin and his children carried out hundreds of these experiments, and he came to a firm conclusion:

> *If we consider these several cases, we can hardly escape from the conclusion that worms show some degree of intelligence in their manner of plugging up their burrows.*

CAN YOU LIVE LIFE UPSIDE-DOWN?

HOW OUR BRAINS INTERPRET WHAT WE SEE

1896
THE STUDY
RESEARCHER:
George Stratton
SUBJECT AREA:
Perception
CONCLUSION:
Our brains carry out a form of perceptual adaptation that allows us to function when what we "see" isn't as it seems.

When you look at something, its image is projected onto your retina upside-down (as it does on the sensor or film in a camera). In the late nineteenth century, the prevailing scientific theories suggested that this must be necessary if we are to "see" things the right way up. However, George Stratton, a professor at Berkeley in California, questioned the current thinking, and wondered whether it is possible to live one's life with the whole visual field upside-down. He set about constructing a pair of mini-binoculars that turned everything he saw upside-down, so that the image would appear on his retina the right way up, or "upright," as he put it.

Turning the world over
He placed two convex lenses of equal refractive power in a tube at a distance equal to the sum of their focal lengths. Looking through the tube turned everything upside-down. He fitted

BELOW: The object you look at casts an upside-down image on your retina. Your brain turns this back the right way up.

together two tubes, one for each eye, and strapped the whole contraption to his head. He was careful to exclude all other light, using black cloth and pads round the edges of his device. He wore it continuously for ten hours, then shut his eyes while he removed it, and put on a blindfold so that he could see nothing. He spent the night in complete darkness.

The next day he repeated the process, wearing his device all day and taking care not to see anything without it. The instrument gave him a clear field of vision and was reasonably comfortable to wear. At first he hoped to use both eyes together, but coping with two separate images was difficult; so he covered the end of the left tube with black paper, and used his right eye alone.

To begin with everything seemed to be upside-down. The room was upside-down; his hands, when raised into sight from below, appeared from above. Yet although these images were clear, they did not at first seem real, like the things we see in normal vision, but felt as if they were "misplaced, false, or illusory images." Stratton observed that his memories of normal vision still continued to be the "standard and criterion of reality" that his brain used to understand what was put before his eyes.

Memory or reality

When attempting to move around while wearing the contraption, Stratton at first blundered and stumbled. It was only when his actions were aided by touch or memory—"as when one moves in the dark"—that he was able to walk or perform hand movements with any degree of success.

Stratton concluded that his problems seemed to consist entirely of the resistance offered by experience, and reasoned that someone whose vision had been upside-down from the very beginning (or who had at least spent considerable time observing the world in this way) would not feel that this was unusual. Therefore he carried on this experiment for several days, and, by the seventh day, he reported feeling more at home in the upside-down scene than ever before, recording that there was by now a "perfect reality in my visual surroundings."

Getting used to the view

Despite the "perfect reality" of the upside-down world that he now inhabited, Stratton was still struck by just how difficult it was to operate in such an environment. Having mastered moving in the "wrong" direction he still found that his perception of depth and distance was flawed: "My hands frequently moved too far or not far enough...." In trying to shake a friend's hand he raised his own too high, or while brushing a speck from his paper he found he didn't move far enough. And he still observed that his hand movements were much less accurate when he looked at them than when he closed his eyes and depended upon touch and memory to guide him.

ABOVE: The upside-down image maybe disorientating, but your brain can still recognize the setting sun on a summer's evening.

Nonetheless, he was, however, gradually getting used to living upside-down, and during his walk that evening he was able to enjoy the beauty of the evening scene for the first time since the beginning of the experiment.

Stratton's overall conclusion was that it does not matter how images appear on your retina; your brain can learn to cope by using what has come to be described as "perceptual adaptation" to match your vision to your sense of touch and spatial awareness.

1898

THE STUDY

RESEARCHER:
Edward Thorndike

SUBJECT AREA:
Animal behavior

CONCLUSION:
There is no evidence that animals use reason or memory to learn.

HOW CLEVER IS YOUR CAT?

THORNDIKE'S PUZZLE-BOX EXPERIMENTS

At the age of 23, Edward Thorndike wrote one of the first accounts of behavioral research, and laid the foundations for generations of others, notably B. F. Skinner (see page 37). He took a hungry cat, put it in a box, and then placed food outside the box where the cat could see it. The cat could only reach the food by operating some mechanism to open the box. Thorndike made 15 puzzle boxes in all, and labeled them from A through to O. Box A, the simplest, only required the cat to press a lever in order to open the box. In another, the cat had to pull a loop of string to open it and in an even more complex box the cat had to press a lever, pull a string, and push down a bar.

Thorndike tried his experiment on several cats, placing them in the same box again and again, and recording how long it took to escape each time. He noted that at first the cats usually tried "to squeeze through any opening," or to claw or bite their way out of the boxes. He also observed that the cats were not paying very much attention to the food, but seemed to simply "strive instinctively to escape from confinement."

However, when the cat was put into the same box again it seemed to become more efficient, "and gradually . . . after many trials, the cat will, when put in the box, immediately claw the button or loop in a definite way." In the simplest box one cat took 160 seconds the first time, but after 24 attempts it could escape in 6 seconds. He drew graphs of the time taken to escape against the number of trials, which not only demonstrated that the cats' speed of escape generally improved, but also showed that the more complex boxes generated more erratic behavior, and the cats took longer to learn their way out.

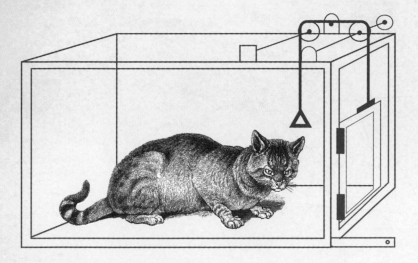

He also found that the cats took their learning with them. Cats that had learned to escape from box A by clawing had a greater tendency, when put into another box, "to claw at things than it instinctively had at the start," and were also less likely to attempt escape by squeeze through holes.

Can cats reason?

Many people in the nineteenth century believed that sophisticated species of animals, such as cats, could learn by association of ideas, and supported this through anecdotes of animals doing apparently clever things. Thorndike was doubtful, writing: "Thousands of cats on thousands of occasions sit helplessly yowling, and no one takes thought of it . . . but let one cat claw at the knob of a door supposedly as a signal to be let out, and straightway this cat becomes the representative of the cat-mind in all the books."

He also explored whether cats might learn through imitation and let one cat watch another escaping from a box. However, when the second cat was put into the box it went through the same gradual process of learning by trial and error.

Thorndike concluded that there was no real evidence that animals could learn by reasoning. The behavior of the cats in his boxes was far from reasoned—"just a mad scramble to get out." Even when they did manage a successful escape the animals did not seem to remember the technique on their next attempt and still took time trying other things. Furthermore if a cat had learned to escape by pulling a loop, it would continue to claw at the air where the loop had been, even though the loop had been taken away. Thorndike noted that when he reached in and took the cat's paw, put it in the loop, and pulled, the cat still failed to learn the trick, and would be clueless next time.

Trials with dogs and chicks

The young scientist also built puzzle boxes for dogs, and pens for chicks. The chicks typically had to step onto a platform, pull a string, or peck a tack, but in one of the most complex trials the chick had to climb a spiral staircase, struggle through a hole, walk over a horizontal ladder, and jump off a ledge. As with the cats, both dogs and chicks improved with practice, but the chicks learned more slowly than either cats or dogs.

Thorndike wrote in his dissertation: "When the crude beginnings of this research have been improved and replaced by more ingenious and adroit experimenters, the results ought to be very valuable." And indeed they were: Thorndike laid the foundations for the new science of behavioral psychology.

DOES THE NAME
PAVLOV RING A BELL?

LEARNED RESPONSES AND
CLASSICAL CONDITIONING

1901
THE STUDY

RESEARCHER:
Ivan Pavlov

SUBJECT AREA:
Animal behavior

CONCLUSION:
Conditioning can create
powerful responses to
otherwise neutral stimuli.

In the late 1890s and early 1900s, the Russian physiologist Ivan Pavlov was leading the field in the exploration of the digestive process. He often used dogs as his subjects and, among his many other observations, he noticed that when their food was delivered by his white-coated assistant, all the animals would begin to drool or salivate.

After many years of study Pavlov knew the body's production of saliva when food, or any foreign body, enters the mouth was a reflex action (an involuntary and often instantaneous response to an external stimulus), which aids digestion or helps to dilute or expel any unwanted matter.

He called this response "psychic secretion" noting: "a similar reflex secretion is evoked when [the food and its container] are placed at a distance from the dog and the receptor organs affected are only those of smell and sight. Even the vessel from which the food has been given is sufficient to evoke an alimentary reflex complete in all its details."

However, he also observed that the dogs soon began to salivate whenever the assistant entered, regardless of whether he brought food: "the secretion may be provoked even by the sight of the person who brought the vessel, or by the sound of his footsteps." Pavlov reasoned that the dogs had learned to associate the assistant, or simply just his white coat, with the arrival of food, and so began to salivate in anticipation whenever they saw him.

Pavlov wondered whether he could make the dogs salivate by using a different signal that bore no relation to the food, so he

arranged for a metronome to start ticking just before their food was brought in. Sure enough, within just a few days, he only had to start the metronome and the dogs would salivate, even if no food appeared. He described one experiment in detail:

> So long as no special stimulus is applied, the salivary glands remain quite inactive. But when the sounds from a beating metronome are allowed to fall upon the ear a salivary secretion begins after nine seconds, and in the course of 45 seconds 11 drops have been secreted. The activity of the salivary gland has thus been called into play by impulses of sound—a stimulus quite alien to food ... The sound of the metronome is the signal for food, and the animal reacts to the signal in the same way as if it were food; no distinction can be observed between the effects produced on the animal by the sounds of the beating metronome and showing real food.

He tried other stimuli, too: an electric buzzer, the smell of vanillin, and probably a bell (although one colleague claimed that Pavlov had never used a bell). He even used an electric shock. In every case the potential stimulus had to be applied a few seconds before the food appeared; if it was applied afterward, by even one second, then it caused no response.

Natural and conditioned reflexes

Pavlov pointed out that food, evokes the salivary reflex in every dog right from birth; this is a natural or unconditioned reflex. Salivation at the sound of the metronome is known as a conditioned response, or a conditioned reflex. He compared this to his newly installed telephone:

> *My residence [is] connected directly with the laboratory by a private line ... or on the other hand a connection may have to be made through the central exchange. But the result in both cases is the same. The only point of distinction between the methods is that the private line provides a permanent and readily available cable, while the other line necessitates a preliminary central connection being established. ... We have a similar state of affairs in reflex action.*

Second-order conditioning

Once the dog has begun responding to a conditioned stimulus, a secondary stimulus can be introduced. Thus a metronome is repeatedly started just before food arrives, until the dog salivates at the sound. Then a bell rings with the metronome, and the dog not only salivates, but also associates the bell with food; so when the bell sounds on its own the dog salivates, even though the bell alone has never signaled the arrival of food.

This entire process has come to be called "classical conditioning," and has been the basis for thousands of subsequent research studies into behavior and learning, including the controversial case of Little Albert (see page 28)

1910

THE STUDY

RESEARCHER:

Mary Cheves West Perky

SUBJECT AREA:

Cognition and perception

CONCLUSION:

When asked to imagine,
our brains find it difficult to
separate our mental images
from real perceptions.

CAN YOU IMAGINE A PERKY TOMATO?

COMPARING PERCEPTION, MEMORY, AND IMAGINATION

In around 1910, the American psychologist Mary Cheves West Perky carried out a series of ingenious experiments to explore how our imaginations function. First of all she set out to compare real images with mental images, or, as she put it, to compare perceptions with images of imagination.

Her subjects were asked to look at a small ground-glass screen, and told to fixate on the white dot in the center while imagining a colored object, such as a tomato or a banana. An unseen projector then began to cast a faint colored glow on the screen—so faint as to be almost imperceptible—while a tomato-shaped stencil was placed between the projector and screen, so that a very faint image of a red tomato appeared. The edges of the stencil were at first softened with a gauze so that there was no sharp image, and the stencil was moved gently from side to side, to give the appearance of fluttering.

The gauze was gradually removed so that the image became sharper, although it was still exceedingly faint. Meanwhile the subject explained what he or she was imagining. The red tomato was followed by a blue book, a deep yellow banana, an orange, a green leaf, and a pale yellow lemon.

Preparing the set-up was difficult. The primitive projector was hard to use and the two assistants required to operated it and move the stencils had to maintain total silence, yet still communicate with the experimenter who sat with the subject. Occasionally things went wrong; a mask slipped, or a shaft of light hit the screen, and the trial had to be abandoned.

22

Real or imaginary?

All 24 subjects reported imagining the expected images, but none of them reported anything until an image was actually visible (to the researchers) on the screen.

All the subjects were asked whether they were "quite sure that they had imagined all these things." They were surprised by the question, sometimes even indignant, and felt certain they had imagined them, saying:

- *I am imagining it all; it's all imagination.*

- *I was making them up in my mind.*

- *The banana is up on end; I must have been thinking of it growing.*

Perky's Experimental Setup

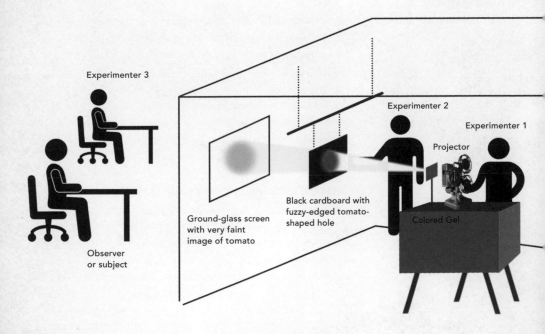

Experimenter 3

Experimenter 2

Experimenter 1

Projector

Black cardboard with fuzzy-edged tomato-shaped hole

Colored Gel

Ground-glass screen with very faint image of tomato

Observer or subject

In fact, many reported some surprise that the banana appeared upright, rather than horizontal as they had first thought of it—yet this seemed to arouse no suspicion. One graduate student even supplied additional context: he saw the tomato painted on a can; the book was a particular book whose title he could read; the lemon was lying on a table.

What this experiment shows is that people find it extremely difficult to distinguish between perception and imagination – between seeing something and imagining it. This is known as the Perky Effect, and remains true today.

Images from imagination and images from memory

In her next series of experiments Perky tried to distinguish between mental images created from imagination and those influenced by memory. She had noticed that when she tried to create a mental picture of something personal, such as her own bedroom or home, she had to reach into her memory, and she tended to move her eyes about when thinking, as if searching for the object. Whereas when she imagined something impersonal, such as a tree or a boat, it had to come from the imagination and she didn't move her eyes. Her brain appeared to operating in a different way to carry out the different tasks.

To test this today, psychologists would use laser eye-trackers, but lasers had not been invented in 1910; so Perky had to invent another way to look for eye movement.

The subjects sat in a dark room and were asked to envisage something personal or something impersonal. They were instructed to keep their eyes on a bright spot on the wall in front of them while they did so. There were several other bright spots just outside their field of view. Out of 426 trials, images from memory appeared in 212, and in 90 percent of these the subjects reported seeing extra bright spots, showing that they must have moved their eyes.

Images from imagination appeared in 214, of which 68 percent showed no eye movement. Eye movements here seemed to be caused by images of animals running across the scene, or by vistas too wide to be seen without eye movement. Occasionally the subjects would "see themselves" in the image: "There was somebody in the boat which I supposed was myself."

Perky also found that sound images caused movement of the larynx when the image was from memory, but not when it was from imagination, while memory smell images produced movements of the nostrils.

Perky wrote: "The mood of memory is that of familiarity or recognition, intrinsically pleasant, the mood of imagination is that of unfamiliarity or novelty." She concluded that memory involves eye-movement and some body movement, while imagination requires steady fixation, but no movement. Also memory images are scrappy, filmy, and give no after-images, while images of imagination are substantial, complete, and sometimes give after-images.

CHAPTER 2: The challenge of behaviorism: 1920—1940

After the work of Thorndike and Pavlov became widely known, psychologists became more interested in the study of behavior, both in humans and other animals. All sorts of remarkable experiments were carried out that shaped the science of psychology (although some, such as John B. Watson's classical conditioning of baby Albert, would be considered cruel by today's standards). B. F. Skinner, arguably the most famous behaviorial psychologist of all, devised ingenious tests for rats and pigeons, to discover how they learn. In Berlin, "Gestalt" psychologists

tried to understand the human ability to make meaningful perceptions in a chaotic world, and in the United States, scientists marched into the factory looking for new ways to improve efficiency and productivity. The psychology of business had begun.

1920
THE STUDY
RESEARCHERS:
John B. Watson and
Rosalie Rayner
SUBJECT AREA:
Behavior
CONCLUSION:
All individual differences in
behavior are due to different
experiences of learning and
conditioning.

WHAT'S THE MATTER WITH LITTLE ALBERT?

EXPLORING CLASSICAL CONDITIONING IN HUMANS

Albert B., or "Little Albert" as he came to be known, was a placid, happy, healthy child, who weighed 21 pounds at the age of nine months, and had lived almost all his life in a hospital, where his mother was a wet nurse.

In 1919, psychologist John B. Watson and his graduate student, Rosalie Rayner, set out to test whether a human would respond to the same kind of conditioning that Pavlov had carried out with his dogs (see page 19). Watson hypothesized that a baby's fear of loud noises is, like a dog's salivation, an innate, reflex response. He reasoned therefore, that, through the principles of classical conditioning, it should be possible to provoke fear in response to otherwise unrelated objects.

Watson and Rayner chose Little Albert as their subject and began by first presenting him with a live white rat, then a rabbit, a dog, and a monkey, as well as a range of other objects. He seemed to want to handle them, but showed no fear at this stage and didn't cry.

Next they made a loud and surprising noise by hitting a steel bar with a hammer just behind his head. Their notes report:

> *The child started violently, his breathing was checked, and the arms were raised in a characteristic manner. On the second stimulation the same thing occurred, and in addition the lips began to pucker and tremble. On the third stimulation the child broke into a sudden crying fit. This is the first time an emotional situation in the laboratory has produced any fear or even crying in Albert.*

Classical conditioning

Then they set out to test whether they could condition fear of an animal, specifically a white rat, by visually presenting it and simultaneously striking a steel bar, and then whether this conditioned fear would be transferred to another animal. Establishing conditional emotional responses as follows:

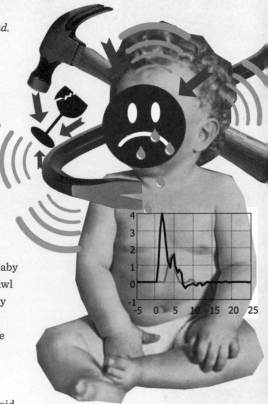

1. White rat suddenly taken from the basket and presented to Albert. He began to reach for rat with left hand. Just as his hand touched the animal the bar was struck immediately behind his head. The infant jumped violently and fell forward, burying his face in the mattress. He did not cry, however.
2. Just as the right hand touched the rat the bar was again struck. Again the infant jumped violently, fell forward, and began to whimper.

They showed him the rat and walloped the iron bar three more times. By this time Albert had begun to whimper when he saw the rat alone. Then they gave him two more treatments with rat and noise, and finally, as soon as the rat was shown alone "the baby began to cry. Almost instantly he ... began to crawl away so rapidly that he was caught with difficulty before reaching the edge of the table." Thus the unconditioned response to the noise had become a conditioned response to the rat.

Generalized response

A few days later, knowing that Albert was still afraid of the rat but otherwise appearing happy and smiling, Watson and Rayner wanted to see if his fear of the rat would be transferred to other furry animals and began by showing him

the rabbit He leaned as far away from the animal as possible and then burst into tears. He was less upset at the sight of the dog, but still began to cry, and was also disturbed by cotton balls.

Researchers continued to condition the child with the rat and the dog, hammering the steel bar as soon as they came close to him. A month after these experiences Albert continued to show signs of distress in response to the rat and the dog, and was still uncomfortable with the rabbit.

This experiment was considered highly controversial at the time and has been heavily criticized since both for the validity of its findings and its morality. It goes without saying that this sort of experiment would never be allowed today, and there is some doubt about whether Albert's mother actually gave formal consent at the time. Watson himself made small note of the questionable ethics behind his venture: "At first there was considerable hesitation upon our part in making the attempt ... A certain responsibility attaches to such a procedure." But quickly he took "comfort" from the notion that "such attachments would arise anyway as soon as the child left the sheltered environment of the nursery for the rough and tumble of the home."

Watson also recorded that he had wanted to try "detachment" or desensitization—the removal of the conditioned emotional responses—but Albert was taken away from the hospital before this could be attempted.

There is no evidence that the conditioning lasted (and no evidence that it remained), since efforts to find Albert were made only recently. The most likely candidate seems to have been one Albert Barger, who died in 1987 before anyone had contacted him.

His niece said he had always disliked dogs.

DO YOU WORRY ABOUT UNFINISHED BUSINESS?

THE ZEIGARNIK EFFECT

1927
THE STUDY
RESEARCHER:
Bluma Zeigarnik
SUBJECT AREA:
Cognition and memory
CONCLUSION:
Incomplete tasks or events
without "closure" are more
easily remembered (or less
easily forgotten) than those
that are completed.

Bluma Zeigarnik was a Lithuanian psychologist working at the Berlin School of experimental psychology in the 1920s. Professor Kurt Lewin roused her curiosity when he observed that a waiter could remember all the orders that had not been paid for, but once they were paid for he promptly forgot them.

Zeigarnik set out to investigate. She gave 164 individual subjects 22 simple tasks, such as writing down the names of cities beginning with L, making clay models, or building boxes out of cardboard. During half of the tasks Zeigarnik interrupted the subjects before they had time to complete them. Afterward she found that subjects remembered 68 percent of the unfinished tasks but only 43 percent of the finished ones.

Some subjects were interrupted when a task was almost finished; these people recalled 90 percent of the unfinished tasks. Our ability to remember unfinished tasks in much greater detail than those we satisfactorily completed, as well as our desire to see it completed, has come to be called the "Zeigarnik effect" and it occurs in many areas of our lives. For example, 200 students questioned after an exam were found to remember many more of the questions they had been unable to answer than those they had answered correctly.

It is evident that we can worry about unfinished business for days, just because it is unfinished. This is why producers of soap operas for television and radio so often conclude each episode with a cliff-hanger; it keeps the audience thinking about the problem until the next program.

One interesting possible consequence of the Zeigarnik effect is that students whose learning is interrupted to do unrelated things—sport or socializing—may actually remember their work better than those who never stop studying; perhaps all work and no play really does make Jack a dull boy. It should be noted, however, that this does not apply to those who interrupt themselves before anything has actually been learned.

The frustration of interruption

Interrupting a task also influences the subject's estimate of how long the task takes, as was shown by a study in 1992. Subjects were asked to unscramble ten three-letter anagrams: GBU, TEP, ARN, FGO, OLG, UNF, TAS, TOL, EAC, UNP. Then they were asked to estimate how much time it had taken. Their guesses proved to be within 10 percent of the actual time.

Next the subjects were asked to solve 20 three-word anagrams—EDB, ANC, YDA, ODR, OTE, UME, ADL, XFO, DLI, XEA, PZI, AEG, ARO, BTI, SYE, NIF, GRA, FTI, DCO, ILE.

Halfway through the task they were asked to estimate how long the task had taken so far. Their estimate was 65 percent longer than the actual time. Then they completed the task and again asked to estimate how long the second half had taken. This time their estimate was 35 percent higher than the real time.

Their first estimate seems to have been so high because they were frustrated at being interrupted, and felt a sense of failure, which appears to have made them think they were slower than they really were.

Can closure help?

Zeigarnik herself concluded that we naturally feel the need to complete tasks, following the Gestalt notion of "closure." That is, we feel a need to complete a task once begun. The finished task is a completed Gestalt, and we can stop thinking about it. Failure to complete the task as instructed sets up tension which keeps us thinking about the incomplete task, and is not resolved until the original need is satisfied. Zeigarnik wrote:

The strength with which such tension systems arise and persist evidently varies greatly between different individuals but remains very nearly constant with the same individual. Strong needs, impatience to gratify them, a childlike and natural approach—the more there is of these, the more will unfinished tasks enjoy in memory a special advantage over those which have been completed.

Furthermore, according to John Gottman, in his 2013 book *What Makes Love Last?*, arguments between lovers that are resolved by confession or discussion do much less harm than denial, or regrettable incidents that go unaddressed, for then the pain persists in active memory and becomes a constant and dangerous irritant. It seems humans seek completion, or "closure" in many areas of their lives.

Contradicted by cash?

However a study at the University of Mississippi in 2006 showed that the Zeigarnik Effect can be quickly undermined when money changes hands. Forty college students were asked to carry out a five-minute task while they would be measured for "hemispheric-activity." They were fitted with fake recording apparatus—a plastic helmet fitted with electronic chips and wires leading to a computer—and left to begin the task. Half the students had been promised $1.50 for taking part; but the other half expected no reward. Halfway through the alloted time the students were informed that the "hemispheric recording" was complete. Forty-two percent of the students expecting payment immediately abandoned the task and walked away with the money. However, of the volunteer half, only 14 percent left before the task was complete.

1932

THE STUDY

RESEARCHER:

Frederic Bartlett

SUBJECT AREA:

Cognition and memory

CONCLUSION:

Remembering is not simply
the recall of static facts,
but an active process,
similar in kind to imagining
and thinking.

ARE YOU GOOD AT TELLING TALES?

THE ACCURACY OF LONG-TERM MEMORY

As part of his long investigation into how memory works,
Frederic Bartlett, the first Professor of Experimental Psychology
at the University of Cambridge, tested people's ability to
remember figures, photographs, and stories. In the 1920s and
1930s he persuaded his subjects to read and retell various short
stories; one of his favorites was the following Canadian folk tale.

"The War of the Ghosts"

*One night two young men from Egulac went down the
river to hunt seals, and while they were there it became
foggy and calm. Then they heard war cries and they
thought: "Maybe this is a war party." They escaped to
the shore and hid behind a log. Now canoes came up
and they heard the noise of paddles and saw one canoe
coming up to them. There were five men in the canoe
and they said: "What do you think? We wish to take
you along. We are going up the river to make war on the
people." One of the young men said "I have no arrows."
"Arrows are in the canoe," they said. "I will not go along.
I might be killed. My relatives do not know where I
have gone. But you," he said, turning to the other, "may
go with them." So one of the young men went but the
other returned home. And the warriors went on up the
river to a town on the other side of Kalama. The people
came down to the water, and they began to fight, and
many were killed. But presently the young man heard
one of the warriors say: "Quick, let us go home; that
Indian has been hit." Now he thought: "Oh, they are*

ghosts." He did not feel sick but they said he had been shot. So the canoes went back to Egulac and the young man went ashore to his house and made a fire. And he told everybody and said: "Behold I accompanied the ghosts and we went to fight. Many of our fellows were killed and many of those who attacked us were killed. They said I was hit and I did not feel sick." He told it all and then he became quiet. When the sun rose he fell down. Something black came out of his mouth. His face became contorted. The people jumped up and cried. He was dead.

Less-than-total recall

Bartlett asked his first subject to read the story, and then tell it from memory to a second subject, who in turn told it from memory to a third, and so on until they reached seven reproductions. The effect was similar to the children's game "Telephone," in which a message is passed from person to person, gradually mutating on the way, as people introduced errors or changes.

Not surprisingly the story changed and lost details as it was told and retold. The subjects were all young British men, and therefore not familiar with the style and content of the story; so they made many mistakes. In particular they steadily shortened the story and left out details. On the other hand the story became more coherent—Bartlett called this "conventionalization"—and it remained a sensible story. The tale became more British, as the subjects inserted words and ideas from their own cultural backgrounds. For example some of the subjects remembered "hunting seals" as "fishing," and "canoes" as "boats." They also forgot and excluded elements that did not make sense to them.

Another of Bartlett's methods was to ask one subject to read the story, retell it, and then retell it again after various periods of time—perhaps half an hour, a week, or three months. This method achieved similar results.

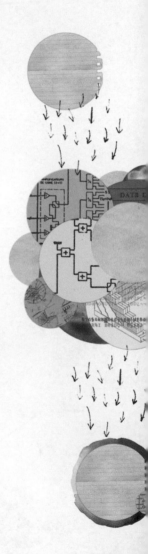

STORY

SCHEMA

MODIFIED STORY

Using schemas

Bartlett suggested that long-term memory consisted of "schemas," where each schema was "a mass of organized experiences," or "an active organization of past reactions, or past experiences." He felt that "the past operates as an organized mass rather than as a group of elements each of which retains its specific character." He therefore concluded that all new incoming information interacts with the old information in the schema, to make a modified schema. Remembering, he suggested, is not a process of picking static facts from a rack, but an active process, not fundamentally different in kind from imagining and thinking. Recalling a memory is therefore a process not of reproduction, but of reconstruction, within what may well be a new cultural context, and constitutes "an effort after meaning." There are bound to be errors in remembering something like the "War of the Ghosts," because the story must inevitably be reinterpreted in the light of the general character of the subjects' previous experience.

Bartlett used tennis as an analogy for his findings. He said that with each stroke of the racket, he did not produce something absolutely new, and he never merely repeated something old. The stroke was literally manufactured out of the living visual and postural schemas of the moment, and their interrelations.

Bartlett's schema approach was not widely accepted at the time, but has come into favor more recently, and in particular was championed by computer scientist Marvin Minsky in the field of artificial intelligence.

HOW DO ANIMALS LEARN?

"OPERANT CONDITIONING" AND REINFORCEMENT OF BEHAVIOR

1938
THE STUDY
RESEARCHER:
Burrhus Frederic Skinner
SUBJECT AREA:
Animal behavior
CONCLUSION:
Positive reinforcement is more effective than punishment in shaping behavior.

After the pioneering work of Thorndike (see page 16) and Pavlov (see page 19), the American psychologist B. F. Skinner took a more scientific approach to animal learning. He thought there was no point in trying to understand what an animal wanted or planned; he preferred to see what it actually did in tests carried out under controlled laboratory conditions.

He observed that humans appear to learn from the consequences of their actions, and that we repeat actions that are rewarded, such as working hard at school. Skinner wondered if animals learn in the same way, and whether by studying animals he might be able to uncover fundamental principles of how people learn.

Operant conditioning

Skinner put a rat in a box that contained a lever. When the rat pressed the lever a food pellet was dispensed. To begin with the rat just ran around the box, but when by chance it pressed the lever it noticed the arrival of the food. This was positive reinforcement in its most direct form. Quite soon the rat learned that whenever it pressed the lever it got food, and it began to press the lever five times a minute.

Skinner described only what he could observe, and never stated that the rats learned to press a lever because they wanted food. He explained that it was the action, not the rat, that was reinforced (or sometimes punished). He called this process operant conditioning, because the rat learned not from any stimulus but from its own actions. Skinner's "operant

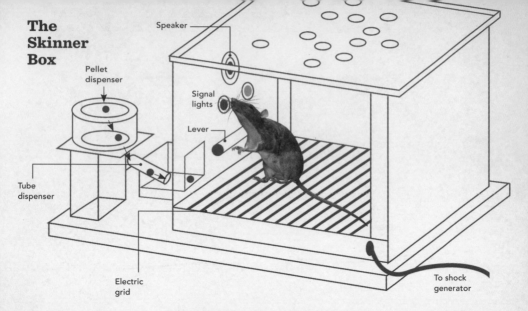

The Skinner Box

Speaker

Pellet dispenser

Signal lights

Lever

Tube dispenser

Electric grid

To shock generator

conditioning" differs from the classical conditioning of Pavlov and Watson in that it operates on the environment, rather than on the reflexive behavior of the subject.

"Chaining"

Skinner boxes were similar to Thorndike's puzzle boxes, but often more elaborate, and linked to automatic recording devices; so he knew exactly how often the rat pressed a lever without having to sit there with a notebook.

In another box, food was dispensed only after a lever was pressed ten times, but the rats soon learned to do this, and indeed afterward they pressed their lever more often than those in an "every-time" box.

Gradually he made his boxes more elaborate, with more difficult things for the rats to do. In some cases he used unpleasant stimuli; as the rat wandered around its box it might suddenly be bombarded by loud noise, but when it accidentally touched a lever, the noise was turned off. In due course it learned to press the lever as soon as it was put into the box. In another box, rats would learn to press a lever as soon as a light came on, because otherwise they would receive an electric shock shortly afterward.

Skinner found that rats could learn to perform a complex sequence of simple actions, as long as they learned them one at a time. A rat could learn, for instance, to turn when a buzzer sounded, and then press a lever after a light came on; food would be dispensed. Skinner called this process "chaining."

Using pigeons in place of rats

Skinner also ran trials with pigeons and found that they could learn to peck at a red spot on the wall for a food pellet to appear. Indeed it would peck even if food appeared sometimes, rather than every time. Skinner pointed out that this was just like a gambler with a one-armed bandit. The gambler has learned that putting a coin into the machine and pulling the lever will occasionally yield a reward; he or she just hopes it will be big enough to compensate for the investment.

Skinner found that pigeons, like rats, could learn to perform complex tasks, such as turning in a circle and then pecking at a target, as long as there was reinforcement for each step.

As Skinner himself said, "The consequences of behavior determine the probability that the behavior will occur again." He was convinced that by using his principles it should be possible to create a Utopian society, where all behavior would be good, and everyone would be happy. In his novel *Walden Two* (1948), he described a wonderful community where people worked only four hours a day, benefited from superb recreation, were environmentally responsible, and enjoyed complete equality between the sexes.

1939

THE STUDY

RESEARCHERS:

Fritz Jules Roethlisberger
and W. J. Dickson

SUBJECT AREA:

Social psychology

CONCLUSION:

Paying attention to the
thoughts and feelings of
workers will increase
productivity.

CAN PSYCHOLOGY INCREASE PRODUCTIVITY?

THE HAWTHORNE EFFECT

The Hawthorne Works, at Cicero, Illinois, was an enormous factory built by Western Electric in 1905. In 1924 their electrical suppliers claimed that better lighting would increase productivity and the managers at Hawthorne commissioned a study to find out whether this was true.

The researchers moved in, measured productivity, divided the workforce into a test group and a control group, and then carefully increased the light levels for the test group. To the surprise of the investigators, productivity shot up in both groups. Then they made the test group's workplace progressively dimmer, and, until they began to complain that they could not see what they were doing, the productivity of both groups rose again. Productivity even went up when they returned the light levels to where they had been in the first place.

Intrigued, the researchers tried changing the work environment in various other ways. The longest-running study focused on the assembly of electrical relays—switching devices used in telephone exchanges. Making a relay required repetitive work, putting together by hand 35 pins, springs, armatures, insulators, coils, and screws. Western Electric produced over 7 million relays a year, and the speed of individual workers determined overall production levels.

The relay test room

The experimenters chose two women and invited them to choose four more to make a team, and set them up in a separate test room. Here an experimenter discussed changes with them, and at times implemented their suggestions.

The experimenters wondered whether the women might get tired during their long working days, and therefore slow down. So they suggested the possibility of two five-minute rest periods during the day; after some discussion the women chose to have them at 10 am and at 2 pm. They enormously appreciated the breaks, although some commented that they were too short.

When the experimenters discovered that the productivity had gone up, they offered ten-minute breaks instead. Some of the women were worried that they would not be able to make up for the time lost, but the experimenters suggested that they would work faster because they would be less tired. The women loved their longer breaks and they actually produced more relays than ever before. When the researchers changed to six five-minute breaks, however, productivity went down.

The researchers tried serving lunch during a 15-minute break in the morning. They shortened the working day by half an hour, and output increased; they shortened it again: output per hour increased, but daily output dropped. Then they put the working day back to where it had been originally, and output reached the highest level ever, up 30 percent.

The final set of tests was to introduce an incentive pay scheme with 14 men in the Bank Wiring factory. The surprising result was that productivity did not increase; the men in the group had established a "norm" for themselves, and kept working at the same rate, in spite of the offer of more pay for more work.

The experimenters noted that almost everything they did—apart from offering more money—produced a temporary increase in productivity. One possibility was that the workers thought they were being monitored individually. Choosing their own colleagues, working as a team, and being treated as special with their own room must have been encouraging to them.

Conclusion

One conclusion was that productivity depends on informal interactions in the work group. In addition, this group had an interested and sympathetic supervisor who provided feedback on their efforts. One of the women, Theresa Layman Zajac, later said "I had no idea there would be so much happening and so many people watching us." Yet another possibility was that the workers wanted to please the experimenters, which is often the case in psychological experiments.

Hawthorne superintendent George Pennock said that:

> ... a relationship of confidence and friendliness has been established with these girls to such an extent that practically no supervision is required. In the absence of any drive or urge whatsoever they can be depended upon to do their best. They say they have no sensation of working faster now than under the previous conditions. They have a feeling that their increased production is in some way related to the distinctly freer, happier, and more pleasant working environment.

What the Hawthorne studies had actually done was show that productivity will increase if the management treat the workers with respect and as human beings, rather than as mere appendages to the machines they operate. These ideas were entirely new in the 1930s, but the Hawthorne experience made it clear that organizations that do not pay attention to personal and cultural values will be less successful than those that do.

HOW DO YOU MANAGE A DEMOCRACY?

EXPLORATIONS IN LEADERSHIP STYLES AND GOOD GOVERNANCE

1939

THE STUDY

RESEARCHERS:
K. Lewin, R. Lippitt,
and R. K. White

SUBJECT AREA:
Social psychology

CONCLUSION:
Effective democracy needs
proactive group management
rather than unlimited
individual freedom.

After the pioneering psychologist Kurt Lewin escaped Nazi Germany in 1933 and fled to America, he wrote about:

> ... *the peculiar mixture of desperate hope, curiosity, and skepticism with which the newly arrived refugee from Fascist Europe looks at the United States. People are fighting for it, people are dying for it. It is the most precious possession we have. Or is it but a word to fool the people? Democracy?*

How could he learn what a real democracy was like, and how to organize it? First, he set up a "lab" that was really more like a kids' den—a space in an attic, with wooden boxes to sit on, surrounded by all sorts of junk—mainly construction equipment—and enclosed by crude sacking walls. It was crowded, undisciplined, unstructured, and fun—just the opposite of a clean white classroom.

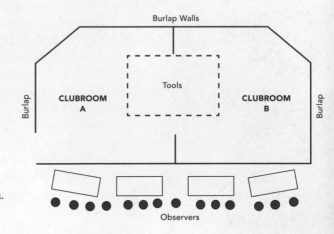

He recruited groups of 10- and 11-year-old children, and divided them into four clubs, each of which would meet once a week. The children were asked, with the help of an adult leader, who was one of the researchers, to make theatrical masks, make furniture and paint signs for the room, carve soap and wood, and build model aircraft. In other words their club room was also their workshop.

Lewin deliberately aimed to set up different types of social climate, by using different styles of leadership—groups of children would experience first one type of leader and then another, over several weeks. A dozen researchers sat in a dark corner and took notes of how the children were reacting to one another, and to the leader, while Lewin himself secretly filmed the proceedings. Interestingly, this was one of the first experiments in social psychology in which the experimenters played central parts, as leaders; before this they had merely been the observers or helpers.

Three styles of leadership

The first leader was strict; he told the children exactly what to do, step by step, so that they rarely knew what the final plan was. He told them which child should do which tasks, and exactly where they should work—mostly in the center of the floor. In his praise or criticism he was direct and personal. He always stood in one place, wore a suit and tie, and remained outside the group.

The second leader set up a "democratic atmosphere" in which the entire club discussed the project in advance, and took decisions about what to do. They chose their own working groups. When they asked for advice, the leader suggested two or three options for them to choose from. He was entirely objective in his praise and criticism of their performance. He was one of the group: took off his jacket, rolled up his sleeves, and moved around the space with the children, although he did little actual construction.

The third leader just sat still, let the children get on with it, and scarcely interfered at all. This "laissez-faire" attitude happened originally by mistake, when a new leader, Ralph White, forgot to steer the children toward democracy, and anarchy set in. As he said later: "The group started to fall apart. There were a couple of kids who were real hell-raisers, and they found a great opportunity to raise hell, which wasn't productive."

The results

In the first regime there was endless trouble. The strict leadership led to a great deal of tension; arguments and fights broke out between the children. They were clearly unhappy, and tended to blame one another for mistakes. After one session they smashed up the masks they had been making. As Lippitt noted, "They couldn't fight the leader, but they could [fight] the masks."

In the democratic atmosphere the children were much happier, less aggressive, and more objective about the work. They were also much more productive and imaginative, doing their work all over the club room.

In the laissez-faire groups, the children rarely concentrated on their tasks, but just wandered around the room. The researchers decided that this style of leadership was also interesting; so they persisted, and the leaders had to work hard at being passive and uninvolved.

When children were moved from one group to another they rapidly shifted to the new regime, and learned how to fit in with the group and the leader.

Lewin concluded that democracy would never come from unlimited individual freedom; it would need strong, proactive group management.

Conclusions

Lewin's experiment showed that democratic behavior can be generated in a small group, which ushered in the concept of focus groups and group therapy. More important, it showed that leadership should be a teachable skill, and need not merely be associated with charisma or military prowess.

CHAPTER 3 : Changing concerns: 1941–1961

After World War II, psychologists widened their focus from human and animal behavior to include the practical ramifications of studying the mind. Questions arose about how psychology could help in the classroom, and researchers devised ways of studying how children think. Scientists examined whether animals could solve problems, and pondered what those answers meant in terms of human interaction. Pure "thinking" was not the only subject matter: emotional and social behavior became relevant realms of psychology. In these contexts,

new questions arose: How do you track emotions as fundamental as a mother's love for her baby? Can we believe in two totally disparate or contradictory realities? Why is conformity so important to us? And is aggression an innate trait?

1948
THE STUDY
RESEARCHER:
Edward C. Tolman
SUBJECT AREA:
Animal behavior
CONCLUSION:
Rats displayed latent
learning and can memorize
details, demonstrating
cognitive behaviors

CAN RATS MAKE MENTAL MAPS?

HIDDEN, LATENT, OR INCIDENTAL LEARNING

The famous behaviorist B. F. Skinner (see page 37) had said that it was not worth considering what animals might think about or want; all you could do was see how they reacted to reinforcement. Berkeley professor Edward Tolman was not so sure. He wanted to find out how much they can think, and what they hold in their memories.

Like Skinner, Tolman and his students built mazes for rats to run through, but they specifically designed mazes that would show thinking—cognitive behavior—by the rats. One of the first of these was a series of narrow passages connected by T-junctions on a horizontal surface (the plan is shown at the top of page 50).

The rats were divided into three groups. Once a day each hungry rat was put in at the bottom left of the maze and had to find its way to the top right. On the way it came to six T-junctions, and had to choose the correct turn each time; so it had six possibilities for going wrong.

Rats in group 1 always found a food pellet at the end of the maze. The result was they learned their way through the maze more quickly each day, and by the seventh day were making no wrong turns at all, as you can see in the graph opposite. (The graphs show the average for each groups of rats.)

Delayed reward

For six days the rats in group 2 found no food at the end; so there was no incentive for them to hurry. They wandered about, making various wrong turns each day, but at the end of the seventh day they found the food, which was there on all subsequent days. On the eighth day they made only one wrong turn, and on the ninth day they went directly to the food with no errors. Group 3 found food at the end of the third day, and afterward found their way to it rapidly.

The point here is that the first group took seven days to work out the direct route to the food. The second and third groups took only two or three days to do it, once they knew the food was waiting. It follows that in their earlier wanderings, they must have formed a mental map (or "cognitive map") of the maze, even though there was no hurry to get to the end. This was hidden learning, since the fact that they had formed a mental map was not obvious until after they found the food and their learning was not revealed until then. It is also called "latent learning," or "incidental learning."

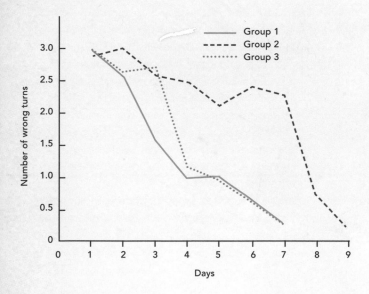

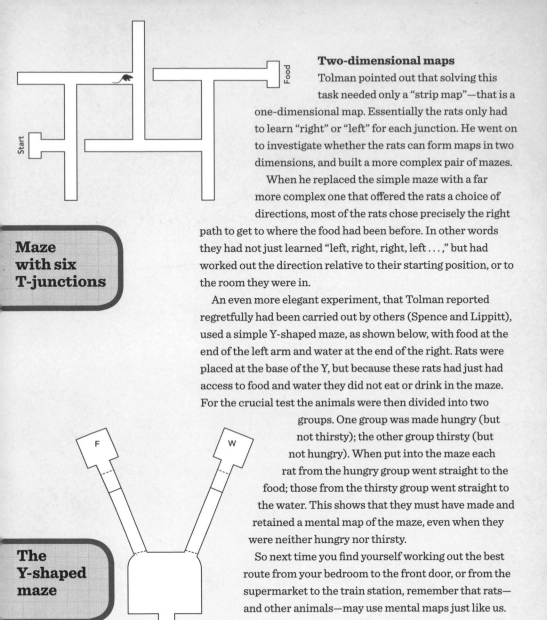

Food

**Maze
with six
T-junctions**

**The
Y-shaped
maze**

F

W

S

Two-dimensional maps

Tolman pointed out that solving this task needed only a "strip map"—that is a one-dimensional map. Essentially the rats only had to learn "right" or "left" for each junction. He went on to investigate whether the rats can form maps in two dimensions, and built a more complex pair of mazes.

When he replaced the simple maze with a far more complex one that offered the rats a choice of directions, most of the rats chose precisely the right path to get to where the food had been before. In other words they had not just learned "left, right, right, left ...," but had worked out the direction relative to their starting position, or to the room they were in.

An even more elegant experiment, that Tolman reported regretfully had been carried out by others (Spence and Lippitt), used a simple Y-shaped maze, as shown below, with food at the end of the left arm and water at the end of the right. Rats were placed at the base of the Y, but because these rats had just had access to food and water they did not eat or drink in the maze. For the crucial test the animals were then divided into two groups. One group was made hungry (but not thirsty); the other group thirsty (but not hungry). When put into the maze each rat from the hungry group went straight to the food; those from the thirsty group went straight to the water. This shows that they must have made and retained a mental map of the maze, even when they were neither hungry nor thirsty.

So next time you find yourself working out the best route from your bedroom to the front door, or from the supermarket to the train station, remember that rats—and other animals—may use mental maps just like us.

WHAT ARE YOU THINKING, CHILD?

PIAGET'S THEORY OF COGNITIVE DEVELOPMENT IN CHILDREN

1952
THE STUDY
RESEARCHER:
Jean Piaget
SUBJECT AREA:
Developmental
psychology
CONCLUSION:
Children think differently
to adults and their learning
gradually develops in
measurable stages.

Jean Piaget was a Swiss psychologist who, after interviewing boys at a school, began to wonder how children think. Most people imagined that children were just like adults, but less good at thinking. He showed they are born with a primitive mental structure that they gradually develop by learning, and that they think in ways that are quite different from adults.

He developed his theories by talking to children, including his own, and by doing experiments to tease out aspects of their thoughts about the world.

Conservation

Young children, he found, don't understand the concept of conservation. He showed them two identical wide glasses containing the same amount of liquid. Then he poured the liquid from one of the glasses into a tall narrow one, so that the liquid there was deeper. Almost without exception, children aged two to seven said there is now more liquid in the thin glass.

Children were fooled when he placed two lines of candies on the table, with equal numbers in each line, but with the candies

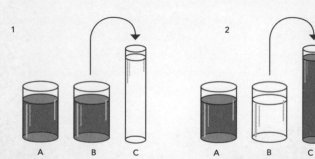

1

2

A   B   C   A   B   C

Piaget's Conservation Task

close together in one, and farther apart in the other. When he asked them which line has more candies he noted:

Children between 2 years, 6 months old and 3 years, 2 months old correctly discriminate the relative number of objects in two rows; between 3 years, 2 months and 4 years, 6 months they indicate a longer row . . . to have "more"; then after 4 years, 6 months they again discriminate correctly.

Piaget defined four stages of development: "sensory-motor" from zero to two years, "preoperational" from two to seven years, and so on. In the first part of the second stage, from two to four years, children begin to use symbols to represent their world. They will draw their families, and although the drawings are not to scale or even remotely like the people, the children don't seem to mind.

In the second part, from about four to seven, children become curious and ask enormous numbers of questions, some of which show signs of reasoning. Some are impossible to answer: I remember my son asking "Why is it a cat?" And while I was wondering what to say, he followed it up with "And what happens if it isn't?"

Piaget suggested that children develop "schemas"—building blocks of intelligent behavior, or actual units of knowledge (see page 36). Even tiny babies have a few schemas, including a sucking schema; they will suck nipples, comforters, or fingers. These schemas are constantly being revised and added to as the child investigates the world, and, as children develop and grow, they develop more schemas to include the new information they acquire. When new information can go straight into an existing schema it is assimilated; when it won't fit then it has to be accommodated by modifying an existing schema into something larger.

Egocentric views?

Small children, he surmised, have an egocentric view of the world, which means they cannot imagine things from someone else's viewpoint. He demonstrated this with a neat experiment, called "The Mountain Scene."

The child was shown a three-dimensional model of three different mountains; one had a cross on it, and one had a tree. A teddy bear or doll was seated on the other side of the table. The child was then shown a series of pictures, and asked which one showed the scene from the doll's point of view. Invariably they chose the one that showed their own view, rather than the doll's.

Piaget wrote: "Far from representing the various scenes which the doll contemplates from different viewpoints, the child always considers his own perspective as absolute and thus attributes it to the doll without suspecting this confusion."

This experiment has been criticized on the grounds that the children may not have understood the question and similar studies with less complex set ups have achieved different results. For example, in 1975 the British developmental psychologist Martin Hughes showed children a model of two intersecting walls, with two model policemen and a model baby. He asked the children to put the baby where he would be hidden from both policemen. The children were aged from 3.5 to 5, and 90 percent of them gave correct answers, even though they needed to understand the points of view of both policemen.

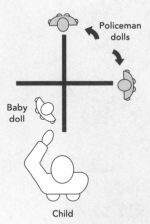

Piaget's work was enormously influential in both developmental psychology and education, and as such his theory and experiments have been closely scrutinized and challenged, and his developmental stages have been subject to many revisions.

THE STUDY

RESEARCHERS:
Morris F. Heller and
Moe Bergman

SUBJECT AREA:
Neuropsychology

CONCLUSION:
Subaudible tinnitus may
be a phenomenon
experienced by everyone
without their knowledge.

WHAT'S THAT NOISE?

IS TINNITUS AN ILLNESS, OR IS IT ALL IN OUR HEADS?

Some people suffer from a ceaseless buzzing or ringing in their ears; it stops them from hearing; it interrupts their sleep; and it can make their lives a misery. This buzzing or ringing is called tinnitus and the sensation can vary from faintly irritating to downright unpleasant.

Before the 1950s it was suggested that there were two types of tinnitus. Vibratory tinnitus, caused by real sounds from a physical source such as muscle activity and nonvibratory tinnitus—an illusion of sound caused by an irritation of the auditory neural elements; that is, it came from inside the brain.

Doctors have suggested all sorts of possible treatments, for the condition including medication with half a dozen different types of drugs; elimination of all drugs and intoxicants; correction of faulty gastrointestinal function or blood-forming organs; dietary control of fluids, salt, and water balance; dental work; medication inside the ear; psychotherapy; or the use of a hearing aid—not to mention a variety of surgical operations.

Is tinnitus an illness or a symptom?

One researcher, E. P. Fowler, who had once asserted that tinnitus is always associated with deafness, altered his view and noted that it was often present in people who had no apparent ear disease. He went on to examine 2,000 patients, and found tinnitus in 86 percent of them.

Morris F. Heller and Moe Bergman, two American doctors specializing in audiology, noted that tinnitus at times appeared to interfere with their patients hearing, but they wondered if, actually, the reverse was true. Perhaps the symptom of tinnitus became more apparent as the patients hearing worsened:

Patients often state that were it not for their head noises, their hearing would be better, and that when the head noises are louder the deafness is more severe. It does not necessarily follow that the tinnitus is always responsible for this. Possibly with increased deafness the head noises are less easily masked and so appear louder subjectively.

Loudness is measured in decibels (dB); very loud noises, such as drills or motorbikes, produce 100dB, while normal conversation is about 70dB and whispering about 50dB. Heller and Bergman estimated that the loudness of tinnitus was only around 5 to 10 dB above the threshold of human hearing—the quietest sound one could possibly hear.

Heller and Bergman wondered if, since tinnitus had been observed in perfectly healthy people, whether it may be an early symptom preceding impaired hearing. And they realized that they could study subaudible tinnitus (tinnitus that people cannot normally hear) by exposing healthy people to an extremely quiet environment.

The soundproof room

Heller and Bergman recruited 80 volunteers from a variety of backgrounds (all healthy adults, males and females, aged from 18 to 60) who had normal hearing, and reported no deafness or tinnitus. Each person was taken into a soundproof chamber, where the ambient noise level was probably between 15 and 18 dB (they could not measure it precisely because the sound-

level meters at the time were not sophisticated enough to detect any noise at all).

The volunteers sat in the soundproof chamber for five minutes and were asked to make notes of any sounds they heard; there was no suggestion that the source of the sound might come from within the subject's head. The researchers also tested 100 hard-of-hearing patients, mostly service veterans.

The results were surprising. Of the hard-of-hearing patients 73 percent reported hearing sounds. Of the other, hearing subjects, 94 percent reported hearing sounds. In all they reported 39 different sounds. Most people reported hearing one sound; some reported two, and a small fraction reported three, four, or five.

These results suggest that the effect of tinnitus is always present for almost everyone, but is normally masked by the ambient noise that floods the environment. In ordinary quiet living conditions the ambient noise is usually more than 35 dB, which seems to be loud enough to hide the tinnitus, which remains subaudible.

Incurable condition

One immediate conclusion that could be drawn from Heller and Bergman's research is that tinnitus cannot be "cured" or eliminated by any treatment, and at best it can only become subaudible. Yet this has not prevented many people from suggesting both causes and preventive measures. For example, one proposed cause is the consumption of coffee and tea. In response to this suggestion, the British experimental psychologist, Lindsay St. Claire recruited 67 volunteers for a 30-day trial in 2010 to find out whether caffeine has any effect on tinnitus. His team observed significant adverse symptoms of caffeine withdrawal, but found no evidence to justify caffeine abstinence as a therapy to alleviate tinnitus.

THE END IS NIGH, OR IS IT?

THE DISCOMFORT OF COGNITIVE DISSONANCE

1956
THE STUDY

RESEARCHERS:
Leon Festinger, Henry Riecken, and Stanley Schachter

SUBJECT AREA:
Cognitive dissonance

CONCLUSION:
Humans find coping with two or more contradictory beliefs highly distressing.

In August 1954, Marian Keech predicted that the world would end in a great flood, just before dawn, on December 21. Mrs. Keech was the leader of a semireligious cult called The Seekers. She claimed to receive messages by "automatic writing," where her hand and the pen seemed to function all by themselves, and words appeared in handwriting quite unlike her own. These messages ranged from descriptions of the environment on other planets, to warnings of war and devastation on Earth, along with promises of amazing joy and salvation for all true believers.

She claimed to have had received the apocalyptic message from the planet Clarion, and added that just before the flood a flying saucer would come to carry The Seekers away to safety.

The believers

The other members of the group—a physician, his wife, and other middle-aged professionals, quit their jobs, left their husbands and wives, and gave away their money and possessions to get ready for the trip. As one put it, "I've given up just about everything. I've cut every tie; I've burned every bridge. I've turned my back on the world. I can't afford to doubt. I have to believe."

Leon Festinger and his colleagues saw the headline in a local newspaper—PROPHECY FROM PLANET CLARION. CALL TO CITY: FLEE THAT FLOOD—and decided to infiltrate the group and follow the action. As social scientists they wanted to observe what psychological processes would happen when the prophecy failed. In October they went to visit Mrs. Keech, and managed to join the cult.

Occasionally they got into difficulties; one evening Mrs. Keech invited researcher Hank Riecken to lead the evening session. He was terrified of refusing, and therefore arousing suspicion, and also of putting his foot in it and ruining everything; so he agreed, and then as the session began, he raised his hand and said, "Let us meditate."

As the fateful day approached, the researchers noted that the group avoided publicity, gave interviews grudgingly, and allowed only true believers into Keech's house.

Preparing for liftoff

On December 20 the group expected a "Guardian" from outer space to call on them at midnight and take them to a waiting spacecraft. In the evening they carefully removed all metal

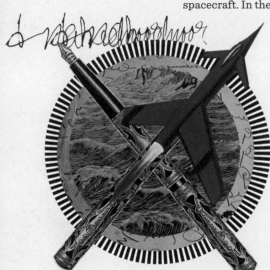

objects, including coins, rings, buttons, zippers, belt buckles, and bra straps.

At 12:10 am there was still no visitor and the group was reduced to horrified silence. At 4 am Mrs. Keech began to cry.

At 4:45 am another message came to Keech by automatic writing. It said that the little group, sitting all night long, had spread so much light that the God of Earth had decided to save the world from ultimate destruction.

Suppose you were a member of the group; what would you do now? Would you replace all those metal things, leave quietly, sneak back home, and hope that your family and your boss would take you back without too much fuss? One member did just that, but the rest of The Seekers did exactly the opposite.

Resolving the conflict

Changing tack completely, they began an urgent campaign to spread their message to as wide an audience as they could. By

6:30 they had phoned newspapers and arranged interviews, trying to bring the whole world into their belief system. They searched outer space for guidance, made various other prophecies, and began to issue pamphlets detailing them. In other words, far from driving them away, the events actually increased their allegiance to the cult.

Festinger said that for this remarkable turnaround, they must have held their beliefs with deep conviction, must have taken important action that could not be undone, must have understood that the prediction has gone completely wrong, and must have had firm support from the others in the group.

He said they had changed their stance because of the mental stress caused by holding two incompatible beliefs at the same time. He called this "cognitive dissonance." We come across it all the time. Suppose your friend Bob has bought a new car, or a new phone. He will probably say it is the best, fastest, most efficient, most economical, and so on. This may all be true, but what is more important is that he has invested time and money in making the purchase, and does not want you or anyone else to suggest it is not altogether perfect.

In the case of The Seekers, Festinger suggested that once the world had not ended, and they were faced with severe cognitive dissonance, the believers found it easier to modify the original prophecy, and to accept the additional belief that the aliens had actually saved the world because of their efforts rather than accept the prophecy, and therefore their deeply held beliefs, was flawed.

1956

THE STUDY

RESEARCHER:

Solomon E. Asch

SUBJECT AREA:

Social psychology

CONCLUSION:

A percentage of subjects
will agree with a group
decision, even if they
believe it to be wrong.

WOULD YOU BUCKLE UNDER PEER PRESSURE?

ASCH'S EXPERIMENTS IN CONFORMITY

Would you always say what you believe to be true, even if several other people say you are wrong? How independent are you?

Groups of individuals often seem to take group decisions: "Let's all go to the restaurant," or "We'll all sing 'Happy birthday to you.'" Sometimes, however, one or two will disagree, and decide to do something else. Behavioral psychologist Solomon Asch wanted to measure how much people are persuaded by the rest of the group.

The experiment

A male college student was invited to join a group of other students for a psychological study. He found them waiting in a passage, and they all went into a classroom, where the new recruit found himself sitting last but one in the row of six or seven others. What he did not know was that they were all stooges, and had been given a strict set of rules to follow. He was the only outsider—the "critical subject."

An experimenter came in and explained that what they had to do was estimate the relative lengths of lines. For each trial he put on a stand a card with three black lines of different lengths, and a separate card with one test line, which was the same length as one of the three. The lines varied from 1 to 10 in (2.5 to 25 cm) long. The job of the team was to say which line was the same length as the test line.

Here was the important part: they had to call out their choice aloud, one by one; so the critical subject was the second last to speak, and had already heard several answers before it was his turn. Each experiment had 18 trials, which in fact consisted of nine trials repeated twice.

All the stooges always gave identical answers; so if the first said it was line B all the others said the same. For the first two trials everyone gave the right answers. In the third trial the stooges deliberately chose the wrong line, and the critical subject, often looking puzzled, had to choose whether to say what he thought was the right answer, or to go along with the majority. This was a tough decision, for he had to speak out in public, and to disagree meant that he was saying that all the others were wrong.

Then they carried on with the subsequent trials; the stooges gave right answers in six of the 18 trials, and wrong answers in 12 of them. Curiously, the critical subjects succumbed, and gave wrong answers, most often in trials four and ten, which involved the same set of lines.

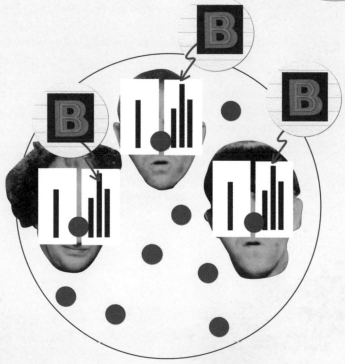

To make sure that choosing the correct line was reasonably easy, Asch ran a series of trials in which a single subject looked at the sets of lines, and wrote down his answers. Without peer pressure, the subjects were accurate more than 99 percent of the time; so the task was not too difficult.

Asch went through dozens of these experiments, and the overall result was that the critical subjects went along with the majority and gave "wrong" answers in 37 percent of the trials. Some critical subjects remained completely independent, and ignored all the others in the group. Others gave in completely, and simply went with the majority every time. Some chose a middle ground, and in 20 percent of the trials gave wrong answers that were not as wildly wrong as those of the majority, but still wrong.

He interviewed them all after the trials, and found that they tried to explain why they had been confused:

- *I thought they were measuring width after a while.*

- *I thought there was some trick to it—an optical illusion.*

- *First I thought something was the matter with me or most of them.*

- *I was sure they were wrong, but not sure I was right.*

The pressure of the group
Late in each interview Asch revealed what had been going on; the victims were all greatly relieved. One even went so far as to say, "The duty of a government is to do the will of the majority, even if you are convinced they are wrong." Others gladly shared their relief:

- *Either these guys were crazy or I was—I hadn't made up my mind which—I was wondering if my judgments really were as poor as they seemed to be, but at the same time I had the feeling that I was seeing them right.*

- *I agreed less because they were right than because I wanted to agree with them. I think it takes a lot of nerve*
to go in opposition to them.

- *When I disagreed I felt outside the group.*

Asch came to a series of conclusions. With only two or three stooges the critical subjects were more likely to be independent, and less swayed by the majority. The pressure of the majority did not build up with time; most of the critical subjects maintained a constant level of independence. So peer pressure really works, although these experiments were only about judging the length of lines; more research would be needed to find out how far it goes. But in general, as one victim put it, "It is hard to be in the minority."

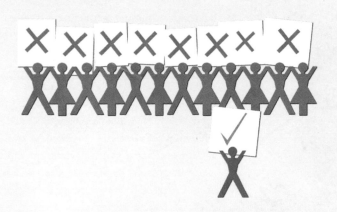

1959

THE STUDY

RESEARCHERS:

H. F. Harlow and
R. R. Zimmermann

SUBJECT AREA:

Developmental psychology

CONCLUSION:

Evidence suggests that
maternal attachment is
based on more than feeding.

HOW DO BABIES FORM ATTACHMENTS?

MATERNAL SEPARATION, DEPENDENCY NEEDS, AND SOCIAL ISOLATION

Why do babies form strong attachments to their mothers? Is it a natural automatic process, or is it something they learn because their mother's feed them? The controversial American psychologist, Harry F. Harlow wrote:

> *Psychologists, sociologists, and anthropologists commonly hold that the infant's love is learned through the association of the mother's face, body, and other physical characteristics with the alleviation of internal biological tensions, particularly hunger and thirst. Traditional psychoanalysts have tended to emphasize the role of attaching and sucking at the breast as the basis for affectional development."*

In other words, either the baby goes to its mother for milk, and learns to associate the food with her face, smell, and feel, and so becomes conditioned to attach to her, or there is an innate evolutionary bond ready to be formed, regardless of the supply of milk. Harlow wanted to find out which of these is right by exploring the effects of mother and infant separation. Realizing that using human babies was impossible, he turned to the rhesus macaque monkeys held at his workplace in the University of Wisconsin Primate Laboratory.

He had to find a way to separate the supply of milk from the warmth and softness of the mother. He noticed that after they had been taken from their mothers, the babies spent some time clinging to their cloth diapers, and this gave him an idea.

Surrogate mothers

He took eight baby monkeys from their mothers between six and 12 hours after birth, and put each one in a cage with two surrogate mothers made of stiff wire mesh, fitted with crude heads. One mother was wrapped in terry toweling cloth, the other was left bare.

In four of the cages the wire mother was equipped with a feeding bottle that dispensed milk; the terry-cloth mother had no milk. In the other four cages only the terry-cloth mother had milk.

Both sets of monkeys drank the same amount of milk and put on the same amount of weight, but the critical observation was that in all eight cages the infants spent most of their time climbing and clinging onto their terry-cloth mothers.

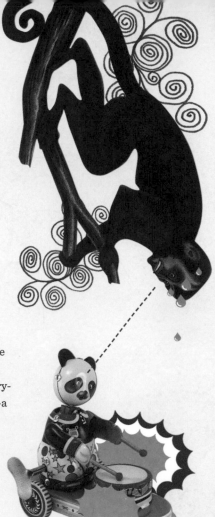

The baby monkeys were left with these surrogate mothers for six months. Generally those whose milk came from the wire mother went briefly to her when they were hungry or thirsty, but spent most of the time with the terry-cloth mother, and formed an affectionate bond with her—a bond that was strong and stable.

The fact that they spent so much time with their soft mothers was evidence that the attachment is not just about food, but something more instinctive. Harlow wondered, however, whether the terry-cloth mothers would provide comfort and security when the babies were frightened. So he presented them with a mechanical bear with a drum that made a loud noise.

Regardless of which mother had provided milk, the terrified babies all cuddled up to the terry-cloth mom. This echoes the behavior of baby monkeys raised by their real mothers. The babies spend many hours every day clinging to their mothers, and run to them for comfort and reassurance when they are frightened.

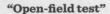

"Open-field test"

He put each baby in novel environments with strange objects, and found that if their surrogate mothers were there, they would cling to her for a while, then go off and explore, and run back to her when they got scared. If she was not there they would just curl up in a corner and stay there, sucking their thumbs.

Human babies who are deprived of affection often have difficulty forming emotional ties later in life. Harlow found similar behavior in his baby monkeys. He took four babies from their mothers and left them without even surrogate mothers to bond with. After eight months he put them into cages with both cloth and wire surrogate mothers, but they failed to form attachments to either surrogate. Harlow concluded that baby monkeys develop normally only if they can cling to a cuddly object during their first few months of life; clinging seems to be a natural, automatic response to stress.

On the other hand they seemed to need social rather than maternal interaction. Another group of four babies were brought up on their own, but put for 20 minutes every day into a cage with the other three. These monkeys grew up with relatively normal emotional and social behavior.

While Harlow's work casts some light by inference on the behavior of human babies (it suggests, for example, that prolonged bodily contact with the mother is beneficial), it has also, however, been heavily criticized as unnecessarily cruel. His babies never became entirely normal; when placed in a cage with a monkey that had been brought up by its mother the "orphan" would huddle miserably in a corner, and always show signs of unhappiness. He also created much anxiety in the mothers that were deprived of their babies; they often became neurotic, and attacked their babies furiously if they were reintroduced.

HOW SHORT IS SHORT-TERM MEMORY LOSS?

THE FAST DECAY OF ICONIC MEMORY

1960

THE STUDY

RESEARCHER:
George Sperling

SUBJECT AREA:
Cognition and memory

CONCLUSION:
Human possess a powerful short-term visual memory.

If you saw a row of random letters, such as **NDRKSQ**, how many could you remember? And what if you saw a whole grid of letters? George Sperling, an American psychologist working at the famous Bell Laboratories in New Jersey, wanted to find out how much can be seen in a brief exposure, and how long can we remember it. We normally see the world in brief glimpses between eye movements called "saccades"; so presumably we must get our whole world view in brief glimpses. Sperling set up a simple and elegant set of experiments to investigate.

One problem was finding a way to show images so briefly. He used a tachistoscope, which allowed him to show a card to the subject for a fraction of a second. Today we would use a computer, but research labs didn't have computers in 1960, and the two-field mirror tachistoscope was considered state-of-the-art equipment.

From 22 in (55 cm) away the subjects were shown 5 x 8 in (12 x 20 cm) cards, each with an array of half-inch-high letters. The tachistoscope was generally set to give a flash of light lasting 50 ms (one twentieth of a second). There were 500 different cards; so no one learned any of the letter patterns, except for a few striking ones, such as **XXX**.

Some cards had three, four, five, six, or seven letters in a line, either normally spaced or bunched together. Others had two or three rows of letters, either spaced or bunched. No vowels were used.

RNFBTS

**KLBJ
YNXP**

Experiment 1

The subject saw one of these arrays for 50 ms, and then had to write down the letters in the correct places on a grid. They were told to guess any they were not sure about. In each run the subjects saw between five and 20 cards, at a pace of their own choosing—generally about three or four per minute.

All the subjects could consistently score 100 percent when there were only three letters. When there were more letters, the subject's scores (immediate-memory spans) varied, but remained almost constant for each subject, ranging from 3.8 to 5.2. The average immediate-memory span was about 4.3 letters. The arrangement of the letters, whether bunched or spaced, and whether in one row or more, did not make much difference.

Experiment 2

Sperling then tried varying the length of the flash, exposing the card for periods from 15 ms up to 500 ms (half a second). Surprisingly, this made no appreciable difference to the scores; the limit stayed the same.

Experiment 3

Sperling noticed that the subjects often said that they had seen more than they could remember afterward, which seemed very odd. Do we really see things and then forget them so quickly? He realized that this meant that the question "What did you see?" asked them to report both what they could remember and what they had forgotten.

He devised a brilliant procedure to find out whether they had really seen more than they could say. He actually showed them far more information than they could possibly report, but asked them to report only some of it.

This time the subjects saw two rows of letters, with either three or four letters in each row. They were told that immediately after the light had gone off, a tone would sound for half a second. If it was a low tone they had to report the lower row; if it was a high tone they had to write down the upper row. Later they saw three rows, and three tones.

high tone	D W R M
medium tone	S K Z T
low tone	Q M C R

The results were astonishing. The subjects scored a higher percentage of correct letters than they had before, and their accuracy increased day by day with practice; after seeing cards with 12 letters, their average score reached 76 percent. In other words they must have "seen" about nine of the 12 letters.

This suggests that at the time of the exposure, and for a few tenths of a second afterward, the subjects had two or three times as much information available as they could later report. Sperling writes "the persistence is that of a rapidly fading, visual image of the stimulus."

This graph shows the number of letters available. The numbers of letters on the card are shown along the bottom axis. The maximum score is the diagonal line. The lower curve shows the immediate memory (experiment 1); the upper curve shows the number of letters available, as revealed by experiment 3.

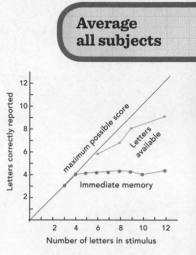

Average all subjects

Iconic memory

Shown a white card immediately after the card with letters, the subjects did much worse, suggesting overload of the visual system, and supporting the idea that the letters must have been held as a persisting visual image.

In other words, Sperling had discovered a quickly forgotten, picturelike memory. Nobody had ever suggested such a thing before. Today it is called "iconic memory," or visual short-term memory (VSTM).

THE STUDY

RESEARCHERS:
A. Bandura, D. Ross,
and S. A. Ross

SUBJECT AREA:
Developmental and social
psychology

CONCLUSION:
Children exposed to
aggressive behavior are
more likely to act in
physically aggressive ways.

IS AGGRESSIVE BEHAVIOR LEARNED?

THE BOBO DOLL EXPERIMENT

Children are bombarded by aggressive behavior on television and in electronic games. By the age of 10 or 11, most children will have seen thousands of murders and hundreds of thousands of acts of violence, all of which tends to be glamorized. Even cartoon characters get swatted, flattened, or thrown off cliffs. Does this endless violence encourage kids to be violent?

That is what Albert Bandura and his colleagues wanted to find out. So they took three groups of children aged between three and six, exposed one group to aggressive behavior by adult role models, one group to normal behavior, and left the control group alone; and then watched to see what the children would do.

The researchers expected that girls would be more likely to imitate a female role model, and boys a man. They also expected boys to be more aggressive, especially after watching an aggressive male role model.

The experiment

Each child was taken into a room, and the role model was then invited to come in and join the game. The child sat in one corner with a table, and material to make pictures using potato prints and stickers. The role model went to the opposite corner, where there was a small table and chair, a Tinkertoy set, a mallet, and a 5-ft (1.5-m) Bobo doll—a life-size inflated character that always bounces up again when pushed over.

In half the trials the models were not aggressive, they sat and assembled the Tinkertoys, ignoring the Bobo doll. In aggressive trials the models worked on the Tinkertoys for a minute, and then spent the rest of the time violently attacking the Bobo doll: punching it, sitting on it and punching it repeatedly on the nose, lifting it up and bashing it on the head with the mallet, then tossing it up in the air and kicking it furiously around the room. They then repeated these actions about three times, shouting "SOCK HIM IN THE NOSE... HIT HIM DOWN... THROW HIM IN THE AIR... KICK HIM... POW."

Ten minutes later the female experimenter came back, and took the child to another building. After two minutes' play in an anteroom they went into the observation room, where she waited, but sat in a corner keeping busy with paperwork at a desk, and avoiding any interaction with the child.

In this experimental room was a variety of toys, from crayons and paper, a ball, dolls, bears, cars, and trucks, to plastic farm animals; but also a mallet and a 3-ft (1 m) inflated Bobo doll. Each child spent 20 minutes in this room, while their behavior was rated by judges, watching through a one-way mirror.

After they had watched the nonaggressive model, the boys were almost invariably more aggressive than the girls, especially with the mallet, and the Bobo doll. Whether the role model had been male or female did not seem to make an enormous amount of difference, and the amount of aggression was much the same as was shown by girls and boys in the control group, who had not witnessed the actions of the role models, although both girls and boys in the latter group were decidedly vicious with the mallet. An interesting feature was that both boys and girls who had watched a nonaggressive man showed less aggressive behavior than those who had not watched anyone.

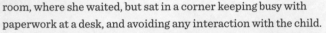

There were much greater changes in the children who had witnessed aggressive behavior. Girls turned out to be considerably more aggressive than boys in their shouting, after seeing a female role model shouting, while boys shouted more after seeing the man shouting.

The Bobo doll

Curiously, in punching the Bobo doll the boys were more aggressive after watching a woman doing it, but the girls were more aggressive after watching a man.

In almost every category the children who had watched aggressive behavior were noticeably more aggressive than those who had not, which confirmed the researchers' predictions that violent behavior would be learned. They concluded that in masculine type behavior such as physical aggression, both boys and girls imitate a man more than a woman. On the other hand for verbal aggression (" HIT HIM DOWN . . . THROW HIM IN THE AIR . . . KICK HIM . . . POW") both boys and girls imitated the model of their own sex more than the other.

There were noticeable sex differences in the children's comments about the aggressive models: "Who is that lady? That's not the way for a lady to behave. . . . She was just acting like a man. I never saw a girl act like that before. She was punching and fighting." And on the other hand, from a girl "That man is a strong fighter, he punched and punched and he could hit Bobo right down to the floor. . . . He's a good fighter, like Daddy."

These experiments have been quoted thousands of times, and yet still the argument goes on: does seeing violence on the screen make children violent? We still do not know the answer.

DO YOU WANNA BE IN MY GANG?

GROUP MENTALITY AND THE GANGS OF ROBBERS CAVE

1961

THE STUDY

RESEARCHERS:
M. Sherif, O. J. Harvey,
B. J. White, W. R. Hood, and
C. W. Sherif

SUBJECT AREA:
Social psychology/conflict
theory

CONCLUSION:
Conflict arises from
competition for resources
rather than individual
differences

What is it about groups or gangs that creates tension, and what can be done to prevent it? All over the world there are problems when rival groups compete for scarce resources. People complain bitterly about immigration, because they think that foreigners will take their jobs. Inner-city gangs compete over drugs and territory. Competition over land, or water, or oil, frequently escalates into war and even genocide.

Social psychologist Muzafer Sherif, born and educated in Turkey, was interested in Realistic Conflict Theory. He decided to set up deliberate conflict between groups, and then see how it could be resolved.

Robbers Cave

He invited two groups of 12-year-old boys to a summer camp at the Robbers Cave State Park in Oklahoma. The 11 white middle-class boys in each group did not know one another; nor did they know about the other group.

For the first week both groups went swimming and hiking, and did baseball practice together, and so established their own cultures. One group called itself the Eagles; the other group the Rattlers, and they stenciled the names on their T-shirts and flags.

Competing for limited resources

Next the researchers arranged for the groups to compete over several days at baseball, tug-of-war, touch football, tent pitching, and a treasure hunt, and promised prizes to the winners—a trophy for the team, and a medal and a four-bladed knife for each

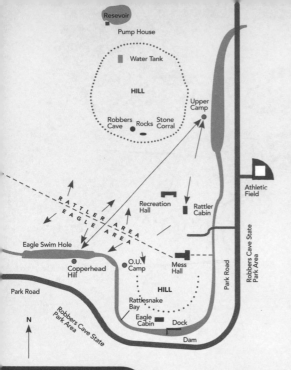

The figure shows a map with the following labels:

Resevoir
Pump House
Water Tank
HILL
Upper Camp
Robbers Cave
Rocks
Stone Corral
RATTLER EAGLE AREA
Recreation Hall
Rattler Cabin
Athletic Field
Eagle Swim Hole
Copperhead Hill
O.U. Camp
Mess Hall
Park Road
Robbers Cave State Park Area
HILL
Park Road
Rattlesnake Bay
Eagle Cabin
Dock
Dam
N
Robbers Cave State Park Area

General layout of the campsite and respective areas of the two camps

member—but nothing to the losers, in order to provoke frustration. When they first heard of these competitions, the Rattlers were totally confident they would win.

The competition was level until the final event, the treasure hunt, which the researchers rigged so that the Eagles won. They were "jubilant at their victory, jumping up and down, hugging each other, making sure in loud tones that everyone present was aware of their victory. On the other hand, the Rattlers were glum, dejected, and remained silently seated on the ground."

One group went for a picnic, but were delayed on the way, and when they arrived found that the others had eaten all their food.

The friction increased. The groups started name calling and taunting each other. The Eagles burned the Rattlers' flag; the Rattlers ransacked the Eagles' cabin, overturning beds and stealing things. Both groups became so heated that the researchers had to physically keep them apart.

Sherif noted that:

> The derogatory attitudes toward one another are not the consequence of preexisting feelings or attitudes. . . . They are not the consequence of ethnic, religious, educational, or other background differentiation among the subjects. The state of friction was produced systematically through the introduction of conditions of rivalry and frustration perceived by the subjects as stemming from the other group.

They then spent two days cooling down, but Sherif found, as he expected, that bringing the groups together was not enough; they kept calling one another names, and at meals they threw food and napkins at one another. The best way to resolve the conflict, he decided, was to present the groups with a problem too large for either to solve alone—Sherif called it a "superordinate goal"—so that they had to work together.

The researchers cut off the water supply at the tank above the camp, then announced that to sort it out would need about 25 people. Members of both groups volunteered. By the time they reached the tank they were thirsty, but could get no water. At this stage they began to cooperate to remove the sacking that had been blocking the pipe (put there by the experimenters).

The next task was over money. The boys were told they could see a movie, but it would cost $15 to get it from town, and the camp could afford only $5. After much discussion and voting, both groups agreed to contribute, and they all enjoyed the movie.

Later both groups went by truck to camp at Cedar Lake, where they were induced to work together to tow a truck which was "stuck," and then both groups agreed to cook food for everyone, on alternate days.

Finally they all went home on the same bus, and even agreed at a rest stop on the way that the $5 prize one group had won should be spent on buying drinks for everyone.

Nearing Oklahoma City, the boys at the front of the bus (mostly high status members from both groups) began to sing "Oklahoma." Everyone in both groups took part, all sitting or standing as close together as possible in the front end of the bus. A few boys exchanged addresses, and many told their best friends that they would meet again.

Conclusions

From the study, Sherif reckoned that because the groups were created to be approximately equal, individual differences are not necessary for intergroup conflict to occur. When the boys were competing for valued prizes, hostile and aggressive attitudes arose because they were competing for resources that only one group could get.

CHAPTER 4: Mind, brain, and other people: 1962—1970

By the mid-60s, psychology was flourishing. The field was becoming a "respectable" science, and noted courses were opening in both universities and high schools worldwide. The growing numbers of psychologists and experimenters explored a widening range of topics—such as how bystanders react to emergencies or the invasion of personal space. The interest in social psychology and human behavior in group situations blossomed, especially after Milgram's influential studies into obedience and subordination to authority.

In addition, the 1960s also saw the beginnings
of new technology, such as the invention of
electroencephalography (EEG), which gave the first looks
inside the living brain. As technology advanced, so did the
opportunities to combine neuroscience and psychology,
leading to a wealth of new directions for the field.

1963

THE STUDY

RESEARCHER:

Stanley Milgram

SUBJECT AREA:

Social psychology

CONCLUSION:

Some subjects will obey
an authority figure who
instructs them to act
against their conscience.

HOW FAR WOULD YOU GO?

THE MILGRAM EXPERIMENT

Yale psychology professor Stanley Milgram wanted to find out how far volunteers would go in obeying authority. He was prompted by C. P. Snow's 1961 comment that "more hideous crimes have been committed in the name of obedience than have ever been committed in the name of rebellion." He was also painfully aware that before and during World War II millions of innocent people were slaughtered on command, in the gas chambers of death camps.

"Teachers" and "learners"

Milgram invited 40 volunteers to take part in a learning experiment, ostensibly to test the effect of punishment on memory. Each volunteer met another person in the lab at Yale University; the procedure was explained to them by an impassive, stern-looking experimenter in a gray lab coat, and they were invited to draw slips of paper from a hat to find out which of them would be "teacher" and which would be "learner." In practice this was rigged; all the slips of paper were marked "teacher"; so the volunteer was always the "teacher."

The "teacher" then saw the "learner" being strapped into a chair and an electrode attached to his wrist. If the "teacher" queried this, the experimenter explained that "Although the shocks can be painful, they cause no permanent damage."

The "teacher" was then taken to another room, and was able to communicate with the "learner" only by sound, using a microphone and headphones.

For the memory test, the "teacher" read out a list of word pairs. Then he repeated the first word, and four possibilities for

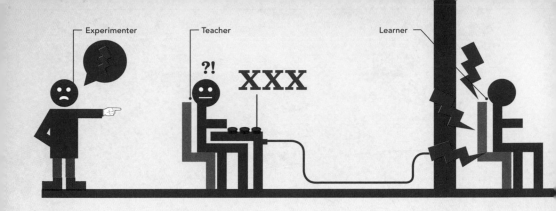

its pair. If the "learner" got it right the "teacher" moved on to the next word in the list. If it was wrong, he was to administer an electric shock, by pressing a switch. There were 30 switches in a line, and the teacher was to start at the beginning and move up one switch at each wrong answer.

How far would they go?

The first shock was only 15 volts ("Slight shock"), but they went up to higher and higher voltages—30 volts, 45 volts, 60 volts, and so on up to 420 volts ("Danger: severe shock") and a maximum of 450 volts (simply marked "XXX"). As each switch was pressed, a pilot light was illuminated in bright red; an electric buzzer sounded; an electric blue light labeled "voltage energizer" flashed; and the voltage meter swung to the right.

In order to convince the "teacher" of the authenticity of the generator before the experiment began, he was given a 45-volt sample shock.

In fact the "learner" was a trained confederate of Milgram's—a 47-year-old accountant—and the "shock generator" was a dummy; he received no real shocks. The "learner" kept answering questions, often wrongly, up to 300 volts; after that he did not speak at all. What the "teacher" did hear, however, was pounding on the wall between them.

At this point the "teacher" usually asked the experimenter for guidance, and was told to wait ten seconds for an answer, and

then deliver the next higher shock. After 315 volts he again heard pounding on the wall, but from this point there was no further response to increasing shocks.

At some point the "teacher" usually asked whether he should go on with the test. The experimenter replied with a series of prods, delivered politely but firmly:

1. Please go on.
2. The experiment requires that you continue.
3. It is absolutely essential that you continue.
4. You have no other choice; you must go on.

How many "teachers" do you think would refuse to carry on the sadistic procedure? You might imagine that most of them would quickly have refused to go on, and a group of psychologists predicted that at the very worst three percent of the volunteers would go right through the sequence. In fact none of the "teachers" stopped below 300 volts, and no fewer than 26 carried right on to the final 450-volt shock.

As each experiment proceeded, all the "teachers" sweated profusely, trembled, stuttered, groaned, dug their fingernails into their palms, and 14 of them broke into nervous laughter. At some point a few of them did actually refuse to carry on.

Responsibility?

The ramifications of this astonishing series of experiments are profound. In the twenty-first century prison guards from the Nazi death camps are still being hunted down and tried for war crimes, but were they just obeying orders? Groups of soldiers of many nationalities have been accused of hideous atrocities—including raping and murdering innocent civilians—but were they just obeying instructions from superior officers? And if so, does this relieve them of responsibility for their actions?

CAN YOU EVER RECOVER FROM BLINDNESS?

LEARNING TO SEE AT THE AGE OF 50

1963
THE STUDY

RESEARCHERS:
R. L. Gregory and
J. G. Wallace

SUBJECT AREA:
Cognition and perception

CONCLUSION:
Sensory experience is not
straightforward.

What would it be like to recover your sight after being blind for 50 years? Born in 1906, S. B. lost the sight of both eyes at the age of ten months after a smallpox vaccination. At the Birmingham Blind School in England he turned out to be bright, and good at mental arithmetic; he learned to recognize capital letters by feeling plastic ones. He was ambitious, and became adept at carpentry, knitting, and boot repairing. When he left school he got a job repairing boots at home.

As R. L. Gregory later discovered, "He was proud of his independence as a blind man. . . . He would go for long cycle rides, holding the shoulder of a friend, and he was fond of gardening, and making things in his garden shed."

Seeing again

To clear his vision he had surgery in December 1958 and in January 1959. When the bandages were removed, the first thing he saw was the surgeon's face. He heard a voice coming from in front of him and to one side: he turned to the source of the sound, and "I saw a dark shape with a bump sticking out and heard a voice, so I felt my nose and guessed the bump was a nose. Then I knew if this was a nose, I was seeing a face."

The surgeon reported that:

After the operation he . . . could recognize faces and ordinary objects (i.e., chairs, bed, table, etc.) immediately. His explanation is that . . . he had a definite and accurate mental image of all things he was able to touch.

Gregory and Wallace first saw S. B. seven weeks after the first operation. He "struck us immediately as a cheerful, rather extrovert and confident, middle-aged individual. . . . He could even tell the time by means of a large clock on the wall. We were so surprised at this that we did not at first believe that he could have been in any sense blind before the operation. However he proceeded to show us a large hunter watch with no glass, and he demonstrated his ability to tell the time very quickly and accurately by touching the hands."

Three days after the operation he saw the moon for the first time. When he was told it was the moon, he expressed surprise at its crescent shape, expecting a "quarter moon" to look like a quarter piece of cake.

Illusions that failed

Gregory showed him some well-known illusions, including the Poggendorf Illusion and reversing depth illusions:

Most people say the diagonal line is not straight, but S. B. simply said "All one line."

Do the Necker cube and the steps look three-dimensional? Can you get them to flip inside out, from one perspective to the other? Can you see the steps from underneath? For most people the Necker cube and the steps will turn "inside-out" as you look at them—the steps can be seen

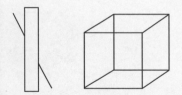

either from above or from below—but S. B. saw no depth, and the figures did not reverse. Asked to draw pictures, S. B. was at first slow and incompetent, but gradually improved. Below are three of his drawings of a bus, done 48 days, six months, and one year after the first operation.

The drawings clearly improved, and he learned to draw capitals and then lower-case letters, but he could not draw the hood of the bus, never having touched it.

One of the features Gregory and Wallace found intriguing was how S. B. transferred information from early touch experience to vision many years later. He could recognize capital letters that he had learned by touch, but could not recognize lower case letters that he had not learned by touch.

In the Science Museum in London, he was fascinated by the Maudslay screw-cutting lathe. From outside the glass case he could not "see" it, but when the case was removed and he was allowed to touch it, "He ran his hands eagerly over the lathe, with his eyes tight shut. Then he stood back a little and opened his eyes and said: 'Now that I've felt it I can see.' He then named many of the parts correctly and explained how they would work."

In his book *Dioptrics* (1637), René Descartes wrote that blind men "feel things with perfect exactness that one might almost say that they see with their hands."

S. B. said "I'd pick up a fork, feel it, and remembering how a fork felt when I was blind I could say: 'This is a fork.' Then I had to learn to remember it the next time I saw it."
A year after the operation, however, he became depressed, and he died on August 2, 1960.

As Gregory wrote, "We have ascertained that vision, although it may prove genuinely useful to the man long blind, is at the same time a potential source of grievous hurt."

1965

THE STUDY

RESEARCHERS:

E. H. Hess

SUBJECT AREA:

Experimental psychology

CONCLUSION:

Study of the eyes can give an
indication as to what
the brain is doing.

ARE YOUR EYES THE WINDOWS TO YOUR SOUL?

PUPIL SIZE AS AN INDICATOR OF INTEREST OR EMOTION

Eckhard Hess wrote, "When we say that someone's eyes are soft, hard, cold, or warm, we are in most instances referring only to a certain aspect of that person's eyes: the size of the pupils."

The pupils are the small black holes in the center of the eyes. Their size is controlled by the autonomic nervous system, and normally varies according to the intensity of the light; so they dilate—get bigger—in a dark room, and shrink to pinpoints in sunlight. Hess hoped that simply by measuring the dilation of the pupils while he gave volunteers things to look at or think about, he might be able to find out something about what was going on in the brain at the time. He devised a clever technique to record pupil size.

Dilated pupils

Hess had shown that most people's pupils dilate when they see something interesting or exciting. The pupils of heterosexual men and women dilate when they see a scantily clad and attractive member of the opposite sex. Homosexuals' pupils dilate when they see an attractive member of the same sex. Women's pupils dilate when they see pictures of babies, or mothers and babies.

In a more subtle study, he showed to a group of 20 men two pictures of the same woman, except that in one version her pupils were enlarged and in the other they were constricted. The men showed marked preference for the picture with enlarged pupils, even though afterward most said the pictures were identical—but one or two said she was soft, prettier, or more feminine, whereas the "other" woman was hard, selfish, or cold. They could not explain their judgments. Hess concluded that pupil size is important in nonverbal communication.

There is a drug called belladonna, which means "beautiful lady." Women used to put drops of it into their eyes, because they thought it made them more beautiful. The active ingredient in belladonna is atropine, which makes pupils dilate. An eyewash containing atropine used to be popular with American women until it was banned by the Food and Drug Administration.

The curious thing is that people don't explicitly know that bigger pupils make people look better and happier, and yet Hess showed that when asked to draw in the pupils on line drawings of sad and happy faces, both adults and children drew bigger pupils on the happy faces.

Mathematical problems

Hess went on to investigate the size of pupils while the subjects were struggling with problems in mathematics. As they thought about each problem, their pupils gradually dilated, until the point when they got the answer, and then the pupils suddenly constricted back to their normal size.

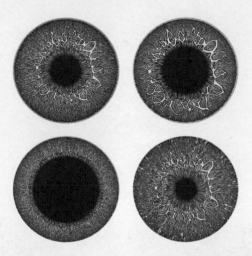

LEFT: Pupils dilating and constricting.

85

He asked volunteers to solve multiplications of increasing difficulty: 7 x 8   9 x 17   11 x 21   16 x 23

While they were tackling the simplest sum, their pupil sizes increased by an average of 4 percent. When they tackled the hardest the increase was a massive 30 percent. In all cases they then dropped back to where they had been before. In other words the amount of dilation seemed to be a measure of the amount of cognitive effort involved.

Later studies showed that pupil dilation during such tasks was greater for average students than for bright students, suggesting that the less brainy ones had to put relatively more effort into solving the problems.

Hess also showed photographs of food to hungry people and to people who had just eaten. The hungry group showed greater pupil dilation; in the full group some pupils actually contracted; they really did not want more food.

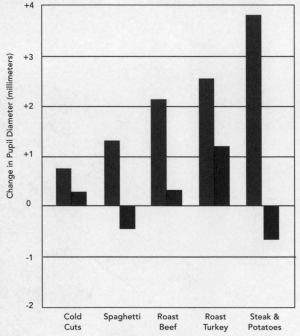

Kahneman and Beatty asked subjects to repeat strings of digits—e.g., 538293. They found that the pupil size increased step by step as subjects heard each successive digit, and then decreased step by step as the subjects repeated each one, until they were the same size as before. The same thing happened when they were asked to call out telephone numbers from memory, although pupil dilation was greater in this memory task.

So even if the eyes are not the windows to the soul, they do tell us something interesting about what the brain is doing.

"ARE YOU SURE, DOCTOR?"

THE HOFLING HOSPITAL STUDY

1966
THE STUDY

RESEARCHERS:
C. K. Hofling,
E. Brotzman,
S. Dalrymple, N. Graves,
and C. M. Pierce

SUBJECT AREA:
Social psychology

CONCLUSION:
Subjects will knowingly
break rules designed to
save lives, if ordered to do
so by an authority figure.

The Milgram experiment (see page 78) of 1963 raised many questions about obedience and the power of authority. A few years later the American psychiatrist Charles K. Hofling conducted a similar test to explore this area further. He and his colleagues were aware that doctors sometimes tread on the toes of nurses, for example by walking into an isolation unit without taking the necessary precautions, or by telling a nurse to do something that contravenes professional standards. The researchers wondered whether this behavior would extend to nurses. Would they knowingly endanger a patient if instructed to do so by a doctor?

They carried out their experiments in 12 departments at a public psychiatric hospital and in ten departments at a private psychiatric hospital. They also asked a control group of nurses and a group of nursing students what they would do in these circumstances.

Astroten

"Dr. Smith from the Psychiatric Department" (actually a stooge) telephoned each one of the 22 night nurses, and asked them whether they had any Astroten in store. Astroten was actually a fake medicine—tablets of harmless glucose that could not cause any damage. Packets of Astroten tablets had been put

in the drug store. "Dr. Smith" then ordered them to give "Mr. Jones" a dose of 20 milligrams of Astroten—he said it was urgent. He said he was running late, and would sign the paperwork when he got to the hospital in ten minutes' time.

The label on the bottle said *Astroten 5mg. Maximum dose 10mg. Do not exceed stated dose.*

Here was the problem for the nurses: Clearly 20mg was much too high a dose; the order was given over the telephone, which was against hospital policy; the medicine was unauthorized; it had not been put on the ward stock list and cleared for use; finally, "Dr. Smith" was unknown to any of the nurses.

Furthermore it was the middle of the night; the nurses were alone on night duty and therefore unable to make contact with anyone else at the hospital.

What would you do?

Suppose you had been one of those nurses. The patient's life might have been at risk. What would you do?

In the control group, 10 of the 12 nurses said they would not have given the medication, and all 21 nursing students said they would have refused. In the real live study, however, 21 out of 22 nurses would have given the medication as ordered, although in practice they were stopped at the bedside by a researcher and doctor, and explained what was going on.

Most of the telephone conversations were short, and most of the nurses did not offer much resistance; none were hostile. Afterward 16 of them felt they should have been more resistant.

In retrospect

Only 11 of those who would have given the medicine admitted that they knew the dosage limit; the other 10 did not notice, but reckoned it must have been safe if the doctor said so. Nearly all of them admitted that they should not have contravened hospital policy: they should not have taken orders over the phone; they should have checked that "Dr. Smith" was a real doctor; and they should not have given drugs that were not authorized.

Most of them said, however, that it was normal practice to obey a doctor's instructions without question; 15 remembered similar incidents, and said that doctors got angry with nurses who resisted their orders.

Hofling and his colleagues concluded that "Nurses will knowingly break hospital rules in a way that endangers a patient's life, if given orders by a doctor."

Rank and Jacobson carried out a similar trial a few years later. Nurses were asked to give a nonlethal overdose of Valium to appropriate patients. In this case the nurses were able to speak to colleagues, and 16 out of 18 refused to administer the drug. This was explained as due mainly to knowledge of the drug's effects and contact with colleagues, but also the result of an increased willingness to challenge doctors' orders, rising self-esteem among nurses, and fear of lawsuits.

In 1995 Smith and MacKie reported that there was a daily 12 percent error rate in U.S. hospitals, and that "Many researchers attribute such problems largely to the unquestioning deference to authority that doctors demand and nurses accept."

1966

THE STUDY

RESEARCHERS:

N. J. Felipe and R. Sommer

SUBJECT AREA:

Social psychology

CONCLUSION:

Uninvited invasions
of our personal space
have a disruptive effect.

ARE YOU A SPACE INVADER?

RESEARCH INTO PERSONAL SPACE

Personal space is that area around you into which you don't like anyone else to enter uninvited. In the 1960s, the American psychologists Felipe and Sommer spent two years sitting very close to complete strangers, to see how long it would be before their subjects moved away.

They were a little worried about where to try their experiments, but eventually settled on Mendocino State Hospital, a facility for those suffering from mental illness: "We had visions of a spatial invasion on a Central Park bench resulting in bodily assault or arrest . . . [but] it seems that almost anything can be done in a mental hospital, provided it is called research."

Getting up-close and personal

The hospital was set in parklike surroundings and there was easy access to the grounds; so patients could easily isolate themselves by finding a deserted area.

Felipe and Sommer carried out their research both inside and outside, choosing subjects who were male, sitting alone, and not doing anything too engrossing, such as reading or playing cards. A male researcher would sit beside the subject, without saying a word, at a distance of about 6 inches (15 cm). If the subject moved his chair or moved further down the bench, the researcher would also move the same amount and maintain the small space between them.

The researcher either just sat there or took notes about what was going on. He also took note of other patients sitting some distance away, as controls. In all they invaded the space of

64 patients, for a maximum of 20 minutes each. Usually the patient would immediately turn away from the researcher, pull in his shoulders, and place his elbows at his sides. Within two minutes, 36 percent of the patients had moved away, while none of the controls had moved. At the end of 20 minutes, 64 percent of the patients had moved. It was also found that writing notes was slightly more effective than not writing in driving victims away.

Intimate Zone
6 to 18 in
(15–45 cm)

Friend Zone
18 in to
4 ft (45 cm
–1.2 m)

Social Zone
4 to 11 ft
(1.2–3.5 m)

In one ward five patients were extremely territorial, and sat in the same chairs day after day. Two of these patients absolutely would not move. The researcher described them as fixed like "the rock of Gibraltar."

Personal space

Influence of gender

Next the researchers decided to carry out further research in large room at a university library—where the students tended to sit as far away from one another as possible.

The researcher, female this time, would go in and deliberately sit down next to one of the female students, completely ignoring her. She then unobtrusively moved closer, so that their chairs were only about 3 in (7.5 cm) apart. Then she leaned over her book, in which she took notes, and tried to maintain a distance of about 12 in (30 cm) between their shoulders. This was sometimes difficult, because the library chairs were wide, and the student would occasionally slide across to the other side of her chair.

If the student moved her chair away, the researcher followed by pushing her chair backward at an angle, and then forward again, under the pretense of adjusting her skirt.

Many of the students quickly drew in their arms, turned away, put their elbows on the table, or made piles of books, purses, and coats as barriers to separate themselves from the researcher. She sat there for a maximum of 30 minutes, at the end of which time 70 percent of the victims had moved away. Only two of the students spoke to the researcher, and only one of the students asked the researcher to move over.

The researcher also tried sitting next to the student, but keeping a normal distance of around 15 in (38 cm) between the chairs, or 2 ft (60 cm) between their shoulders. She also tried sitting one or two spaces away from the student, or on the opposite side of the table. None of these invasions of space had much effect.

Conclusions

Felipe and Sommer concluded that invasions of personal space have a disruptive effect, ranging from discomfort to flight, and also quoted the Australian behavioral scientist, Glen McBride, who had observed that when the dominant bird in a flock approaches, the other birds will look away and move aside to make extra space for it.

The intensity of the subject's reaction was influenced by many factors including territoriality, the dominance-submission relationship between invader and subject, and the "attribution of sexual motives to the intruder" (even though intruder and subject were always of the same gender).

However, they also noted that notions of personal space are culture-dependent: Japanese people and people from Latin countries stand closer together than Americans, for example.

WHAT HAPPENS WHEN A BRAIN IS CUT IN HALF?

CONSCIOUSNESS AND FUNCTIONAL HEMISPHERECTOMY

1967
THE STUDY
RESEARCHERS:
S. Gazzaniga and
Roger W. Sperry
SUBJECT AREA:
Neuropsychology
CONCLUSION:
Splitting the brain
seems to create two
separate spheres of
consciousness.

During the 1960s, some patients with severe epilepsy were treated with a drastic operation, a hemispherectomy, during which surgeons would cut through the corpus callosum (the neural fibers that connect the left and right hemispheres of the brain) to prevent seizures from affecting both sides.

After the operation, patients did not respond to touching the left side of their bodies. When an object was placed in the left hand they said it wasn't there. Astonishingly, however, they generally recovered well, and were able to lead normal lives. Their IQ, speech, and problem-solving abilities were not much changed.

However, the American psychologists Michael Gazzaniga and Roger Sperry came up with some experiments to demonstrate the depth of changes that had really taken place.

When a patient held something (say a spoon) in his right hand, he could say what it was and describe it. When he held it in his left hand, he could not describe it, but given a collection of similar objects (knives, forks, etc.) he was able to match it with another spoon.

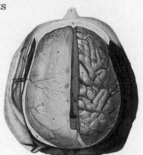

RIGHT: Surgeons cut through the corpus callosum (the blue area) to separate the two sides of the brain.

Patients were seated in front of a screen and told to fixate on a spot in the center. This fixation was important, for as long as

93

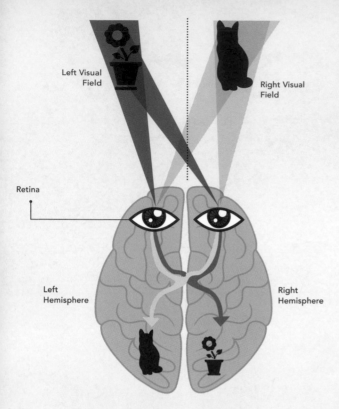

Left Visual
Field

Right Visual
Field

Retina

Left
Hemisphere

Right
Hemisphere

All the visual information from the left side of your
visual field (from both eyes; not just the left) goes
to the right hemisphere, and all the information from
the right visual field goes to the left hemisphere.

they did not shift their gaze, anything flashed on the left side of the screen would go to the right hemisphere, and anything of the right side of the screen would go to the left hemisphere.

When the patient fixed his gaze on the center of a screen and spots of light were flashed briefly across it, he said lights had been flashed only on the right side of the screen. This suggested that the right hemisphere was blind, but oddly, if the patient was told to point at the lights on the left, he could do so. Could the man see the lights, or not?

It seemed that both hemispheres could receive the visual information, but only the left hemisphere could report it with speech.

Sending information to only one side of the brain

The researchers flashed either a picture or written information to only one hemisphere, or placed an object out of sight in one of the patient's hands, so that the information went mainly to the opposite hemisphere. When the picture or word or touch of the object went to the left hemisphere, the patient could describe it normally. When information went to the right hemisphere, however, from something seen in the left visual field or held in the left hand, there was no spoken or written response—or occasionally a wild guess.

When shown a picture of an object (again, such as a spoon) on the left side, however, they were able with the left hand to pick out a spoon from a hidden group of objects—or a fork if there was no spoon in the group—but they still could not say what it was.

The word HEART was flashed across the center of the screen. Asked what they had seen, the patients said "ART," but asked to point with the left hand at cards marked HE and ART, they pointed at HE.

Cross-cueing

Sometimes the researchers spotted "cross-cueing" from one hemisphere to the other. When they flashed either a red or green light to the right hemisphere, the patient would simply guess, because the right hemisphere does not control speech. When he got it wrong, however, he would frown, shake his head, and say he had got it wrong, and meant the other color. Apparently the right hemisphere saw one color, but heard the other, and therefore precipitated a frown and a shake of the head; so the left hemisphere knew it had guessed wrongly. The right hemisphere is not always inferior. Asked to draw a cube, the patients could do it with their left hands, but not with the right; so in this case the right hemisphere was better at control.

Conclusion

Gazzaniga and Sperry came to the conclusion that separation of the hemispheres creates two independent spheres of consciousness within a single brain, but even today no one is quite sure whether that is true, nor what it means.

1968
THE STUDY

RESEARCHERS:
John Darley and
Bibb Latané

SUBJECT AREA:
Social psychology

CONCLUSION:
People in groups
are less responsive
to those in need than
individuals.

WHY DO BYSTANDERS STAND BY?

INDIVIDUAL APATHY IN EMERGENCY SITUATIONS

In March, 1964, a young woman called Kitty Genovese was stabbed to death in a New York City street. The attack went on for more than half an hour. At least 38 people saw what was happening, but not one of them intervened. No one even called the police. Why did no one help?

Perhaps it was apathy or indifference; possibly not wanting to get involved, or even fear of the attacker. One possibility was the simple fact that other people were watching; so someone might already have called the police, or be on their way to help the woman.

American social psychologists John Darley and Bibb Latané were fascinated by the witnesses responses in the face of such a horrible event, and set out to investigate the factors that had stopped people intervening.

The seizure experiment

Darley and Latané asked university students to participate in a discussion about personal problems. Each would be part of a discussion group of varying size, and to avoid embarrassment each of the participants would be in separate rooms communicating through microphones and headphones. The subject was unaware that all the other voices they heard were prerecorded. Some subjects believed they were having one-on-one conversations and some believed they were in groups with up to five others. The subjects were told that the experimenter would wait in the corridor outside the room while the conversation happened.

The first prerecorded voice would confess to the group that he was having difficulty adjusting to life in the city, and that he was also prone to serious seizures. Depending on the size of the "group," other voices were then heard and the subject also spoke. Then the first person spoke again reporting that he was having a seizure, and becoming louder and more incoherent. There was a choking sound, and then deathly quiet.

Almost all the subjects were convinced the fit was real. Of those that thought they were the only person to have heard the attack, 100 percent reported the fit, with 85 percent running to the corridor before the "victim" had stopped speaking.

Of the subjects that thought they were in a group of six, only 62 percent of them reported the fit at all. However these students did not appear to be apathetic or indifferent—as was previously assumed of unresponsive bystanders—they were visibly shaken with trembling hands and sweating palms.

Not so safe in numbers

The research team demonstrated that not only were people in groups less likely to respond to an emergency than individuals, but the response was inversely proportional to the number of witnesses. The greater the number of bystanders, the less likely it was that any one would help.

Darley and Latané concluded that when only one person witnesses an emergency, any help must come from that person and the pressure is on to do something. When there are more witnesses that pressure is dispersed and individuals assume that someone else will take action; or they may worry that their intervention will hinder the efforts of those more qualified.

The scientific context of the situation is also significant, since, as in the Milgram experiment (see page 78), subjects also reported concerns about disrupting or halting the test.

1968

THE STUDY

RESEARCHERS:

Robert Rosenthal and
Lenore Jacobson

SUBJECT AREA:

Social psychology

CONCLUSION:

High expectations can
lead to better results

CAN RESULTS IMPROVE BY SIMPLY EXPECTING THEM TO?

THE PYGMALION EFFECT AND THE POWER OF SELF-FULFILLING PROPHECIES

There are many anecdotes about self-fulfilling prophecies; for example in a group of young men who went bowling, when they "knew" that Matt was going to bowl well one evening he bowled brilliantly, but when they "knew" that Jack was going to lose his touch the next night he could not do anything right. But could there be any scientific evidence to support the superstition?

In 1963 Lenore Jacobson was the principal of an elementary school in San Francisco; she approached Harvard psychologist Robert Rosenthal after reading an article of his. Together they wondered whether something as important as a child's success at school could be affected by a teacher's expectations.

In the schoolroom

They went into a public elementary school, which they called Oak School, where the teachers divided each grade into three tracks: fast, medium, and slow; in general there were more boys and more Mexican children in the slow track. The teachers assigned the tracks on the basis of each child's reading ability and performance in tests.

The researchers tested 350 children using what they grandiloquently described as the "Harvard Test of Inflected Acquisition" and told the teachers that this was an assessment designed to predict "spurting" or "blooming" in children.

The test was actually Flanagan's Test Of General Ability (TOGA), and tested IQ in terms of verbal ability and reasoning. For example, at one level children were shown pictures of a suit jacket, a flower, an envelope, an apple, and a glass of water, and were asked to mark with a crayon "the thing you can eat."

Picking the "spurters"

The researchers did not tell the teachers the results, and instead chose one fifth of the students, from slow, medium, and fast tracks, at random and told each teacher which of the children in their classes were predicted by the "Harvard Test" to put on a spurt in the following year, and to do better than the others in the class. They also made sure to ask the teachers not to mention the test to either the children or their parents.

Results

A year later they gave all the children the same IQ test. All six grades showed an improvement in IQ; the average gain was more than eight points, but the spurters did much better than their peers. They gained on average 12.2 points, 3.8 more than the rest. The effect was almost entirely confined to grades 1 and 2, where 21 percent of spurters gained more than 30 IQ points, compared with 5 percent of nonspurters.

Grade	Control		"Spurters"		
	Number	Increase	Number	Increase	Advantage
1	48	+12.0	7	+27.4	+15.4
2	47	+7.0	12	+16.5	+9.5
3	40	+5.0	14	+5.0	0
4	49	+2.2	12	+5.6	+3.4
5	26	+17.5	9	+17.4	-0.1
6	45	+10.7	11	+10.0	-0.7
Total	255	+8.4	65	+12.2	+3.8

The fact that the effect was confined to the first two grades may possibly have been because teachers have most influence over the youngest children. They are more malleable and more capable of change, or they may not yet have an established reputation in the school.

There was no significant difference between children in different tracks; that is those in the slow and medium tracks did just as well as those in the fast track. Girls did slightly better than boys in the reasoning test: among the spurters, the girls showed a large advantage of 17.9 points over the nonspurters, while the boys actually did worse than the average.

Conclusions

Rosenthal and Jacobson observed what has come to be described as the "Pygmalion effect." When teachers expected certain children to show greater development, those children did just that; a "self-fulfilling prophecy was in evidence."

But why was there any effect at all? It's possible that the teachers had a different attitude toward the spurters or gave them more attention, subconsciously behaving in ways that would encourage the students to achieve.

Curiously, this study was allegedly inspired by a performing horse. At the start of the twentieth century the horse, known as Clever Hans, became famous for his apparent ability to read, spell, and carry out simple mental arithmetic. When challenged to calculate 3+4, for example, Clever Hans would tap his hoof seven times.

The psychologist Oskar Pfungst investigated the case thoroughly and concluded that the animal was probably being guided by the subconscious responses of his audience. When he reached the right answer their reactions would change, and Hans would know he had done enough tapping.

WHAT DO BABIES DO IN "STRANGE SITUATIONS"?

SEPARATION ANXIETY IN BABIES

1970

THE STUDY

RESEARCHERS:
Mary D. Salter Ainsworth
and Silvia M. Bell

SUBJECT AREA:
Developmental
psychology

CONCLUSION:
Infants need their
mothers as a secure base
from which to explore
the world.

Harry Harlow's controversial experiments (see page 64) into maternal attachment had shown that baby monkeys would explore their surroundings in the presence of a soft, motherlike figure, but in the absence of such a figure they became miserable and withdrawn.

Mary Ainsworth and Silvia Bell wanted to find out whether human babies behave in similar ways; so they set up a "strange situation" in their laboratory. Their room had a large space in the middle, and three chairs—one heaped with toys at the far end of the room and two others, near the door, for mother and a female stranger. The baby was put down in the middle of the triangle formed by the chairs. Then they proceeded with eight episodes, identical for each baby:

Episode 1 (M, B, O). Mother (M), with an observer (O), carried the baby (B) into the room, and then O left.

Episode 2 (M, B; three minutes). M put B down, then sat quietly in her chair, participating only if B sought her attention.

Episode 3 (S, M, B). A stranger (S) entered, sat quietly for one minute, conversed with M for one minute, and then gradually approached B, showing him a toy. After three minutes M left the room unobtrusively.

Episode 4 (S, B; three minutes). If B was happily engaged in play, S was nonparticipant. If B was inactive, she tried to interest him in the toys. If B was distressed, she tried to distract or comfort him. If he could not be comforted, the episode was then curtailed.

Episode 5 (M, B). M entered, paused in the doorway to give B an opportunity for a spontaneous response. S then left unobtrusively. Once B was again settled in play with the toys M left again, after pausing to say "bye-bye."

Episode 6 (B alone; three minutes). The baby was left alone, unless he was so distressed that the episode had to be curtailed.

Episode 7 (S, B; three minutes). S entered and behaved as in episode 4, unless distress prompted curtailment.

Episode 8 (M, B). M returned, S left, and after the reunion had been observed, the situation was terminated.

They carried out these experiments with 56 babies, each 11 months old, all "family-reared infants of white, middle-class parents." Observers watched through a one-way mirror, and took notes.

Exploratory behavior

The researchers were particularly interested in how much the baby crawled about ("locomotion"), played with the toys ("manipulation"), and looked at the toys and around the room. When the stranger came in for episode 3 there was a sharp drop in all forms of exploratory behavior. Looking and playing increased again when the mother came back in, but the stranger failed to increase them in episodes 4 and 7; indeed the exploratory behavior dropped to its lowest level in 7, although it is possible they were tired by then, and bored of the room. During episode 2 the baby spent a lot of time looking at the toys, glancing only occasionally at its mother, to make sure she was still there. In episode 3, however, the baby spent more time looking at the stranger.

Crying, clinging, and resisting contact

There was little crying in episode 2, suggesting that the strange situation itself was not very alarming. There was some crying in episode 4, when the mother left, less in episode 5, and much more in episode 6, which the stranger was unable to reduce in 7, suggesting that it was mother's absence that was most distressing, rather than just being alone.

In episodes 2 and 3 the baby made only slight efforts to cling to its mother, but was much more enthusiastic after separation and reunion in episodes 5 and especially 8.

Some babies resisted contact, especially with the stranger. This may have been caused by fear of her, but Ainsworth and Bell think it more likely it was because the baby was angry at the mother's departure.

Conclusions

Babies are generally attached to their mothers. When mother is in the room, the baby will approach new things and explore them, which is what happened in this study; the babies were not terrified by the strange situation; they did not cling to their mothers. When the mother leaves, the baby explores less and shows more signs of attachment behavior, including crying and searching.

As the researchers wrote, "Provided that there is no threat of separation, the infant is likely to be able to use his mother as a secure base from which to explore, manifesting no alarm in even a strange situation, as long as she is present."

CHAPTER 5: The cognitive revolution: 1971–1980

In his 1967 book, *Cognitive Psychology*, the German scientist Ulric Gustav Neisser questioned the paradigms of behaviorism and started other psychologists thinking.

They began to realize that to find out what makes people tick they needed to investigate what goes on in the mind. Before long the label "cognitive psychology" began to embrace perception, language, attention, and memory, as well as thinking. This was not entirely new, but the time

now seemed ripe for cognition to creep into every area of
psychology, and to change the way research was done.

Peter Wason's card tricks teased out the way we think
about what is true; Elizabeth Loftus's work on false memory
provoked decades of further research; while Daniel
Kahneman and Amos Tversky showed us why we make bad
decisions.

1971

THE STUDY

RESEARCHER:

Philip Zimbardo

SUBJECT AREA:

Social psychology

CONCLUSION:

The harsh prison situation caused participants to behave in cruel or violent ways, rather than their individual personalities.

CAN GOOD PEOPLE TURN BAD?

SITUATIONAL INFLUENCES ON BEHAVIOR AND THE STANFORD PRISON EXPERIMENT

Philip Zimbardo was a student at James Monroe High School at the same time as his fellow New Yorker Stanley Milgram (see page 78). After completing his PhD he taught at Yale, NYU, and Columbia, before joining the faculty at Stanford, where he was to carry out the experiment for which he is now best known.

Zimbardo was interested in claims of brutality and violence reported in prisons and wanted to explore the question of whether prisoners are inherently violent and those who become prison guards are naturally authoritarian, and perhaps sadistic; or whether these traits evolve and come to the fore simply because of the prison environment.

Building a makeshift prison

He advertised in local newspapers for male volunteers for a psychological study of prison life. He interviewed 70 applicants, and chose 24 college students who were healthy, middle-class boys. They were told that the study would last for one or two weeks, and that they would be paid $15 a day. Half were randomly chosen to be prisoners and the other half guards.

Meanwhile the experimenters held consultations with experts, including a man who had spent 17 years in prison, and constructed a "prison" in the basement of Stanford University's Psychology Department. Three rooms, each just big enough for three beds, were fitted with heavyweight doors made from steel bars and given cell numbers. The corridor became the "exercise yard," and a small closet became "The Hole"—a solitary confinement cell.

If they wanted to go to the bathroom, the prisoners would have to ask permission, and then be blindfolded and led down the hallway. The rooms were bugged for sound and there was a small hole through which proceedings could be videotaped.

Each "prisoner" was picked up at home, charged, warned of his legal rights, spread-eagled against the police car, searched, and handcuffed, while startled neighbors watched, goggle-eyed. At the "jail," he was searched, stripped naked, and then deloused with a spray—a procedure designed to humiliate. He was then given a uniform: a stocking cap, and a smock to wear at all times, without underwear, and with his prison ID number stenciled on front and back. A heavy chain was bolted around his right ankle.

Humiliation

Real male prisoners don't wear dresses, but they do feel humiliated. Zimbardo aimed to produce similar effects quickly. The stocking cap was a substitute for having all the hair shaved off, which is what happens in real prisons. The chain was used to remind them of the oppressiveness of the environment, and tended to keep them awake at night.

The guards were issued with khaki uniforms, dark mirror glasses, whistles, and police batons, but were given no specific instructions or training.

The experiment began with nine prisoners, and nine guards who worked in eight-hour shifts, three on each shift. At 2:30 am on the first night the prisoners were woken by blasting whistles

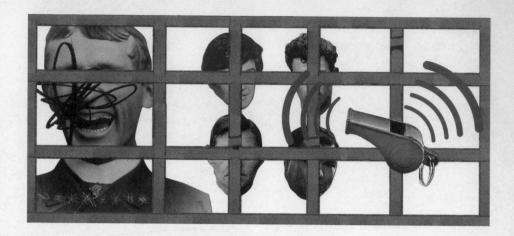

for the first of many "counts." Some of the prisoners were not yet fully into their roles, and would not accept discipline from the guards, who retaliated by making them do push-ups, which were made more difficult by standing on the prisoners' backs.

Prisoner rebellion

The first day went quietly, but on the morning of the second day the prisoners rebelled. They ripped off their stocking caps, and barricaded themselves inside their cells by jamming their beds against the doors. The guards subdued the rebels with icy jets of carbon dioxide from fire extinguishers; they stripped the prisoners naked, took their beds away, and put the ringleaders into solitary confinement.

The guards then decided on psychological warfare to regain control. They took three prisoners who had been least rebellious, put them in rooms with beds and gave them special food, while the others watched. This set the other prisoners against the favored three, and caused most of their frustration and anger to be targeted at one another, instead of at the guards.

Meanwhile the guards became more heavy-handed, and denied even the smallest of privileges; sometimes they refused prisoners permission to go to the bathroom, and left them a bucket to use in their cells. The stench became overpowering.

Within 36 hours one of the prisoners showed acute emotional disturbance. Even though the experimenters had begun behaving like prison warders, and initially did not believe the prisoner was in real distress, they eventually allowed him to leave the experiment.

As the days passed, the guards became gradually more brutal and sadistic, especially at night, when they thought no one was watching. Meanwhile the prisoners, who had at first fought with the guards, gradually broke down emotionally. One developed a rash all over his body. By the end of the study they were, as Zimbardo put it, "disintegrated, both as a group and as individuals. There was no longer any group unity; just a bunch of isolated individuals hanging on."

Termination

The situation became so bad that Zimbardo had to stop the experiment on the sixth day. All the prisoners were happy about this, but not the guards. Two months later one of the prisoners wrote: "I began to feel that I was losing my identity . . . because it was a prison to me; it still is a prison to me."

Zimbardo wrote:

> After observing our simulated prison for only six days, we could understand how prisons dehumanize people, turning them into objects and instilling in them feelings of hopelessness. And as for guards, we realized how ordinary people could be readily transformed from the good Dr. Jekyll to the evil Mr. Hyde.
>
> The question now is how to change our institutions so that they promote human values rather than destroy them.

1971

THE STUDY

RESEARCHERS:

Peter Wason and
Diana Shapiro

SUBJECT AREA:

Cognition, decision-
making

CONCLUSION:

We struggle with abstract
problems, but the same
problem becomes easy
when expressed in
concrete terms

CAN YOU PICK THE LOGICAL ANSWER?

WASON'S SELECTION TASK: ABSTRACT REASONING IN CONCRETE TERMS

Try this logic problem:

Every card is colored on one side and has a number on the other. All blue cards should have an even number on the back. Which of these cards would you have to turn over to find out whether that is true?

Beware; at least 70 percent of people get this wrong. Which cards would you turn over?

Peter Wason was interested in how people tackle logical problems, and first introduced some like this in 1966. He explained how to approach it in terms of pure logic, which you may or may not find helpful.

In this example, p is the blueness of the card, and q is the evenness of the number; so p is true for the first card and false for the second, and q is true for the fourth card, but false for the third.

Therefore you have to turn over the blue card, to see whether it has an even number on the back. You also have to turn over the 3 card, because that is an example of q being false; 3 is not an even number. Turning over the 8 card does not help, for if it is blue that is fine, but if it is pink (or yellow, or any other color) that is still fine; it does not contravene the rule.

So the correct cards to turn over are blue and 3.

Wason and Shapiro gave students a total of 24 tests like this one. There were only seven correct answers (29 percent). The students were too concerned with verifying the rule, and ignored the possibility of falsification. In other words, they ignored the chance to falsify the rule by turning over the q false card.

The researchers wondered whether the problem might be easier if it was related to the real world, and devised what they called "thematic" problems. They divided 32 undergraduate students into two groups. Those in the abstract group were given a task like the one above: four cards had a letter on one side and a number of the other. They were showing D, K, 3, and 7. The rule was "Every card with a D on one side has a 3 on the other." Which cards do you have to turn over to decide whether it was true or false?

Can you solve it? The answer is at the end of this entry.

Those in the thematic group were told that the experimenter had made four journeys on particular days. She claimed that every time she went to Manchester she traveled by car. Four cards represented her journeys; each had a town on one side and a mode of transport on the other:

Which cards would they have to turn over to verify her claim?

Manchester	Leeds	TRAIN	CAR

Results

The abstract group averaged only two correct (12.5 percent). The thematic group did much better, with ten correct (62.5 percent). The researchers concluded that the thematic problem is easier because it deals with concrete material rather than abstract letters and numbers, and also has a relationship between the words; they are all about travel, and situations that could happen in real life.

But the easiest of all is a situation that happens all the time when you go out drinking. Suppose you are in a bar, where no one under the age of 21 is allowed to drink beer. Each card represents one drinker:

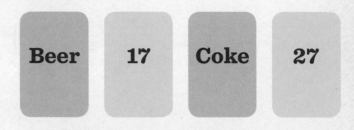

Which cards do you have to turn over to find out whether these four are obeying the law? You should find this one easy.

The conclusion seems to be that we can solve such problems easily when they involve social compliance. This might be because we are more familiar with social situations, or because our brains have evolved to solve social problems, rather than abstract ones.

The correct answers are D and 7, Manchester and train, and Beer and 17.

CAN PSYCHIATRISTS TELL IF YOU'RE SANE?

THE ROSENHAN EXPERIMENT AND "BEING SANE IN INSANE PLACES"

1973
THE STUDY

RESEARCHER:
David L. Rosenhan

SUBJECT AREA:
Social psychology

CONCLUSION:
Professionals in some
psychiatric hospitals
cannot distinguish
the sane from the
insane and exhibit
dangerous levels of
dehumanization.

In 1973, American psychologist David L. Rosenhan published "On Being Sane in Insane Places" which detailed his study into the validity of psychiatric diagnoses. Rosenhan persuaded eight perfectly sane people to admit themselves to different psychiatric hospitals across the United States. These "pseudopatients" comprised a psychology student, three academic psychologists, a pediatrician, a psychiatrist, a painter, and a housewife; three female, five male. They all used false names, and those working in mental health gave different occupations.

Hearing voices

First the pseudopatients called the hospital for an appointment. The only symptom they described was that they heard voices. They said the voices were often unclear, but seemed to say "empty," "hollow," and "thud." Otherwise they told the truth about their lives, families, and relationships.

They were all admitted immediately, which was worrying, but thereafter they gave no further signs of any abnormality. When asked by staff how they were feeling, they said they were fine, and no longer heard voices. They were all keen to be discharged, and were described in nurses' reports as "friendly," "cooperative," and "exhibited no abnormal indications." They accepted medication, although they did not swallow it. In total they were given 2,100 pills of a variety of drugs, but flushed them down the toilet, where they often found pills deposited by real patients. And yet in spite of their displays of sanity, the

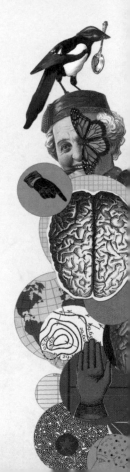

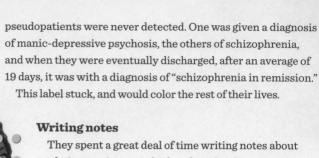

pseudopatients were never detected. One was given a diagnosis of manic-depressive psychosis, the others of schizophrenia, and when they were eventually discharged, after an average of 19 days, it was with a diagnosis of "schizophrenia in remission." This label stuck, and would color the rest of their lives.

Writing notes

They spent a great deal of time writing notes about their experiences, which at first they kept secret. Soon, however, it became apparent that the staff were not interested and never looked at the notes; so often they wrote them openly, in the day room. One nurse noted every day in the pseudopatient's case notes "Indulges in writing behavior." This was clearly perceived as a symptom of schizophrenia.

In each hospital, the staff were strictly segregated from the patients. Staff had their own living space, including dining facilities, bathrooms, and assembly places. The pseudopatients called these glassed quarters "the cage," and noted that the staff emerged on average just 11.3 percent of the time.

When they did come out they were extremely unwilling to indulge in conversation. If a pseudopatient approached with a question such as "Pardon me, Dr. X, when am I likely to be discharged?" the most common response was "Good morning, Dave. How are you today?" and walked on without waiting for an answer.

Rosenhan writes vividly about depersonalization: "Powerlessness was evident everywhere. The patient is deprived of many of his legal rights by dint of his psychiatric commitment. . . . Personal privacy is minimal. Patient quarters and possessions can be entered and examined by any staff member, for whatever reason . . . His personal hygiene and waste evacuation are often monitored. The water closets may have no doors."

Rumbled by patients

Even though they were never spotted as frauds by the staff, they were found out by other patients. During the first three hospitalizations, one third of the patients on the admissions ward were suspicious: some, having watched the note-taking, even said things like, "You're not crazy. You're a journalist, or a professor. You're checking up on the hospital."

As Rosenhan writes:

> *The fact that the patients often recognized normality when staff did not raises important questions. Failure to detect sanity during the course of hospitalization may be due to the fact that physicians ... are more inclined to call a healthy person sick ... than a sick person healthy ... it is clearly more dangerous to misdiagnose illness than health. Better to err on the side of caution, to suspect illness even among the healthy.*

> *Whenever the ratio of what is known to what needs to be known approaches zero, we tend to invent 'knowledge' and assume that we understand more than we actually do. We seem unable to acknowledge that we simply don't know. The needs for diagnosis and remediation of behavioral and emotional problems are enormous. But ... we continue to label patients 'schizophrenic,' 'manic-depressive,' and 'insane,' as if in those words we had captured the essence of understanding. The facts of the matter are that we have known for a long time that diagnoses are often not useful or reliable, but we have nevertheless continued to use them. We now know that we cannot distinguish insanity from sanity.*

1973

THE STUDY

RESEARCHERS:

Mark R. Lepper, David
Greene, and
Richard E. Nisbett

SUBJECT AREA:

Social psychology

CONCLUSION:

Rewards can undermine
a child's natural interest
in certain activities.

ARE CHILDREN PUT OFF BY BRIBES?

THE TROUBLE WITH GOLD STARS

Children in school are often given gold stars, and other extrinsic rewards, and there is a danger that these rewards may actually reduce enthusiasm: "I am doing this arithmetic in order to get a star"—rather than because it's interesting or fun.

Lepper and his colleagues set out to test this theory in a kindergarten on the Stanford University campus. They chose a group of white, middle-class children who were known to be interested in drawing, and divided them randomly into three groups. Each of the children in Group A was told in advance that he or she would receive a reward in the shape of a handsome Good Player Award certificate embellished with a gold star, a red ribbon, and the names of the child and the school. Each of the children in Group B were given the same type of award, but did not know about it until after they had finished drawing. Children in Group C received no award.

The experiment phase one

For each trial a single child was brought into the room and invited by an experimenter to do some drawing with a set of multicolored fluorescent pens, which were not normally available to the children. If the child was in Group A the experimenter produced a sample award, and said the child would get one for his or her drawing. Children from other groups were simply invited to start drawing.

After six minutes the experimenter interrupted, and for children from Groups A and B presented the Good Player Award, after writing the child's name and school on it. Together they then pinned the award on a special "Honor Roll" board "so that everyone will know what a good player you are."

The experiment phase two

One week later, the researchers began the second phase of the experiment. First, teachers placed on a small hexagonal table the set of special pens, and white paper. The classroom also contained a number of other activities, including building blocks, easels, housekeeping equipment, and sometimes playdough. When the group of children came in they were free to do anything they liked. For the first hour, the activities at the hexagonal table were carefully monitored by observers from behind a one-way mirror.

In the end a total of 51 children—19 boys and 32 girls— completed the experiment. Of these 18 were in Group A, 18 in Group B, and 15 in Group C.

Prediction of outcome

The researchers predicted that the promise of a reward would make the children less interested in the task. This was precisely borne out by the results. The children in Group A, who had been told in advance that they would get an award, were afterward much less interested in drawing with the colored pens; in fact they spent only about half as much time drawing as the others. There was no significant difference between girls and boys.

Children in both Groups B and C showed slightly more interest in drawing with colored pens than they had before the experiment began.

The pictures drawn by the children during the experimental trials were rated for quality, on a scale of 1 to 5, by three judges who did not know which groups the artists had been in. The average scores were Group A, 2.18; Group B, 2.85; Group C, 2.69. In other words, having been told they would get an extrinsic

reward, the children in Group A drew pictures that were significantly less good than the others.

How to incentivize

The researchers concluded that their findings:

> ... have important practical implications for situations in which extrinsic incentives are used to enhance or maintain children's interest in activities of some initial interest to the child. Such situations, we would suggest, occur frequently in traditional classrooms where systems of extrinsic rewards—whether grades, gold stars, or the awarding of special privileges—are applied as a matter of course to an entire class of children.

> Many of the activities we ask children to attempt in school, in fact, are of intrinsic interest to at least some of the children; one effect of presenting these activities within a system of extrinsic incentives, the present study suggests, is to undermine the intrinsic interest in these activities of at least those children who had some interest to begin with.

HOW ACCURATE ARE YOUR MEMORIES?

FALSE MEMORIES AND THE MISINFORMATION EFFECT

1974

THE STUDY

RESEARCHER:
Elizabeth F. Loftus

SUBJECT AREA:
Memory

CONCLUSION:
Our memories of events
can be affected by
information we receive
after the fact.

Do you think your memories are accurate and unchanging? If so, you are probably wrong. When someone says "I saw it with my own eyes," people are inclined to believe it. However, between the time you witness an event and the time you recount it to someone else, your memory may change a great deal, especially if an interested person asks you leading questions.

Professor Elizabeth Loftus found that when she staged an event, and then asked individuals about it afterward, they often all gave different accounts. This is common after road accidents; the versions of the event given by eyewitnesses frequently differ.

Eyewitnesses in court

Eyewitness error can be serious in courts of law. In one case in 1973, 17 witnesses identified a man as the person who shot a police officer. Later it turned out that the man they had identified had been nowhere near the scene of the crime.

Loftus explains that "When we experience an event, we do not simply file a memory, and then on some later occasion retrieve it and read off what we've stored. Rather, at the time of recall or recognition, we reconstruct the event, using information from many sources. These include both the original perception of the event and inferences drawn later, after the fact. Over a period of time information from these sources may integrate, so that a witness becomes unable to say how he knows a specific detail. He has only a single, unified memory."

In other words, the brain takes in what it actually experiences of the event, and makes up a plausible story to account for

what seems to have happened. Later, if other information or suggestions come in, the brain may reconstruct the memory to fit the new input. Loftus had noticed that the form of questions asked of witnesses seemed to change their memories, and she set up an experiment to find out how easily this could happen. She showed 100 students a short film depicting a multiple car accident.

Leading questions

After seeing the film the students filled out a questionnaire which included six critical questions: three about items that had appeared in the film, and three others about items that had not.

For half the subjects the critical questions were of the form "Did you see *a* broken headlight?" For the other half the critical questions were of the form "Did you see *the* broken headlight?" The second of these questions implies that there was a broken headlight, whether or not one actually appeared in the film. Witnesses who received questions with "the" were much more likely to report having seen something that had not really appeared in the film: 15 percent in the "the" group said "yes" when asked about a nonexistent item, while only 7 percent of the "a" group said "yes." In other words, just changing from "a" to "the" seemed actually to have altered the memories of 8 percent of the students. Alternatively 38 percent in the "a" group answered "Don't know," as opposed to 13 percent in the "the" group.

To find out whether other small changes in the question could affect quantitative judgments she showed 45 subjects seven films of traffic accidents. For some of the viewers of one film, she asked in a questionnaire "about how fast were the cars going when they *hit/ smashed/ collided/ bumped/ contacted* each other?" The resulting estimates varied considerably.

Average speed estimates for various verbs	
smashed	40.8 mph
collided	39.3 mph
bumped	38.1 mph
hit	34.0 mph
contacted	31.8 mph

Memory adjustment

Another group of students were shown a similar film and asked how fast the cars were going when they "hit" or "smashed into" each other. One week later, they were asked whether they had seen any broken glass, although there had been none in the film. Twice as many of the students who had been asked a question with "smashed" said they had seen broken glass as those who had been asked a question with "hit." In other words, their memories of the film had apparently been altered just because they had been asked a question with a slight change of words.

Loftus concludes:

> *Eyewitnesses are inaccurate in estimating not only speed but also time and distance. Yet in courts of law they must make quantitative judgments all the time. Accident investigators, police officers, lawyers, reporters, and others who must interrogate eyewitnesses would do well to keep in mind the subtle suggestibility that words carry with them. When you question an eyewitness, what he saw may not be what you get.*

False memories

The phenomenon of a person's recall becoming less accurate because of post-event suggestion or information has come to be known as the "misinformation effect," and Loftus's work led to decades of research into the condition of "false memory." There is considerable danger that interrogators, not only in courts of law and in police stations but also in the armed services, can plant false memories, either by accident or by design.

1974

THE STUDY

RESEARCHERS:

Amos Tversky and
Daniel Kahneman

SUBJECT AREA:

Cognition, decision-making

CONCLUSION:

When an outcome is
unknown, cognitive bias
can cause us to make
poor decisions.

HOW DO YOU MAKE TRICKY DECISIONS?

"HEURISTICS" AND ASSESSING POTENTIAL RISK

Most people find decisions difficult when they don't know the exact outcomes, and they often get them wrong. Israeli-born psychologists Daniel Kahneman and Amos Tversky began collaborating by looking at contradictions in human behavior. For example, people may drive across town to save $5 on a $15 phone but not drive across town to save $5 on a $125 coat.

Heuristics

The researchers found that when people have to make judgments about uncertain futures they tend to use "heuristics," which are mental shortcuts that rely on simple, efficient rules, often focusing on one aspect of a problem while ignoring others.

For example, suppose you are told that "Steve is very shy and withdrawn, invariably helpful; a meek and tidy soul, he has a need for order and structure, and a passion for detail." You are also told that Steve may be a farmer, a salesman, an airline pilot, a librarian, or a doctor. Which do you think is the most likely?

You might be tempted to say librarian, but there are many more farmers than librarians, and therefore Steve is in fact more likely to be a farmer than a librarian, in spite of his personal characteristics. This is the "representativeness heuristic."

In an experiment, a group of students were told about one man in a group of 100 professionals: "Dick is married with no children. A man of high ability and high motivation, he promises to be quite successful in his field. He is well-liked by his colleagues."

Half the students were told that the group comprised 70 engineers and 30 lawyers; the other half were told it was

30 engineers and 70 lawyers. When asked whether Dick was likely to be a lawyer or an engineer, they all said it was 50-50. They ignored the fact that he was much more likely to be one of the larger group: the chances were 70-30, one way or the other.

What are the chances?

Consider the letter K. In a typical chunk of English prose, is the letter K more likely to occur as the first letter in a word or as the third letter? What do you think?

This question was posed to 152 subjects, and 105 (69 percent) said it was more likely to be the first letter. In fact K typically turns up twice as often in position three as in position one. The problem is that thinking of words beginning with K is easy, while thinking of words with K as the third letter is much harder. The same is true of L, N, R, and V. This is called the "availability heuristic," because it relies on the immediate examples that come to mind.

Regression toward the mean

Imagine that a large group of children has been examined with two equivalent versions of an aptitude test. Suppose you choose the ten best on the first version, then you will probably find they did worse on the second version. On the other hand if you choose the ten children who did worst on the first version, you will find they did better on the second. This is called "regression to the mean," and was first discussed by Francis Galton in the nineteenth century.

The ten best may really be better than everyone else in the class, but they may well have done slightly better on this test just by luck; they are likely to be closer to the average, or mean. The consequence is that the ten best are likely to move back again, and the ten worst to move forward.

The researchers point out that ignoring this can lead to dangerous consequences:

In a discussion of flight training, experienced instructors noted that praise for an exceptionally smooth landing is typically followed by a poorer landing on the next try, while harsh criticism after a rough landing is usually followed by an improvement on the next try. The instructors concluded that verbal rewards are bad for learning, while verbal punishments are good, which is contrary to accepted psychological doctrine. This conclusion is unwarranted because of the presence of a regression toward the mean.

Probabilities of Death in US From Various Causes (percent)		
Cause	Subject Estimates	Real probability
Heart Disease	22	34
Cancer	18	23
Other Natural Causes	33	35
All Natural Causes	73	92
Accident	32	5
Homicide	10	1
Other Unnatural Causes	11	2
All Unnatural Causes	53	8

How will you die?

The researchers asked 120 Stanford graduates how likely they thought they were to die from various causes. Listed in this table are the averages of the answers they gave.

They slightly underestimated the probabilities for natural causes, and vastly overestimated the probabilities for unnatural causes. It looks as though they spent too much time worrying about accidents and murder, and perhaps not enough worrying about their health.

Tversky and Kahneman conclude: "Analysis of the heuristics that a person uses . . . may tell us whether his judgment is likely to be too high or too low. We believe that such analyses could be used to reduce the prevalence of errors in human judgment under uncertainty."

Following the extensive work of Tversky and Kahneman there has been an enormous amount of research on human biases.

CAN SHEER TERROR BE SEXY?

HEIGHTENED SEXUAL ATTRACTION UNDER CONDITIONS OF HIGH ANXIETY

1974

THE STUDY

RESEARCHER:
Donald G. Dutton
and Arthur P. Aron

SUBJECT AREA:
Experimental psychology

CONCLUSION:
There is a definite
connection between the
body's response to fear
and arousal.

If you were frightened, would you find potential partners more attractive? Can you tell the difference between sexual arousal and sheer terror?

There is some evidence to suggest that sexual arousal is associated with strong emotion, or even increased by strong emotion; that is why people take their partners to dangerous funfair rides and scary movies. In fact we may simply be unable to distinguish the separate emotions.

The Capilano bridges

Researchers Donald Dutton and Arthur Aron devised a clever way to investigate the relationship between fear and sexual arousal. They found two bridges over the Capilano River in North Vancouver, Canada. One—the control bridge—was a wide, solid, cedarwood bridge with high hand rails, 10 feet (3 m) above a small stream. The other—the experimental bridge—was the Capilano Canyon Suspension Bridge. This was a long, narrow bridge, made of wooden boards hanging on wire cables, 230 feet (70 m) above a raging torrent. The handrails were low, and the bridge swayed and wobbled when pedestrians attempted to cross it. Many of those who crossed the suspension bridge walked slowly and carefully, hanging on to the handrail.

The subjects were men who were judged to be between 18 and 35, and happened to be walking across one of the bridges, without a female companion.

As the subject walked across the bridge, he was approached by an interviewer, who asked him, as part of a psychology experiment, to fill in a short questionnaire. The first page had basic questions about age, sex, education, previous visits to the bridge, and so on. On the second page they were asked to write a short dramatic story based on a picture of a young woman covering her face with one hand and reaching out with the other.

The stories were later scored for sexual content, ranging from 1 for no sexual content, through 3 for a kiss, to 5 for any mention of intercourse.

Sexy interviewer

When the subjects had finished their questionnaires, the interviewers thanked them and offered to explain the experiment in detail when there was more time. They offered their phone numbers, and invited the men to call if they wanted to talk further.

What differences would you expect from the sex of the interviewer? The interviewers were students—some male, some female. Most of the men who were approached agreed to take part, especially with a female interviewer.

The results showed first that men were more inclined to take part if the interviewer was female; second that being interviewed on the frightening suspension bridge seemed to have made

Interviewer	Number taking part	No. accepting phone no.	No. who phoned	Sexiness of story
M — control bridge	22/42	6/22	1	0.61
M — scary bridge	23/51	7/23	2	0.80
F — control bridge	22/33	16/22	2	1.41
F — scary bridge	23/33	18/23	9	2.47

the female interviewer appear more attractive. The men had definitely been aroused. Not only did they write sexier stories than the others, but many more of them called her afterward.

In other words there is a definite connection between fear and arousal, or to put it another way, you cannot tell whether the adrenaline coursing through you is caused by sexual arousal or sheer terror.

1975

THE STUDY

RESEARCHERS:
William R. Miller
and Martin Seligman

SUBJECT AREA:
Behavioral psychology

CONCLUSION:
A perceived absence
of control over negative
events may lead to
clinical depression.

CAN DOGS GET DEPRESSED?

LEARNED HELPLESSNESS AND DEPRESSION

As a research student working with the experimental psychologist Richard L. Solomon, Martin Seligman began working with dogs in an extension of the work of Pavlov (page 19). He placed a dog in a box that was divided in half by a low barrier, chest-high to the dog. After a while, the dog would get a brief electric shock, and then another and another. If the dog jumped over the barrier into the other half of the box, the shocks stopped. Then the shocks started there instead, and to stop them the dog had to jump back to the first side.

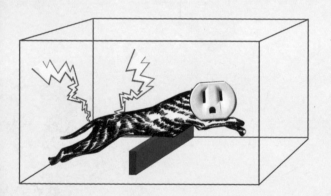

Soon the dogs learned to jump the barrier as soon as they felt the first shock. This was known as operant conditioning.

Seligman then exposed another group of dogs to intermittent electric shocks that they could not avoid. When he put one of these dogs into the divided box, they never did learn to jump over the barrier. They just stood or lay around, waiting for the shocks to stop. A third group of dogs, which had never experienced shocks, quickly learned to jump the barrier.

Learned helplessness

Seligman concluded that the uncontrollable shocks had produced "learned helplessness" in the second group of dogs;

they had learned that they could not control of the shocks, whatever they did; so why bother to do anything? Even when they took the barrier away and put food on the other side, the dogs did not bother to move.

As Seligman wrote:

> *When an experimenter goes to the home cage and attempts to remove a nonhelpless dog, it does not comply eagerly; it barks, runs to the back of the cage, and resists handling. In contrast, helpless dogs seem to wilt; they passively sink to the bottom of the cage, occasionally even rolling over and adopting a submissive posture; they do not resist.*

People too?

Seligman and Miller found similar effects with people. In one experiment they asked people to do mental arithmetic while bombarding them with distracting noise. When some of them found they could turn off the noise, their performance improved, even though they often did not bother to turn it off. The point was that they knew they could turn it off; so they no longer felt helpless.

The researchers worked also with depressed people, and noticed that they sometimes showed the same sort of behavior as the helpless dogs: weariness, sleeplessness, foreboding of disaster, mental numbness, and so on. They made two suggestions: 1. some of the characteristics of depression are the effect of learned helplessness, and 2. the depressed people *believed* they were helpless.

How do depressed people explain this?

Miller and Seligman suggested they adopt "a depressive explanatory style"—beliefs such as "The problem is that I am incompetent. I'll never be any use. I'm hopeless at everything." Because they *believe* they are helpless, they are even more likely to get depressed.

They found that depressed patients in the hospital were more likely to give this sort of explanation than schizophrenics or normal medical students. Also, ordinary college students who got grades lower than they had hoped for—a B instead of an A, or a D instead of a C, were merely disappointed, but depressive students adopted a depressive explanatory style.

People are likely to succumb to "hopelessness depression" if highly desirable things do not happen, or if highly unpleasant things do happen, and there is nothing that they can do about it.

Varieties of helplessness

Suppose a group of people are exposed to uncontrollable noise, but an experimenter tells them that it is controllable. They cannot find out how, and may come to believe either that it really is uncontrollable, or that they simply lack the ability to control it.

The researchers went on to describe universal and personal helplessness. Suppose a child contracts leukemia, and his father does everything he can to save the boy's life. Nothing helps, and he comes to believe there is nothing he can do, and nor can anyone else. He finally gives up, and shows signs of behavioral helplessness and depression. This is *universal helplessness*.

Suppose a student tries hard at mathematics, studies endlessly, takes extra courses, hires tutors, but nothing helps; she still fails her exams. She comes to believe she is stupid and gives up—and any mathematical question from estimating shopping bills to filling in a tax return will always be a nightmare. This is *personal helplessness*.

Conclusion

One ramification, as Miller and Seligman point out, is that people who fail in exams or in business, while others succeed, will acquire lower self-esteem than those who think that success is just a matter of luck. Also, a poor student who fails an exam that others pass will have lower self-esteem than if the others fail as well, for then he will think that the results were out of anyone's control.

CAN YOU LISTEN WITH YOUR EYES?

THE IMPORTANCE OF LIP-READING

1976
THE STUDY

RESEARCHERS:
Harry McGurk and
John MacDonald

SUBJECT AREA:
Perception

CONCLUSION:
We listen with our eyes
as well as our ears.

Lip-reading sometimes makes you hear the wrong thing. When you talk to someone on the phone you have to rely on sound alone, but when you are speaking face to face you probably notice the other person's lips as you hear the voice. Lip-reading is helpful to most people, especially those whose hearing is impaired; indeed profoundly deaf people rely on lip-reading.

Harry McGurk noticed, however, one curious area of speech in which lip-reading actually hinders hearing. He saw a film of a young woman speaking to camera. She was saying "ba . . . ba" but the sound of this had been put exactly into sync with the vision of her lips saying "ga . . . ga."

What do you think you would hear? What McGurk heard was "da . . . da"—until he shut his eyes, when he heard "ba . . . ba," but when he looked at the screen again he heard "da . . . da" once more. Colleagues working with him had the same experiences.

When the process was reversed, and the sound was "ga . . . ga" while the lips were saying "ba . . . ba," what they heard was "bagba" or "gaba."

Making a film

Intrigued by these curious observations, McGurk decided to investigate further, using some clever lab experiments.

He set out to confirm and generalize the discovery, by filming in close-up a woman saying "ba...ba" three times; then "ga...ga," followed by "pa...pa," and finally "ka...ka" each three times in a row. Then he carefully edited the film to make four separate recordings, as shown.

Then he showed his edited film to 103 people: 21 preschool children, 3 or 4 years old, 28 elementary-school children, 7 or 8 years old, and 54 adults, mostly male. Each one watched the film alone, and said what they heard; then they listened to the sound without seeing the lips, and again said what they had heard.

The results were intriguing. Without the distraction of the lips, hearing was accurate: 91 percent correct for the young children, 97 percent for the older children, and 99 percent for the adults.

Recording	1	2	3	4
Voice	ba...ba	ga...ga	pa...pa	ka...ka
Lips	ga...ga	ba...ba	ka...ka	pa...pa

When they watched the lips while listening, they heard the "wrong" syllables 59 percent, 52 percent, and 92 percent of the time.

He defined a "fused" response as one where information from sound and lips was transformed into a new sound, different from either component; for example "ba...ba" and "ga...ga" turning into "da...da." When the sound and lips produced a modified version of one or the other—for example "ga...ga" and "ba...ba" turning into "bagba"—he called it a "combination."

First of all, the results show that a majority of people experience closely similar effects: 98 percent of adults heard the ba/ga combination as da and 81 percent of adults heard pa/ka as ta. The children clearly relied more on hearing than on sight, but nevertheless more than 50 percent of them heard the same fused results.

Adults were more influenced by the lips, and when they relied on one sense it was sight, whereas for the children it was sound.

Conclusion

McGurk points out that for hearing, vowels carry information for the consonants that immediately precede them, and concluded tentatively:

> *If we speculate that the acoustic waveform for 'ba' contains features in common with that for 'da' but not with 'ga,' then a tentative explanation for one set of the above illusions is suggested. Thus in a 'ba' voice 'ga' lips presentation there is visual information for 'ga' and 'da' and auditory information with features common to 'da' and 'ba.' By responding to the common information in both modalities, a subject would arrive at the unifying percept 'da.'*

These experiments remind us how much we—especially adults—rely on vision and lip-reading without realizing it, and should serve as a warning that we may often be confused or fooled by films or videos in which the sound recording is less than perfect.

1978

THE STUDY

RESEARCHER:

Edoardo Bisiach

SUBJECT AREA:

Perception

CONCLUSION:

Brain damage can lead
to a one-sided view of
the world.

HOW CAN YOU LOSE HALF THE WORLD?

HEMIFIELD AND UNILATERAL NEGLECT

Some people can apparently see only half the world in front of them. After a stroke, many people are left paralyzed down one side of their bodies, or unable to speak properly, but a few have hemifield neglect, or unilateral neglect. This means that they do not seem to admit the existence of one side of their field of view.

If the damage is to the right side of the brain, which is most common, the left-hand side of the world seems to disappear. Men with unilateral neglect typically shave only the right side of their faces, and women make up their right side only. They eat only food on the right side of their plates; someone else has to come and turn the plate round before they can finish the meal.

When asked to draw a picture, they will squash everything on to the right side. So a clock face may show only the right side, or the numbers may all be there, pushed over to the right, while a flower may have all the petals on one side.

Patients sometimes bump into objects, or door frames, with their left side, because they neglect them.

What is so curious about this is that they have not really lost the vision of the left field. The information comes in, but they cannot process it. It seems as though they simply neglect it, or do not have their attention drawn to it.

Patients asked to read a word, may read only the right half of the word, and may make the rest up. So invited to read the word PEANUT they may say "nut" or "walnut." If you shake the left hand of such a patient and ask what it is, he may answer "It's a hand." But when asked whose hand he may say "I don't know. It's not mine; it must be yours." So the brain confabulates, or makes up a story to fit what he sees as the facts.

134

Other experiments have shown that the right hemisphere can take in emotional messages, without being able to explain them. Marshall and Halligan showed patients a picture of two houses that were more-or-less identical, except that one had smoke and flames pouring out of the left-hand side.

The patients said the houses were identical, but when asked which house they would rather live in they chose the non-burning one. They had clearly taken in the emotional message of one house on fire, though they had neglected to mention it to their vision system.

Bisiach showed that the neglect is deeper than just in vision, with an elegant demonstration. He works with patients in Milan, Italy, where everyone knows the spectacular *duomo*, or cathedral, and the grand square or piazza in front of it. First he asked them to imagine themselves standing facing the *duomo*, and to describe the scene.

They described all the buildings on the right-hand side of the piazza, but none on the left. Then he asked them to imagine standing with their backs to the *duomo*, and sure enough they described all the buildings on the other side of the piazza. Therefore they had the information about all the buildings, but from any one viewpoint ignored those on the left.

What is more curious is that they did this not while standing in the piazza, but while imagining it; so clearly this one-sided view of the world was not confined to vision, but affected every part of their imagery. Presumably the views of the whole square had been stored in memory before the stroke, but only one side could be retrieved afterward.

CHAPTER 6: Into consciousness: 1981—

Until the 1980s the word "consciousness" was barely admitted in scientific discussions of the mind. How could anyone investigate consciousness? What is more, when the subject did creep in, psychologists had to confront "the hard problem"—that mind and body seem separate, but must be related, or may even be the same thing.

It may feel as though there is an "I" inside, looking out through the eyes to see the world, but we know this is not the case. Instead there is just an enormous mass of neurons, connected in billions of ways, processing information. And it was this bundle of pathways that scientists now began to investigate.

THE STUDY

RESEARCHERS:

Benjamin Libet, Curtis A.
Gleason, Elwood W. Wright,
and Dennis K. Pearl

SUBJECT AREA:

Consciousness

CONCLUSION:

Free will may be a myth,
but we are still responsible
for our actions.

ARE YOU REALLY IN CONTROL?

THE NEUROSCIENCE OF FREE WILL

We all think we have conscious control over our actions, but is it true? In the 1980s, American neuropsychologist Benjamin Libet and his colleagues studied five right-handed college students, who sat in a partially reclining position on a lounge chair with their right arms out in front of them.

When they were comfortable, the trial began with a get-ready tone, at which they were instructed to spend the next second or two relaxing the muscles in the head, neck, and forearm. Then, whenever they felt like doing so, they were to perform a quick, abrupt flip of the fingers or the wrist, and to do so spontaneously: "to let the urge to act appear on its own at any time without any preplanning or concentration on when to act." In other words they should flip their wrists whenever they felt like it, of their own free will, and do so 40 times.

Meanwhile the researchers wanted to measure three things:

1. the time at which the movement began; this was recorded by electrodes on the forearm;

2. the onset of the "Readiness Potential," which is a negative potential shift that slowly builds up a second or more before the action. The command from the brain goes to the muscles in the wrist shortly before they move. The Readiness Potential is the preparation for that command, and they were able to measure it using electrodes on the scalp;

3. The moment of decision—"the appearance of ... conscious awareness of 'wanting' to perform a given self-initiated movement." But this could only be subjective—only the subject could know when it happened. How could anyone possibly measure it?

The moment of decision

If the subjects were asked to shout "Now," there would be a delay before the shout came out—so that would not work—and reflex time would also delay any mechanical action, such as pressing a button.

So what the researchers did was to put a screen in front of the subject and arrange for a spot of light to move around it in a circle about once every 2.5 seconds, like a hand on a clock face. The screen was marked with radial lines and numbers 1—12, as they appear on a clock face. The actual time taken for the spot to move from one number to the next was therefore about 43 milliseconds.

As the subjects decided to move their wrists, they called out the "time" shown by the spot. This turned out to be highly reliable; subjects were consistently precise when asked to call the times of slight shocks delivered (at random intervals) to the back of the hand, and the small biases they displayed in this were used to correct the reported times for their decisions.

Readiness potential

The results showed a considerable range of times, but on average the Readiness Potential started about one second (1,000ms) before the muscles actually moved. The decision to act also preceded the

actual movement. In every single one of several hundred trials, however, the decision came well after the onset of the Readiness Potential. The average difference was about 350ms.

In other words the brain initiated the action about a third of a second before the subject "decided" to act.

Libet and his colleagues wrote that this study:

> ... invites the extrapolation that other relatively 'spontaneous' voluntary acts, performed without conscious deliberation or planning, may also be initiated by cerebral activities proceeding unconsciously. These considerations would appear to introduce certain constraints on the potential of the individual for exerting conscious initiation and control over his voluntary acts.

Libet's results suggest that our conscious decisions may not be the cause of our actions; it's as though we do something spontaneously and then afterward decide that we meant to do it. They even suggest that we may have no free will.

In 1985 Libet reported further experiments in which subjects had been instructed to veto the action after taking the decision to act. This time the muscles did not move. In other words, we do have time to exercise the veto, and so stop an action before it happens.

Conclusions

Libet's conclusions noted that it was:

> ... important to emphasize that the present experimental findings and analysis do not exclude the potential for 'philosophically real' individual responsibility and free will. Although the volitional process may be initiated by unconscious cerebral activities, conscious control of the actual motor performance of voluntary acts definitely remains possible. The findings should therefore be taken not as being antagonistic to free will

but rather as affecting the view of how free will might operate. The concept of conscious veto or blockade of the motor performance of specific intentions to act is in general accord with certain religious and humanistic views of ethical behavior and individual responsibility. 'Self-control' of the acting out of one's intentions is commonly advocated; in the present terms this would operate by conscious selection or control of whether the unconsciously initiated final volitional process will be implemented in action. Many ethical strictures, such as most of the Ten Commandments, are injunctions not to act in certain ways.

THE STUDY

RESEARCHERS:
Diane C. Berry and
Donald E. Broadbent

SUBJECT AREA:
Cognition, decision-making

CONCLUSION:
Practice, training, and
thinking aloud is the best
combination.

DOES PRACTICE MAKE PERFECT?

"THE SUGAR FACTORY TASK"

After you have solved a problem, can you always explain how you did it? Berry and Broadbent wanted to investigate how people tackle complicated mental tasks. Does their performance improve with practice, or with training? And can they afterward explain their methods?

This is a simplified account of one set of their experiments.

The sugar factory

They set up a computer simulation about managing an imaginary sugar factory. The problem appeared to be simple. The factory started with 600 workers producing 6,000 tons of sugar, and the subjects' task was to get sugar production up to 9,000 tons, and if possible keep it there for their entire run, by varying the number of workers in the factory.

In fact the computer was following a deceitful algorithm, but the subjects did not know this, and had to operate by guesswork and intuition.

The difficulties

The computer's algorithm meant that a workforce of say 800 would not always produce the same amount of sugar. This meant that the subjects had to be in control all the time; even if they hit 9,000 tons early on, putting in the same workforce again would almost always change the output.

In each run they could make only a fixed number of "tries" or key presses; their score was the number of tries after which production was between 8,000 and 10,000 tons. If they had chosen workforce numbers completely at random, they would have scored 3.4. If they did better than this they must have learned something about how to control sugar production.

Do you think they would do better than this? And would they improve with practice?

There were five groups of subjects. The first group, A, had only one run of 30 tries. The second group B had two runs of 30 tries, one after the other. The third group, C, had two runs of 20 tries, but after the first run they were given careful and explicit training in how to tackle the problem. The fourth group, D, were given no training, but encouraged to think aloud while they were doing the second run, in the hope of explaining what they were doing. The fifth group, E, were trained after the first run, and encouraged to think aloud while they were doing the second run.

The results

The average scores for Groups A and B are shown in the table.

On average the subjects did much better than chance (3.4). Clearly they improved with practice, since Group B scored almost twice as much in their second run; so they must have learned something about how to solve the problem.

	Score in 30 tries (max 30)	
group	Run 1	Run 2
A	8.7	-
B	8.6	16.2

Explaining how they did it

After finishing their runs, subjects were asked to fill in questionnaires about how they had tackled the problem.

These questionnaires were then assessed and marked on a scale of 1 (bad) to 5 (good). Both groups A and B scored only 1.7 out of 5. In other words they could not explain what they had done. Group B were not significantly better than Group A, even though they had greatly improved at performing the task. What is more, subjects who had not even done the task scored 1.6 on the questionnaire, which was almost as good as those who had.

The subjects must have learned something, but they could not put it into words. Many claimed during their debriefing interviews that they were operating on the basis of "some sort of intuition," making responses because they "felt right."

The results for groups C, D, and E were even more interesting The scores are lower because the subjects had 20 tries instead of 30, but again they show that practice improved production, since the scores in run 2 were all higher than those in run 1. Training, however, seemed to have almost no effect on Group C, since their score in run 2 was hardly any better than Group D's.

The training did, however, have a spectacular effect on their ability to answer questions: Groups C and E scored roughly twice as much on the questionnaire as Groups A, B, or D; so they did understand how they were meant to tackle the task.

Thinking aloud by itself did not have much effect either on performance or on answering questions, but when combined with the training it produced a spectacular increase during the second run.

Conclusions

1. Questionnaires cannot assess a person's performance.
2. Verbal training may not improve performance.
3. Even though people may get better at a task with practice, they may not be able to explain why.

Some people think of intuition as a mysterious force or inexplicable ability, but this experiment shows that it is perfectly normal to be able to learn a skill without knowing what you have learned. When you sense something wrong, or when you make a choice without knowing why, you are using just this kind of intuition.

	Score in 20 tries		
Group	Run 1	Run 2	Questionnaire
C (trained)	4.7	7.0	3.6
D (thinking aloud)	4.5	6.7	1.6
E (trained and thinking aloud)	5.2	13.3	3.4

HOW DO AUTISTIC CHILDREN SEE THE WORLD?

THE "THEORY OF MIND"

1985
THE STUDY
RESEARCHERS:
Simon Baron-Cohen, Alan
M. Leslie, and Uta Frith
SUBJECT AREA:
Developmental psychology
CONCLUSION:
Autistic children
cannot "see" other
minds.

Autism is rare; it affects about four children in 10,000. Autistic children are unusual in many ways, but in particular they suffer from poor communication, both verbal and nonverbal. This is one of the main reasons why they have trouble in coping with the social environment, and they do not develop social relationships. Autistic children are said to "treat people and objects alike."

Baron-Cohen and his colleagues wondered whether they can understand that other people want, feel, or believe things, which is called having a "theory of mind (TOM)." In most children this begins to appear when they are three or four. Autistic children, even those with high IQs, do not enjoy pretend play. So the researchers predicted that autistic children would not have a theory of mind.

They compared 20 autistic children with 14 children with Down's Syndrome and 27 non-autistic preschool children aged between 3.5 and 6. The average mental age of the autistic children was higher than that of both the Down's group and the non–special needs group. The average IQs were 82 for the autistic group and 64 for the Down's group.

The experiment

With each of the children in turn they acted out a little story with two dolls, Sally and Anne. First they named the dolls, and asked the children which was Sally and which was Anne. All 61 children got this right.

Then they played a game, using the dolls. Sally put a marble into her basket. Then she went away, and Anne took the marble and hid it in her box. When Sally came back, the experimenter asked the critical Belief Question: "Where will Sally look for her marble?" If the children point to the basket, then they pass the belief question by appreciating that Sally now has a false belief. If, however, they point to the box, then they fail the question. They know where the marble has been hidden, but cannot appreciate that Sally does not know.

They then repeated the whole procedure, but this time the marble ended in the experimenter's pocket.

After each scene, the experimenter asked two more crucial questions: the "reality" question "Where is the marble really?" and the memory question "Where was the marble in the beginning?" All the children, without exception, got these right; they knew where the marble was now, and they remembered where it had started.

When asked "Where will Sally look for her marble?" the children varied. Among non–special needs children, 85 percent got it right—they pointed to her basket—and so did 86 percent of the Down's children, but only 20 percent of the autistic children did so.

The four autistic children who got the answer right did not appear to be much different from the others in the group; they had average ages and mental ages. Every single child answered the control questions correctly; so they all understood what was going on; they knew (and believed) that the marble was put somewhere else after Sally left.

The non–special needs and Down's children answered the belief question by pointing to Sally's basket. The researchers concluded that:

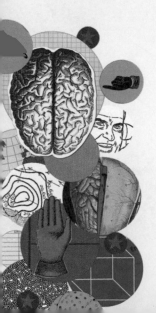

Thus they must have appreciated that their own
knowledge of where the marble actually was, and the
knowledge that could be attributed to the doll, were
different. That is, they predicted the doll's behavior on the
basis of the doll's belief.

The autistic group, on the other hand, answered by
pointing consistently to where the marble really was. They
did not merely point to a 'wrong' location, but rather to the
actual location of the marble.... We therefore conclude
that the autistic children did not appreciate the difference
between their own and the doll's knowledge.

Our results strongly support the hypothesis that autistic
children as a group fail to employ a theory of mind. We
explain this as an inability to represent mental states.
As a result of this, autistic subjects are unable to impute
beliefs to others and are thus at a grave disadvantage when
having to predict the behavior of other people.

Thus we have demonstrated a cognitive deficit that is
largely independent of general intellectual level and has
the potential to explain both lack of pretend play and
social impairment by virtue of a circumscribed cognitive
failure.

On the other hand, the researchers thought that perhaps the
four autistic children who did get the belief question right—and
they all did so in both versions of the story—may employ a theory
of mind, might be able to take part in pretend play, and might
have slightly less trouble in forming social relationships.

The Sally-Anne test has been widely used as a way to study
theory of mind in children and its relationship to social
interaction and empathy.

1988

THE STUDY

RESEARCHER:

Randolph C. Byrd

SUBJECT AREA:

Social psychology

CONCLUSION:

Byrd's data
seemed to show prayer
had a beneficial effect.

CAN PRAYER HELP HEAL THE SICK?

STUDIES ON INTERCESSORY PRAYER

If you became seriously ill, would you like other people to pray for you? Randolph Byrd points out that "Praying for help and healing is a fundamental concept in practically all societies, though the object to which those prayers are directed varies among the religions of the world. In western culture, the idea of praying for the benefit of others (intercessory prayer) to the Judeo-Christian God is widely accepted and practiced."

The Bible has no doubts: "And Abraham prayed unto God: and God healed Abimelech, and his wife, and his maidservants" (Genesis 20:17); "The father of Publius lay sick of fever and dysentery: unto whom Paul entered in, and prayed, and laying his hands on him healed him" (Acts 28:8).

More recently, in 1872, Francis Galton reported on the effectiveness of prayer on clergy, and found no positive effects. He also pointed out that in England, even though thousands of people prayed every Sunday for the health of the royal family, none of the kings or queens had actually lived significantly longer than their prominent contemporaries.

But there was no hard scientific evidence to show whether prayer did any good; so Byrd set out to investigate. In the coronary care unit at San Francisco General Hospital he invited 450 patients to consent to being prayed for, while they were in the hospital; 393 gave their consent.

He divided the patients randomly into two groups, 192 to get intercessory prayer and 201 not to be prayed for. Neither he, nor the patients, nor doctors and staff knew who was in which group.

The intercessory prayers

The intercessors were all "born-again Christians . . . with an active Christian life as manifested by daily devotional prayer and active Christian Fellowship with a local church." Each patient in group 1 was allotted between three and seven intercessors, who were given the patient's first name, diagnosis, and general condition.

The prayer was done outside the hospital every day until the patient was discharged. "Each intercessor was asked to pray daily for a rapid recovery and for prevention of complications and death."

The hospital record provided details of no less than 30 different conditions that affected these patients, but overall there was no significant difference between the two groups when they were admitted to hospital.

The hospital course after entry "was considered to be *good* if . . . events occurred that only minimally increased the patient's morbidity or risk of death. The course was considered *intermediate* if there were higher levels of morbidity and a moderate risk of death. The course of patients who had the highest morbidity and risk of death, or who died during the study was graded as *bad*."

The results

In the prayer group 85 percent of the patients were considered to have a *good* hospital course, as opposed to 73 percent in the control group. An *intermediate* grade was given in 1 percent of the prayer group, and 5 percent of the controls. A *bad* hospital cause was observed in 14 percent of the prayer group, and 22 percent of the controls.

Conclusions

Byrd concludes that the prayer group "had less congestive heart failure, required less diuretic and antibiotic therapy, had fewer episodes of pneumonia, had fewer cardiac arrests, and were less frequently intubated and ventilated . . . " So "the prayer group had an overall better outcome. . . . Based on thus data there seem to be an effect, and that effect was presumed to be beneficial."

On the other hand, some reviewers were skeptical, on the grounds that there were positive outcomes in only six of the 30 different conditions affecting the patients, and one critic compared Byrd's results with the "sharpshooter fallacy"— "searching through the data until a significant effect is found, then drawing the bull's-eye."

Quality of hospital course		
Score	Prayer group (192 patients)	Control group (201 patients)
Good	163	147
Intermediate	2	10
Bad	27	44

Many later experiments, attempting to reproduce Byrd's results, have produced either very little positive evidence, or none. In 2006 Herbert Benson reported the enormous STEP project (Study of the Therapeutic Effects of Intercessory Prayer) with 1,802 coronary artery bypass patients.

They were divided into three groups: Group A (604 patients) were told that they might or might not be prayed for—and in fact were prayed for. Group B (597) were also told that they might or might not be prayed for—and in fact were not prayed for. Group C (601) were told that they would be prayed for, and were. Prayer started the day before their operations and continued for 14 days.

Benson found that within a month major complications and sometimes death happened to 52 percent of Group A, 51 percent of Group B, and 59 percent of Group C. In other words people who knew that prayers were being said for them actually came off worse than the others. Perhaps this was "performance anxiety"; they were seriously stressed by thinking they might fail to respond to the prayers.

DO YOU NEVER FORGET A FACE?

THE EFFECTS OF PROSOPAGNOSIA AFTER A STROKE

1993
THE STUDY
RESEARCHERS:
Jane E. McNeil and
Elizabeth K. Warrington
SUBJECT AREA:
Social psychology
CONCLUSION:
Sometimes sheep are
easier to recognize
than humans.

Can you imagine being able to recognize sheep faces more easily than human faces? After a stroke, some people develop prosopagnosia (also known as "face blindness"), a condition in which people are unable to recognize other people's faces. A few people also suffer prosopagnosia all their lives, and always have trouble with face recognition.

The lack of recognition sometimes extends to other things. There was a prosopagnosic bird-watcher who could no longer recognize birds, a farmer who could no longer recognize his own cows, another who could recognize his cows and dogs, but not human faces, and so on. McNeil and Warrington worked with a patient, W. J., who after a series of strokes had acquired severe prosopagnosia, and could not recognize human faces.

When they tested him, W. J. was able to identify only two out of 12 well-known faces, and even then he had to work them out carefully. He could not judge age, sex, or expression on faces in photographs. On the other hand he was 95 percent accurate in naming famous buildings, breeds of dog, makes of car, and flowers.

He also claimed that he could recognize his sheep by their faces.

Sebastian

Mr. Pickles

Lady

151

Recognizing sheep?

For two years he had owned a flock of sheep, which were identified by number tags in their ears. When they showed him close-up photographs of 16 of his sheep (with ear tags omitted from the pictures) he was able to identify eight of them, but clearly knew some of the others, as in several cases he would say things like, "I know that sheep very well; she's the one that had three lambs last year, but I can't remember her number." Clearly he was better at recognizing the faces of sheep than of humans.

In order to avoid the problem of having to remember numbers, the researchers arranged a different recognition test. They showed, at three-second intervals, eight photographs of sheep's faces, asking only whether or not they were pleasant sheep. Then they showed 16 photographs, which were the same eight plus eight different sheep, but mixed up in a random order, and asked in each case whether the sheep in the photograph was one of the first eight, or not. They gave the same test to two other people with flocks of sheep and five other farmers, who commented that the sheep all looked the same, and were incredibly difficult to recognize.

W. J. was hopeless with human faces; he was probably guessing, since he scored just 50 percent, which is the score you would expect by chance. Clearly, however, he was much better than the others at recognizing sheep.

As a further test, they went through the same procedure again, but using photographs of a different breed of sheep, with which W. J. was not familiar. The results were similar, though less dramatic.

	W. J.	Farmers	Sheep-owners
	Accuracy of recognition (percent)		
Familiar sheep	87	66	59
Unfamiliar sheep	81	69	63
Human faces	50	89	100

The researchers then tried another test. They showed six photographs of faces of unfamiliar sheep, and gave each one a plausible name. Then they showed the photographs in random order and asked the subjects to provide the correct name when the photograph appeared. They did the same thing with six human faces and names. Once again, W. J. was poor with human faces, but outclassed the controls when it came to sheep.

	WJ	Farmers	Sheep-owners
	Accuracy of recognition (percent)		
Human faces	23	71	78
Sheep	57	41	55

The researchers were baffled by W. J. He had bought the sheep after his strokes, and therefore had learned to recognize their faces at the same time as he was unable to recognize human faces. They discuss how he might have done this:

> It is possible that he developed a sheep 'prototype,' which enables the effective encoding of sheep facial features. What is quite surprising, however, is the extent to which his abilities appear to generalize to other visually dissimilar breeds of sheep. Perhaps the more remarkable finding is that W. J. has been totally unable to overcome his prosopagnosia. . . . He seems unable to utilize the sorts of strategies he has learned to use for sheep.

THE STUDY

RESEARCHERS:
Daryl J. Bemm and
Charles Honorton

SUBJECT AREA:
Perception

CONCLUSION:
Honorton's tests show that
mind-reading might
be possible, but his results
have yet to be matched.

IS THERE ANY
SENSE IN ESP?

FINDING EVIDENCE FOR
EXTRASENSORY PERCEPTION

Daryl J. Bemm and Charles Honorton knew they were fighting an uphill battle with this research. Most academic psychologists don't think paranormal phenomena exists at all. And even believers have no idea how it might work.

Extrasensory Perception (ESP)

The first serious research into ESP was conducted in the 1930s at Duke University in North Carolina by J. B. Rhine and his wife Louisa. The Rhines set out to investigate various apparently paranormal phenomena. In particular Rhine used a special pack of cards designed by a colleague, Karl Zener.

There were 25 cards in the pack, five of each type. The pack was shuffled and cut, and then a subject had to guess each card before it was dealt. By pure chance you should guess right one time in five, or 20 percent. Some of Rhine's subjects achieved much higher scores, but the effects were hard to replicate, and there may have been "sensory leakage," (where an experimenter involuntarily reveals the answer) or even cheating. Nevertheless in 1934 Rhine wrote a book called *Extra-Sensory Perception*, which was where the term ESP came from.

Ganzfeld and autoganzfeld

Bemm and Honorton's basic plan was that a "sender" should watch a short film clip—the "target"—and try to send ideas and images from the film to a "receiver." Both sender and receiver were isolated, and no contact was possible between them.

After half an hour, an experimenter went into the receiver's room and showed four possible film clips. The receiver then had to guess the target on the basis of which one most closely resembled the ideas and images that had arrived during that half hour.

There had been suggestions that psychic activity was more likely to occur during relaxation and meditation, so Honorton had invented a method that he called the *Ganzfeld* (German for "entire field"). The receiver lay comfortably on a reclining chair. White noise, like waves on a sea shore, was played into their headphones. Over the eyes were half ping-pong balls, lit with warm red light. These conditions were meant to put the receiver in an ideal state to receive incoming messages from the sender, which would be telepathy, or to see the film clip directly, which would be clairvoyance.

While the sender was "sending," the receiver talked continuously about the ideas and images that turned up, and all this was recorded for later analysis.

People had argued over the original Ganzfeld experiments; there might have been sensory leakage or cheating. Honorton set up the elaborate autoganzfeld in order to try to silence the critics, since the automatic procedure should eliminate any problems. He used a computer, which chose the film clip from a bank of 80 clips, and showed it repeatedly to the sender.

At the end of the session an experimenter removed the ping-pong balls, switched off red light and white noise, and switched on a TV set, where the computer showed, in random order, the film clip, along with three others, for the receiver to guess which had been the target. They were allowed to watch the clips and adjust the ratings as often as they liked. Their final ratings were saved on the computer, and then the sender came in to discuss the results. The experimenter who sat with the receiver did not know until this point what the target was. With his elaborate security arrangements, Honorton reckoned he had prevented both sensory leakage and fraud.

The results

By chance alone they should have got it right one quarter of the time, since there were four possible film clips, and any of them could have been the target. Thus any score significantly above 25 percent could count as evidence for psychic phenomena.

This large study had 240 participants, most of whom were strong believers in ESP. They took part in 329 sessions, and scored 106 hits, which corresponds to a score of 32 percent— much better than the expected 25 percent.

In order to also see if artistic people are be better suited to psychic communication than the average the researchers recruited ten male and ten female students from the Juilliard School—eight music students, ten drama students, and two dancers. This group did one session each, and achieved an extraordinary score of 50 percent.

Conclusions

Bemm and Honorton believed that they had, for the first time, scientifically proved the existence of ESP. Their paper was one of the first paranormal papers accepted for publication by a mainstream psychology journal.

Sadly, however, because of lack of funding, Honorton's lab was closed down before this study could be repeated, and Honorton himself died nine days before the paper was accepted for publication. No one else has been able to produce such positive results.

WHY YOU CAN'T ALWAYS SPOT THE DIFFERENCE?

THE PECULIAR PHENOMENON OF CHANGE BLINDNESS

1995

THE STUDY

RESEARCHERS:
Daniel J. Simons and
Daniel T. Levin

SUBJECT AREA:
Perception

CONCLUSION:
We sometimes fail to
see what's right in front
of our nose.

Sometimes it takes only the smallest distraction to make someone unaware of substantial changes in the scene they are looking at. "Change blindness" was first reported by the English psychologist Susan Blackmore and her colleagues in 1995. They showed that when you look at two pictures showing almost the same scene, you are unlikely to notice a substantial change from one to the other, provided the two versions are separated by a flash frame, or a blank frame, or when they are in slightly different places on the screen. These experiments were all of two-dimensional pictures.

Later that same year, the American experimental psychologists Simons and Levin wanted to try the same test in three dimensions using short films. In one film an actor walked through an empty classroom and sat in a chair. The footage cut to a close up and a different actor completed the action. Even though the actors were easy to tell apart only 33 percent of the 40 participants reported noticing the change.

In the real world

Simons and Levin decided to take this idea into the real world, and see whether change blindness would still work when the participant was actively engaged with an experimenter.

One experimenter waited on the campus of Cornell University holding a map, and then approached an unsuspecting pedestrian to ask for directions to the library. After the experimenter and

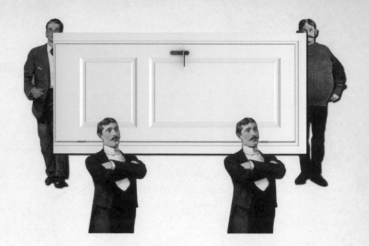

the pedestrian had been talking for 10 or 15 seconds, two other experimenters carrying a door walked toward them along the sidewalk, and rudely pushed their way between them.

As the door passed, the first experimenter grabbed the back of the door and walked on, while one of the others let go of the door, stayed behind, and carried on asking directions.

The second experimenter had a copy of the same map, but was dressed in different clothes. When the directions were complete the experimenter said "We're doing a study as part of the psychology department . . . of the sorts of things people pay attention to in the real world. Did you notice anything unusual at all when that door passed by a minute ago?"

Any subjects who had not noticed the change were then asked directly "Did you notice that I'm not the same person who approached you to ask for directions?"

There were 15 pedestrians in this experiment, male and female, ranging in age from 20 to 65. When asked whether they had noticed anything unusual, most of them said they thought the people with the door were rude, but eight of them—more than half—had not noticed the switch. They simply carried on the conversation, and were surprised to learn that the direction-seeker had been replaced by someone else in mid flow.

Interestingly, those who did notice the change were all in the same age range (20–30) as the experimenters; older people were less likely to notice. The researchers speculate that this might be because the younger pedestrians, recognizing people of their own age group (the in-group) were more likely to expend effort in noting their features.

Experiment 2

In order to test this idea, they carried out the same procedure again, tackling unsuspecting passersby, but this time the experimenters dressed as builders—there was a building site nearby—and wore strikingly different clothes.

They approached only young (20–30) pedestrians, and this time only four out of 12 noticed the switch.

Even though the two experimenters wore strikingly different clothes, the pedestrians probably saw them as "builders"—an out-group—and therefore not worth close inspection. The researchers wrote "One subject said that she had just seen a construction worker and had not coded the properties of the individual. . . . Even though the experimenter was the center of attention, she did not code the visual details and compare them across views. Instead, she formed a representation of the category."

Simons and Levin point out that these experiments build on the work of Loftus (page 119) and Bartlett (page 34), but show in addition that people often do not notice changes in the scene even when they are actively engaged and the change is to the central object in the scene. After hearing about the experiment, 50 introductory psychology students all insisted that they would have spotted the substitution. This has been called "change-blindness blindness."

We certainly don't expect to be fooled in this way—but we often are. These and many other experiments should make us wonder how much of what is going on around us we simply never see.

1998

THE STUDY

RESEARCHERS:
Marcello Costantini and
Patrick Haggard

SUBJECT AREA:
Perception

CONCLUSION:
We can sometimes
misperceive our own
bodies.

COULD YOU CONFUSE A FAKE HAND WITH YOUR OWN?

"THE RUBBER HAND ILLUSION"

Lay your hand on a table in front of you, with a fake hand (try using an inflated rubber glove) beside it in the same orientation. Hide your hand and get someone to stroke both the rubber hand and your own, at the same time and in the same way, and you may suddenly feel that the rubber hand is your hand. This illusion was first invented by two psychiatrists from the University of Pittsburgh, Matthew Botvinick and Jonathan Cohen, in 1998.

We all feel and see our bodies, and have a sense of body ownership. Psychologists talk about "body schema" and body image. Body schema is the model of your body that you can feel with your eyes shut; it allows you to walk around without bumping into obstacles, because you know where your limbs are; this is part of what's known as "proprioception." Your body image is a conscious idea of your body, including what it looks like from the outside. Together with body schema, it forms a coherent base for self-consciousness.

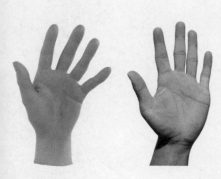
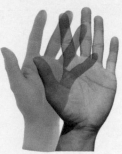

The American researchers, Marcello Costantini and Patrick Haggard, wanted to take the idea further and find out whether our experience of our body come mainly from the inside (schema) or from the outside (image)?

The experiment

The participant sat at a table and put her arm on it, in front of the shoulder, with the hand laid flat, palm down. (There were 13 male and 13 female participants, with an average age of 28.) She could see the rubber hand, but not her own. The rubber hand was lined up with hers, and 12 in (30 cm) away from it.

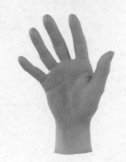

Her hand and the rubber hand were stroked exactly in sync and in the same place, with 1 mm paintbrushes, under computer control. Because she saw and felt the stroking, a typical participant was confused by the conflicting signals of sight and touch, and would say this made the rubber hand "feel like my own hand."

One consequence was that she thought her own unseen hand was closer to the rubber hand than it actually was. This is called "proprioceptive drift," because she felt she knew the position of her hand by proprioception—part of the body schema. Participants were asked to measure the distance between the rubber hand and where they felt their own hands to be, which provided the experimenters with a measure of the power of the illusion. In the strongest cases the participants felt as if their hands were a few inches closer to the rubber hand than they really were; this was the amount of proprioceptive drift.

The researchers investigated what happens as they varied the angles of the stroking and the hands. There were two groups of participants. To begin with they lined up the rubber hand precisely with the real hand, and stroked both down the back to the middle finger. This is the baseline condition.

In one group they manipulated the real hand, while in the second group they manipulated the rubber hand. They expected the illusion to disappear sooner in the second group, since you can detect a change of angle easily by sight, but not so easily by proprioception.

The first manipulation was to vary the angle at which the paintbrush stroked the hand. Next they rotated the hand, but kept the stroking the same with respect to the hand, or in "hand-centered space." Finally they rotated the hand, while the stroking remained in the same direction as it had been in the baseline condition. There would be a mismatch if the participant felt the stroking in a hand-centered space, but no mismatch if she felt it in an external egocentric space.

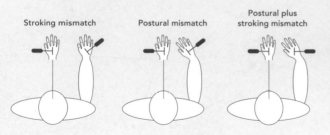

Stroking mismatch Postural mismatch Postural plus stroking mismatch

Keeping up the illusion

In the proprioceptive group—rotating the real hand—the illusion changed a little at 10 degrees, significantly more at 20 degrees and 30 degrees. In the visual group, however—rotating the rubber hand—even a 10 degrees mismatch destroyed the illusion altogether. In other words, changing the angle of the rubber hand, which the participant could see, produced a much greater effect than changing the position of the real hand.

In the first picture above, only the stroking is off line, relative to both the participant and the hand. In the second, the hand and the stroking are off line relative to the participant, but the stroking in not off line relative to the hand. In the third picture, the hand is off line, and the stroking is off line relative to the hand, although not relative to the observer.

Since this third condition gave much the most dramatic change in the R.H.I., the researchers concluded that the stroking is felt in a hand-centered space.

The researchers conclude that "the brain maintains an internal body representation, with its own characteristic spatial organization based on proprioception. This representation uses a frame of reference based on the specific part of the body that is stimulated." In other words, when your hand is stroked, you feel the sensation in hand-centered space, not body-centered space.

WHY CAN'T YOU TICKLE YOURSELF?

FINDING ANSWERS TO A TICKLISH QUESTION

2000
THE STUDY

RESEARCHERS:
Sarah-Jayne Blakemore,
Daniel Wolpert, and
Chris Frith

SUBJECT AREA:
Neuropsychology

CONCLUSION:
There is a surprising
link between
ticklishness and
schizophrenia.

We can easily tell the difference between sensations caused by our own movements and those caused by other things; the difference between pushing and being pushed is obvious. The researchers suggest this is because when we do something, the brain sends messages to the muscles to act, and at the same time provides advance warning, known as an "efference copy," that the relevant action is going to happen; so we are not surprised to see our arms go out to push. If we don't get that warning, on the other hand, we are surprised.

When you turn your head or your eyes to look at something, the warning means that you can work out where in the world is the thing you are looking at.

With both eyes open, try pressing gently with a finger on your eyelid near the outside corner of your eye. In the view from that eye, the world begins to swivel. This is because the brain has not received the warning that the eye was about to move, and can't compute the image it receives.

In normal, everyday movements therefore there is forward planning, which is backed up as the movement happens by further messages, confirming the movement happens as planned, or adjusting it if necessary. The forward planning and the confirming feedback can be used to damp down, or attenuate, the sensations caused by the movement.

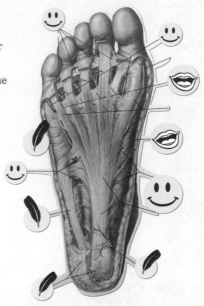

Tickling yourself

Tickling yourself does not really work. You can still feel it, but it is not very tickly. The researchers suggested that the forward planning warns you not only that there is going to be soft movement across your skin, but exactly when and where it will happen—and automatically damps down the sensation.

To test their hypothesis, the researchers asked 16 subjects to hold out their right hands for tickling, and to rate the ticklishness of a piece of soft foam moved 0.6 in (1.5 cm) in a figure-eight pattern across the palm twice every second. The foam was attached to a robotic arm.

First the robot did all the action, which felt very tickly— 3.5 on the Tickle Rating Rank. Then the subjects moved the foam themselves, by using their left hands to move a knob on the arm of a second robot, which transferred the movement to the first; so the subject had complete control of the movement.

The clever part of the experiment was that the researchers were able to introduce variations, either by delaying the movement, so that it was just the same as that caused by the subject, but fractions of a second later, or by changing the direction of the movement, so that when the subject moved say north-south (N-S), the foam went a bit clockwise, through NE-SW to E-W.

The results

The subjects said that tickling themselves (via the robot arms) was much less tickly (about 2.1) than being tickled by the robot, but the sensations became gradually stronger as the self-tickling was progressively delayed or twisted. By the time the movement was delayed by 300ms (nearly one third of a second), or skewed by 90 degrees from north-south to east-west, the tickling was almost as good as if the robot was doing it. These results strongly support the theory that when you try to tickle yourself, the sensation is damped down by the advance warning of your own movements. Without any delay or skewing, the ticklishness is reduced by nearly 50 percent, but the more delay or skewing there is, the less accurate is the advance warning, and the more tickly it becomes.

Connection with schizophrenia

Lack of this forward planning and confirmation may possibly be involved in schizophrenia. A common symptom is hearing voices. This might be the result of internally generated voices or thoughts for which there are no warning messages. Another common symptom is called "passivity phenomena." For example schizophrenics may feel some of their own actions are caused by other people: "My fingers pick up the pen, but I don't control them. What they do is nothing to do with me." Again, that may be what it feels like if there is no advance warning.

Can there really be a connection between ticklishness and schizophrenia? To find out, the researchers tried tickling tests on patients with schizophrenia, bipolar affective disorder, or depression, dividing them into two groups. The 15 patients of group A all had auditory hallucinations (hearing voices) and/or passivity experiences. The 23 patients of group B had neither of these symptoms. Group C comprised 15 non-patients.

All these people held out their right hands, and either were tickled by the experimenter or tickled themselves with the other hand. All those in groups B and C said that when they tickled themselves it was much less intense, tickly, and pleasant. Patients in group A, however, said that tickling themselves was just as effective as being tickled by the experimenter.

This suggests that hearing voices and passive experiences may well be associated with lack of that vital advance warning of movement.

2001

THE STUDY

RESEARCHERS:

Researchers: V. S.
Ramachandran
and E. H. Hubbard

SUBJECT AREA:

Perception

CONCLUSION:

For some people the
senses are all connected.

CAN YOU TASTE THE NUMBER 7?

**THE EXTRAORDINARY EFFECTS
OF SYNESTHESIA**

A small fraction of people, perhaps one in a thousand,
experience curious mixtures of senses; they hear numbers as
distinct musical tones, taste letters, and see days of the week as
colored. This is called synesthesia often runs in families, is more
common among women, left-handers, artists, and poets.

Is it a genuine phenomenon, or just imagination?

The condition was first described by Francis Galton in 1880, but
for more than a century, scientists and refused to take the idea
seriously for a variety of reasons:

1. "Synesthetes" are just crazy. The phenomenon is simply the
    result of a hyperactive imagination.
2. They are just remembering childhood memories such as
    seeing colored numbers in books or playing with colored
    refrigerator magnets.
3. They are just engaging in vague tangential speech or just
    being metaphorical, just as you and I might say "bitter cold"
    or "sharp cheese."
4. They are "potheads" or "acid junkies" who have been on
    drugs. This idea is not entirely absurd, since LSD users
    often do report synesthesia both during the high as well as
    long after.

Ramachandran ("Rama") and Hubbard took the claims more
seriously, and decided to investigate. They used some elegant
experiments to investigate whether synesthesia depends on
the vision system rather than imagination, or memory. Here
are two of them:

How synesthetes see

They showed people displays of square 2s and 5s like the one on the right. Can you see a triangle of 2s within the display?

If you are not a synesthete this may take several seconds, but if 2s and 5s are different colors for you (as in the display on the next page) then the triangle should pop out instantly. This is exactly what happened with the synesthetes. They could not have been making this up.

Synesthetes say that a letter or numeral printed in the "wrong" color is ugly. "Also they often report 'odd' or weird colors they cannot see in the real world but see only in association with numbers. We even saw a color-blind subject recently who saw certain colors only upon seeing numbers."

The researchers noted that color processing in the brain, in both humans and monkeys, happens in an area called the fusiform gyrus, right next to the area in which visual letters and numbers are processed. The most common form of synesthesia is to see letters and numbers as colored. They therefore propose that synesthesia is caused by cross-wiring in the brain between these two processes.

They further propose that since synesthesia runs in families, "a single gene mutation causes an excess of cross-connections or defective pruning of connections between different brain areas. Consequently, every time there is activation of neurons representing numbers, there may be a corresponding activation of color neurons."

Top-down influence

Rama and Hubbard showed synesthetes the Roman numeral IV. They saw the colors for I and V when it looked like letters, but not the color expected for the number four.

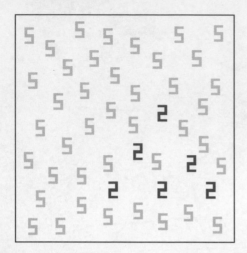

The implication of these observations is that synesthesia can be modulated by top-down influence; there is clearly some processing before the color appears.

Conclusion

At the end of a long, speculative paper, Rama and Hubbard write:

Synesthesia has always been regarded as somewhat spooky. Even though it has been known for over 100 years, it has often been thought of as a curiosity . . . our psychophysical experiments were the first to prove conclusively that synesthesia is a genuine sensory phenomenon.

They speculate that studying synesthesia may help us understand the neural basis of metaphor and creativity:

Perhaps the same mutation that causes cross-wiring in the fusiform, if expressed very diffusely, can lead to more extensive cross-wiring in their brains. [This could] explain the higher incidence of synesthesia in artists, poets, and novelists (whose brains may be more cross-wired, giving them greater opportunity for metaphors).

And finally they suggest a novel synesthetic theory of the origin of language.

IS ASTRAL TRAVEL OUT OF THIS WORLD?

THE SCIENCE BEHIND OUT-OF-BODY EXPERIENCES

2007

THE STUDY

RESEARCHERS:
Bigna Lenggenhager, Tej Tadi, Thomas Metzinger, and Olaf Blanke

SUBJECT AREA:
Perception

CONCLUSION:
There is no conclusive evidence to prove out-of-body experiences exist.

Apparently one person in ten has at least one out-of-body experience (OBE) during their lifetime. In typical OBEs, people feel that they have left their body and can see the world from a location outside it—often looking down from a position near the ceiling. OBEs can form a dramatic part of a near-death experience, for example when patients having a heart attack describe watching medical staff trying to resuscitate them. But OBEs are far more common in people who are perfectly well, most often happening during deep relaxation or on the verge of sleep. Occasionally they happen in such scary situations as giving a lecture or appearing on stage. One woman even reported having an OBE during her driving test, yet she carried on driving.

Some people are terrified that they will not be able to return to their body, when in fact it is much harder to maintain the experience, and it usually lasts only a few seconds or a few minutes. Others enjoy the experience so much that they try to induce it, for example by hovering in the hypnagogic state just before falling asleep, or by taking drugs such as ketamine which can induce something like an OBE.

Allegedly Thomas Alva Edison used to get into this hypnagogic state deliberately while wrestling with invention problems. He would sit in a chair, holding a bucket, and put a silver dollar on his head. When he nodded off, the dollar fell into the bucket and awoke his mind, while his body slept. Sylvan Muldoon, OBE pioneer, used to drift off to sleep with his forearm vertical; he hoped that when it fell he might go into an OBE.

Psychological or paranormal?

The big question for OBEs is whether anything actually leaves the body during the experience. People of many cultures have theories about souls or spirits that can separate from the physical body, and even survive after death. One such is the doctrine of astral projection, according to which we may have "subtle bodies" including an astral body that can project beyond its physical home and into the astral planes. Yet there is no reliable evidence for the existence of astral bodies or planes. Also many people who have OBEs claim to be able to see at a distance, but there is no good evidence that this is true either.

In 2002 a Swiss neurosurgeon discovered a spot in the brain where OBEs can be induced, and suddenly there was a burst of scientific research on OBEs. Some involved brain research; others used virtual reality, like this experiment by Bigna Lenggenhager and her colleagues in Zurich. They wanted to try to induce OBEs artificially, by using an extension of the rubber-hand illusion (see page 160), in which vision dominates proprioception. In order to do this, they invited participants to enter into a world of virtual reality.

Experiment A

The participant wore a head-mounted display unit (HMD), and stood in the middle of the room. A video camera was mounted on a tripod 2.2 yd (2 m) behind, and the image of the participant's back was transmitted to the HMD. The result was that the participants could see a three-dimensional image of their own backs apparently standing 2.2 yd (2 m) in front of them. An experimenter then stroked their backs for one minute, and the sensation of feeling their backs being stroked, while watching the same thing, induced the participants to believe that they were seeing themselves standing over two yards away. When the virtual stroking was arranged to be out of sync with the actual stroking the effect was much reduced.

Immediately after the stroking had finished the participants were blindfolded and asked to move back to their original position. As predicted, they moved forward—toward where the virtual body had been. This "proprioceptive drift" averaged 9.5 in (24 cm), if the stroking had been in sync, but only half as much when it had not.

Experiments B & C

The experimenters then put a fake body in front of the camera, and stood the participants two meters away to the side. Then they stroked both the participant's body and the fake body. Participants therefore saw, apparently standing 2.2 yd (2 m) in front of them, a fake body being stroked at the same time as their own backs. Provided that the stroking was exactly in sync, they came to feel that they were looking at themselves, as when they had been looking at the image of their own bodies. Furthermore when blindfolded, they stepped forward and showed if anything slightly more proprioceptive drift.

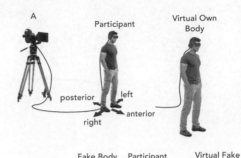

When the fake body was replaced by a box, however, the participants were no longer persuaded that it was their own body they were looking at, and showed virtually no drift.

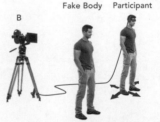

Conclusions

The researchers write: "Illusory self-localization to a position outside one's body shows that bodily self-consciousness and selfhood can be dissociated from one's physical body position."

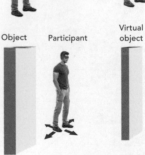

On the other hand they acknowledge that because the participants did not feel disembodied, and maintained their original perspective—as opposed for example to looking down on themselves from above—the experiments induced only some parts of a typical OBE.

Index

Glossary

chaining—reinforcing individual responses in a sequence to form a complex pattern of behavior.

change blindness—a phenomenon that occurs when a change in scene is introduced and not noticed.

cognitive dissonance—mental stress caused by holding two contradictory beliefs at the same time, or by encountering information that contradicts an existing belief.

cognitive psychology—the study of mental processes such as attention, language use, memory, perception, creativity, and problem solving.

corpus callosum—a wide band of nerve fibers that connects the left and right sides of the brain and transfers information between them.

EEG—electroencephalogram, recording electrical activity along the scalp.

extrinsic reward—an expected reward for something done, that does not lead to greater satisfaction.

ingroup—a small exclusive group of people with a common interest.

Gestalt—an organized whole that is seen to be more than the sum of its parts.

heuristic—a shortcut method of solving problems, which may disregard some information, and may not give the right or best answer.

intrinsic reward—reward for a job that comes from satisfaction of a job well done, and a sense of achievement.

kinesthesis—the ability to feel movements of the limbs and body.

operant conditioning—learning through reinforcements and punishments for particular actions.

outgroup—a group of people to which you do not belong.

proprioception—the sense of how the various parts of your body are positioned.

proprioceptive drift—the idea that your body or its parts have moved, or is in the wrong place.

reflex action—instinctive response to a stimulus.

saccade—a brief rapid movement of the eye between fixation points.

schema—a pattern of thought, behavior, or experiences that organizes categories of information and the relationships among them.

theory of mind (TOM)—the ability to understand that others have beliefs that are different from yours.

Acknowledgments

Deciding which experiments should go in this book was tricky, but I had help from half a dozen psychologists—including my wife Sue Blackmore—not to mention a philosopher, a solicitor, and the postman. Then the fun started. In almost every case I was able to go back to the original account and read about the work at first hand, which was a great privilege.

I have tried to explain each experiment in simple language, without unnecessary jargon. Some of the researchers had a delightful way of putting things, while others wrote in the most impenetrable style. I have avoided technical statistics, although I do sometimes say the researchers found a significant result, meaning that it could not have happened simply by chance.

One thing that struck me repeatedly was the importance of ingenuity. A good scientist has to be focused—as Darwin was on his earthworms—but also needs to be ingenious, and imaginative.

To take a modern example, Simon Baron-Cohen's Sally-Anne experiment was simple—it needed no expensive equipment or complicated procedure—and yet it delivered intriguing information about the "theory of mind." Similarly, Sarah-Jayne Blakemore's research on tickling generated important information about schizophrenia.

While writing this book I have learned something about experimental psychology, and a lot about human nature. I have enjoyed writing it, and I hope you enjoy reading it.

Sources

Chapter 1 Darwin, Charles. *The Formation of Vegetable Mould through the Action of Worms, with Observations of their Habits* (London: Murray, 1881).

Stratton, George M. "Some preliminary experiments on vision without inversion of the retinal image." *Psychological Review* 3, no. 6 (1896): 611.

Thorndike, E. L. "Animal intelligence: An experimental study of the associative processes in animals," *Psychological Review: Monograph Supplements,* Jun 1898, 2 (4): i—109.

Pavlov, I. P. "Conditioned Reflexes: an Investigation of the Physiological Activity of the Cerebral Cortex," trans. G. V. Anrep (London: Oxford University Press, 1927).

Perky, Cheves West. "An experimental study of imagination." *The American Journal of Psychology* (1910): 422—452.

Chapter 2 Watson, John B., and Rosalie Rayner. "Conditioned emotional reactions." *Journal of Experimental Psychology* 3, no. 1 (1920): 1.

Zeigarnik, Bluma. "Über das Behalten von erledigten und unerledigten Handlungen," *Psychologische Forschung,* 9 (1927): 1—85.

Bartlett, Frederic C. *Remembering: A Study in Experimental and Social Psychology* (Cambridge: Cambridge University Press, 1932).

Skinner, Burrhus Frederic.

The Behavior of Organisms: An Experimental Analysis (New York: Appleton-Century, 1938).

Roethlisberger, F. J., and W. J. Dickson. "Management and the worker" (Cambridge MA: Harvard University Press, 1939).

Lewin, Kurt, Ronald Lippitt, and Ralph K. White. "Patterns of aggressive behavior in experimentally created 'social climates'." *The Journal of Social Psychology* 10, no. 2 (1939): 269—299.

Chapter 3 Tolman, Edward C. "Cognitive maps in rats and men." *Psychological Review* 55, no. 4 (1948): 189.

Piaget, Jean. *The origins of intelligence in children.* (New York:

International Universities Press, 1952).

Heller, M. F., and M. Bergman. "Tinnitus aurium in normally hearing persons." *Ann Otol Rhinol Laryngol* 62, no. 1 (1953): 73—83.

Festinger, Leon, Henry W. Riecken, and Stanley Schachter. *When Prophecy Fails: A Social and Psychological Study of a Modern Group that Predicted the Destruction of the World* (Minneapolis: University of Minnesota Press, 1956).

Asch, Solomon E. "Studies of independence and conformity: a minority of one against a unanimous majority." *Psychological Monographs: General and Applied* 70, No. 9, (1956): 1—70.

Harlow, Harry F., and Robert R. Zimmermann. "The development of affectional responses in infant monkeys." *Proceedings of the American Philosophical Society* (1958): 501—509.

Sperling, George. "The information available in brief visual presentations." *Psychological monographs: General and applied* 74, no. 11 (1960): 1.

Bandura, Albert, Dorothea Ross, and Sheila A. Ross. "Transmission of aggression through imitation of aggressive models." *The Journal of Abnormal and Social Psychology* 63, no. 3 (1961): 575.

Sherif, Muzafer, Oliver J Harvey, Jack White, William R. Hood, and Carolyn W. Sherif. *Intergroup Conflict and Cooperation: The Robbers Cave Experiment*, Vol. 10 (Norman, OK: University Book Exchange, 1961).

Chapter 4 Milgram, Stanley. "Behavioral study of obedience." *The Journal of Abnormal and Social Psychology* 67, no. 4 (1963): 371.

Gregory, R. L., and J. G. Wallace. "Recovery from early blindness." *Experimental Psychology Society Monograph* 2 (1963): 65—129.

Hess, Eckhard H. "Attitude and pupil size." *Scientific American,* 212, (1965): 46—54.

Hofling, Charles K., Eveline Brotzman, Sarah Dalrymple, Nancy Graves, and Chester M. Pierce. "An experimental study in nurse-physician relationships." *The Journal of Nervous and Mental Disease* 143, no. 2 (1966): 171—180.

Gazzaniga, Michael S. "The split brain in man." *Scientific American,* 217, no. 2 (1967): 24—29.

Darley, John M., and Bibb Latane. "Bystander intervention in emergencies: diffusion of responsibility." *Journal of Personality and Social Psychology* 8, no. 4 (1968): 377—383.

Rosenthal, Robert, and Lenore Jacobson. "Pygmalion in the classroom." *The Urban Review* 3, no. 1 (1968): 16—20.

Ainsworth, Mary, D. Salter, and Silvia M. Bell. "Attachment, exploration, and separation: Illustrated by the behavior of one-year-olds in a strange situation." *Child Development* (1970): 49—67.

Chapter 5 Zimbardo, Philip. *Stanford prison experiment.* Stanford University, 1971.

Wason, Peter C., and Diana Shapiro. "Natural and contrived experience in a reasoning problem." *The Quarterly Journal of Experimental Psychology* 23, no. 1 (1971): 63—71.

Rosenhan, David L. "On being sane in insane places." *Science* 179, no. 4070 (1973): 250—258.

Lepper, Mark R., David Greene, and Richard E. Nisbett. "Undermining children's intrinsic interest with extrinsic reward: A test of the 'overjustification' hypothesis." *Journal of Personality and Social Psychology* 28, no. 1 (1973): 129.

Loftus, Elizabeth F. "Reconstructing memory: The incredible eyewitness." *Jurimetrics J.* 15 (1974): 188.

Tversky, Amos, and Daniel Kahneman. "Judgment under uncertainty: Heuristics and biases." *Science* 185, no. 4157 (1974): 1124—1131.

Dutton, Donald G., and Arthur P. Aron. "Some evidence for heightened sexual attraction under conditions of high anxiety." *Journal of Personality and Social Psychology* 30, no. 4 (1974): 510.

Miller, William R., and Martin E. Seligman. "Depression and learned helplessness in man." *Journal of Abnormal Psychology* 84, no. 3 (1975): 228.

McGurk, Harry, and John MacDonald. "Hearing lips and seeing voices." *Nature* 264 (1976): 746—748.

Bisiach, Edoardo, and Claudio Luzzatti. "Unilateral neglect of representational space." *Cortex*, 14, No. 1 (1978): 129—133.

Chapter 6 Libet, Benjamin, Curtis A. Gleason, Elwood W. Wright, and Dennis K. Pearl. "Time of conscious intention to act in relation to onset of cerebral activity (readiness-potential)." *Brain* 106, no. 3 (1983): 623—642.

Berry, Dianne C., and Donald E. Broadbent. "On the relationship between task performance and associated verbalizable knowledge." *The Quarterly Journal of Experimental Psychology* 36, no. 2 (1984): 209—231.

Baron-Cohen, Simon, Alan M. Leslie, and Uta Frith. "Does the autistic child have a 'theory of mind'?" *Cognition* 21, no. 1 (1985): 37—46.

Byrd, Randolph C. "Positive therapeutic effects of intercessory prayer in a coronary care unit population." *Southern Medical Journal* 81, no. 7 (1988): 826—829.

McNeil, Jane E., and Elizabeth K. Warrington. "Prosopagnosia: A face-specific disorder." *The Quarterly Journal of Experimental Psychology* 46, no. 1 (1993): 1—10.

Bem, Daryl J., and Charles Honorton. "Does psi exist? Replicable evidence for an anomalous process of information transfer." *Psychological Bulletin* 115, no. 1 (1994): 4—18.

Simons, Daniel J., and Daniel T. Levin. "Failure to detect changes to people during a real-world interaction." *Psychonomic Bulletin & Review* 5, no. 4 (1998): 644—649.

Botvinick, Matthew, and Jonathan Cohen. "Rubber hands 'feel' touch that eyes see." *Nature* 391, no. 6669 (1998): 756—756.

Costantini, Marcello, and Patrick Haggard. "The rubber hand illusion: sensitivity and reference frame for body ownership." *Consciousness and Cognition* 16, no. 2 (2007): 229—240.

Blakemore, Sarah-Jayne, Daniel Wolpert, and Chris Frith. "Why can't you tickle yourself?" *Neuroreport* 11, no. 11 (2000): R11—R16.

Ramachandran, Vilayanur S., and Edward M. Hubbard. "Synaesthesia—a window into perception, thought and language." *Journal of Consciousness Studies* 8, no. 12 (2001): 3—34.

Lenggenhager, Bigna, Tej Tadi, Thomas Metzinger, and Olaf Blanke. "Video ergo sum: manipulating bodily self-consciousness." *Science* 317, no. 5841 (2007): 1096—1099.